Threat Mitigation and Detection of Cyber Warfare and Terrorism Activities

Maximiliano E. Korstanje
University of Palermo, Argentina

A volume in the Advances in Information Security, Privacy, and Ethics (AISPE) Book Series

IGI GLOBAL
DISSEMINATOR OF KNOWLEDGE
www.igi-global.com

Published in the United States of America by
IGI Global
Information Science Reference (an imprint of IGI Global)
701 E. Chocolate Avenue
Hershey PA 17033
Tel: 717-533-8845
Fax: 717-533-8661
E-mail: cust@igi-global.com
Web site: http://www.igi-global.com

Library of Congress Cataloging-in-Publication Data

Names: Korstanje, Maximiliano, editor of compilation.
Title: Threat mitigation and detection of cyber warfare and terrorism activities / Maximiliano E. Korstanje, editor.
Description: Hershey, PA : Information Science Reference, [2016]
Identifiers: LCCN 2016048907| ISBN 9781522519386 (hardcover) | ISBN 9781522519393 (ebook)
Subjects: LCSH: Information warfare--Prevention. | Cyberspace--Security measures. | Cyberterrorism--Prevention. | Computer crimes--Prevention. | Internet--Political aspects.
Classification: LCC U163 .T49 2016 | DDC 363.325--dc23 LC record available at https://lccn.loc.gov/2016048907

This book is published in the IGI Global book series Advances in Information Security, Privacy, and Ethics (AISPE) (ISSN: 1948-9730; eISSN: 1948-9749)

British Cataloguing in Publication Data
A Cataloguing in Publication record for this book is available from the British Library.

Advances in Information Security, Privacy, and Ethics (AISPE) Book Series

ISSN:1948-9730
EISSN:1948-9749

Mission

As digital technologies become more pervasive in everyday life and the Internet is utilized in ever increasing ways by both private and public entities, concern over digital threats becomes more prevalent.

The **Advances in Information Security, Privacy, & Ethics (AISPE) Book Series** provides cutting-edge research on the protection and misuse of information and technology across various industries and settings. Comprised of scholarly research on topics such as identity management, cryptography, system security, authentication, and data protection, this book series is ideal for reference by IT professionals, academicians, and upper-level students.

Coverage

- Security Classifications
- Risk Management
- Data Storage of Minors
- Global Privacy Concerns
- Network Security Services
- Telecommunications Regulations
- Cyberethics
- Information Security Standards
- Cookies
- Access Control

IGI Global is currently accepting manuscripts for publication within this series. To submit a proposal for a volume in this series, please contact our Acquisition Editors at Acquisitions@igi-global.com or visit: http://www.igi-global.com/publish/.

The Advances in Information Security, Privacy, and Ethics (AISPE) Book Series (ISSN 1948-9730) is published by IGI Global, 701 E. Chocolate Avenue, Hershey, PA 17033-1240, USA, www.igi-global.com. This series is composed of titles available for purchase individually; each title is edited to be contextually exclusive from any other title within the series. For pricing and ordering information please visit http://www.igi-global.com/book-series/advances-information-security-privacy-ethics/37157. Postmaster: Send all address changes to above address.

Titles in this Series

For a list of additional titles in this series, please visit: www.igi-global.com

Online Banking Security Measures and Data Protection
Shadi A. Aljawarneh (Jordan University of Science and Technology, Jordan)
Information Science Reference • copyright 2017 • 312pp • H/C (ISBN: 9781522508649) • US $215.00 (our price)

Developing Next-Generation Countermeasures for Homeland Security Threat Prevention
Maurice Dawson (University of Missouri-St. Louis, USA) Dakshina Ranjan Kisku (National Institute of Technology, India) Phalguni Gupta (National Institute of Technical Teachers' Training & Research, India) Jamuna Kanta Sing (Jadavpur University, India) and Weifeng Li (Tsinghua University, China)
Information Science Reference • copyright 2017 • 428pp • H/C (ISBN: 9781522507031) • US $210.00 (our price)

Security Solutions for Hyperconnectivity and the Internet of Things
Maurice Dawson (University of Missouri-St. Louis, USA) Mohamed Eltayeb (Colorado Technical University, USA) and Marwan Omar (Saint Leo University, USA)
Information Science Reference • copyright 2017 • 347pp • H/C (ISBN: 9781522507413) • US $215.00 (our price)

Managing Security Issues and the Hidden Dangers of Wearable Technologies
Andrew Marrington (Zayed University, UAE) Don Kerr (University of the Sunshine Coast, Australia) and John Gammack (Zayed University, UAE)
Information Science Reference • copyright 2017 • 345pp • H/C (ISBN: 9781522510161) • US $200.00 (our price)

Security Management in Mobile Cloud Computing
Kashif Munir (University of Hafr Al-Batin, Saudi Arabia)
Information Science Reference • copyright 2017 • 248pp • H/C (ISBN: 9781522506027) • US $150.00 (our price)

Cryptographic Solutions for Secure Online Banking and Commerce
Kannan Balasubramanian (Mepco Schlenk Engineering College, India) K. Mala (Mepco Schlenk Engineering College, India) and M. Rajakani (Mepco Schlenk Engineering College, India)
Information Science Reference • copyright 2016 • 375pp • H/C (ISBN: 9781522502739) • US $200.00 (our price)

Table of Contents

Detailed Table of Contents

Business over the internet such as banking and several online services are growing rapidly. Similarly, social media web portals are also getting more and more involved in our daily life. Since these applications are popular and consist of personal and valuable data, they attract malicious attacks to their vulnerable points. The weakness can also be faced in all businesses and institutions that do not care the necessary security steps. The web parameter tampering is one of the major attacks which is based on the modification of parameters. In order to prevent the parameter tampering, a novel and simple mechanism is implemented by verifying the validity. The mechanism is based on a deterministic finite state machine. Beside this static method, the system also has run time validation which leads for the usage of hybrid analysis approach. As an evaluation, performance assessment of the algorithm is done for real time attacks targeting a web site.

With the continuing evolution of cyber threats, it is only a matter of time before an organisation will suffer a major breach or there is an incident of national significance. This necessitates monitoring to detect possible incidents and mechanisms to respond and recover from breaches. This chapter provides an overview of structures to aid in threat detection and incident recovery. Security Operation Centres (SOCs), Computer Security Incident Response Teams (CSIRTs), and Security Intelligence Centres (SICs) will be covered, and the differences, benefits and limitations will be discussed. Guidance for the implementation of these security capabilities within organisations will be provided.

Paper aimed to consider of approaches to big data (social network content) utilization for understanding of social behavior in the conflict zones, and analysis of dynamics of illegal armed groups. Analysis directed to identify of underage militants. The probabilistic and stochastic methods of analysis and classification of number, composition and dynamics of illegal armed groups in active conflict areas are proposed. Data of armed conflict – antiterrorist operation in Donbas (Eastern Ukraine in the period 2014-2015) is used for analysis. The numerical distribution of age, gender composition, origin, social status and nationality of child militants among illegal armed groups has been calculated. Conclusions on the applicability of described method in criminological practice, as well as about the possibilities of interpretation of obtaining results in the context of study of terrorism are proposed.

Over the last two decades, public confidence and trust in Government has declined visibly in several liberal democracies, giving way instead to disillusionment with current political institutions, actors, and practices; rendering obsolete or inappropriate much of traditional democratic politics. Simultaneously, digital technologies have created huge opportunities for public bodies and agencies. In analysing the No. 10 Downing Street ePetitions Initiative based in the United Kingdom, this chapter engages with issues related to the innovative use of digital network technology by Government to involve citizens in policy processes and to buffer national security within existing democratic frameworks. The work examines whether the application of new digital platforms to participatory democracy in the Government 2.0 era leads eventually to radical transformations in government functioning and the body politic, or merely to modest, unspectacular political reform and to the emergence of technology-based obsessive-compulsive pathologies and trolling behaviours amongst individuals in society.

The discourse of the Matrix not only wakes up philosophy from the slumber they are, but reflects the rise of fears which are proper of a culture, that manipulates anxieties to fix policies otherwise would be rejected. Basically, Neo is subject to an ethical dilemma which means that we are free to make the decision to live in a fake without suffering or in "the real of the dessert" as Morpheus put it. At this extent, if technology is used to protect humankind, Matrix evinces under some circumstances, it can oppress human beings when they renounce to their "free-will". This chapter not only explored the limitations of the society of risks within liberal thought but taking into consideration the role of authenticity serves ideologically to legitimate suffering. Indeed, accepting to live as Neo did, may lead to oppression or liberation.

Preface

The "scourge" of terrorism inscribes into the logic of illegal violence, though one must accept there is a lot of subtypes which merits to be discussed. Cyber-terrorism not only seems to be an emergent form of terrorism that escapes to the definition of illegal violence but the paradox lies in the fact sometimes the nation-state, which was originally designed to protect citizens, violates the constitutional rights in the name of security. Cyber-espionage, cyber-terrorism and cyber-crime are concepts inextricably intertwined. Snowden´s case reveals two important assumptions. On one hand, the discourse of surveillance appeals to what Foucault denominated as "the economy of discipline", which means that any perceived risks allude to a much deeper and preexisting economic matrix (Foucault, 2003). On another, while states devote considerable time and efforts struggling against terrorism to make from this society a better place, important constitutional rights are being vulnerated (Korstanje, 2015). In addition, unlike classic terrorism, cyber-terrorism operates from a virtual landscape which is very hard to grasp.

In this ever-changing and globalized World, the steps of nations to prevent future terrorist attacks are at snail pace. This does not mean nation-states are inefficient to keep the social order, but terrorism operates in the dark side of uncertainness. Since nobody knows where the next blow will take place, specialists have many limitations to forecast the next attack. The forces of securities are clearly identified, but not terrorists. This point opens the doors to a clear sentiment of paranoia and panic, which are the main goals of terrorism. As Luke Howie (2010) puts it, terrorists do not want to kill a lot of peoples, they like a lot of people watching what they have accomplished. This suggests that terrorists and media are two sides of the same coin. The challenge for democratic societies consists in developed democratic solutions to prevent undemocratic expression of violence. In doing so, they should not to lose the sight that terrorism and cyber-terrorism shares the same commonality respecting to trust. Both act to undermine the necessary trust in order for lay people to legitimate their loyalties to authorities.

THE CHALLENGE

Let`s explain first my recent traumatic experience with a Trojan virus which introduced in my personal Computer corrupted all existent Word files. This begs a more than interesting point, how does it happen?

Originally, I received an email from a colleague at Palermo University saying "please review the attached file for me". Confident his email was genuine; I double clicked the document opening the doors to a virus that affected many of documents. Doubtless it started a real nightmare. Since my recent backup copy dates back January, not only I lost a great part of my academic production from March on, but also it expressed a frightening sensation since my private life was somehow vulnerated.

The efficacy of virus depends upon its capacity to contaminate networks taking advance of social ties among members. I would never accept a spam from an anonymous source unless by a colleague. As this backdrop, in fact, I did not lose nothing, my house was there, my Pc was there, my family had health. Then, though I lost everything, I have lost nothing. This point suggests two important things. First and foremost, we work on a society which is based in abstract information, produced and disseminated through a virtual platform. What lies there is very important for capitalist societies, though it cannot be physically touched nor grasped. As society of information, words are exchangeable commodities which delineate the borders between reality and virtuality. A person may lose all its savings in a day when a hacker empties the bank account. This intersection of reality with virtuality seems to be one of aspects that facilitated the expansion of capitalism, which are not shared by other human organization as hunter-gatherers, or other ancient civilization. As Naisbitt puts it, the means of production, as well as the profits, tend to be abstracted over the recent decades (Naisbitt, 1984). Secondly, specialists have unclear the limits between cyber-terrorism and terrorism (Verton & Brownlow, 2003). In this respect, some conceptual differences should be in depth discussed.

While terrorists look to cause political instability by the administration of fear, in which case, they serve to adjoin to spectorship through the media (Eid, 2014; Howie, 2010; Korstanje, 2015, 2016), cyber terrorists work in secrecy, never vindicating their acts of vandalism unless if necessary. No less true is that terrorists are in quest of producing a culture of witnessing to produce a symbolic blow, whereas cyber-terrorists have diffuse goals. Some of them seek criminal acts, while others introduce harms to extract vital information from government agencies. Likely, I felt the same terror once I have been confirmed by experts the whole portion of my word files were encrypted or deleted, than audience while bombing or shooting. At the same time, thoughts struck me that there were higher probabilities an attack of this calibre may be very well repeated, waking up a sentiment of anxiety inside me, I bought some updated antivirus software and ensured the necessary back-up copies. Of course, this thesis suggests that surprise factor is the secondary point in common between terrorism and cyber-attacks. As stated, it is important to discuss, with invited academics, cyber-terrorism as a much wider theme which alternates contrasting position towards the formation of an all-encompassing model.

THE SEARCH FOR SOLUTION

Though at some extent, terrorism and cyber-terrorism are problems which remain unresolved up to date, the examples of FARC in Colombia, ETA in Spain and other classic cells as IRA in United Kingdom evince there are many possibilities to bring this issue to a safer port. Professor James Piazza argues that one of the aspects that determine the radicalization of some groups depends how their original claims are addressed by state. Whenever the group is pressed to operate from clandestinity or has marginal participation in the democratic system, there are higher likelihoods to adopt violent tactics that may very well leads to terrorism (Piazza, 2008). Other studies as McCauley and Moskalenko (2008) and Moskalenko and McCauley (2011) suggest that the radicalization of terrorists depend on the articulation of different stages that oscillate from self-esteem to be accepted by peers, to physical isolation in small groups where opposition to leaders are limited. The various facet of radicalization not only are emotionally in consonant with candidate´s inner-life, but also takes advantage of those people who had a bad image of society and world to instill a radicalized hate. In this vein, Korstanje (2015) has developed an interesting model to understand that the core of terrorism has nothing to do with violence nor religion, but it associates to extortion. One of the problems policy makers, authorities and governments are facing today respecting to terrorism depend upon they do not understand to what they struggle. Terrorism should be defined as a dialectics of hate, which is triggered by a group of insurgents and state. Since state never manages accurate information on where the next attack will take room, sometimes torture surfaces. The oxygen of terrorism and cyber-terrorism not only is previously determined by the generated fear, but as Mahmoud Eid observed, the leading role played by media covering terrorism-related news. Using the fear of citizenry to extortionate a state that somehow is unable, unfit or indifferent to terrorist´ claims is a key factor that explains "the individual radicalization". These groups recruit youth people, who are seduced by peers or relatives to balance their frustrations feeling they are part of something exceptional, important, and historic. The best form to struggle against terrorism combines further education, with less appetite for witnessing news containing cruelty, violence and death. As civilization, our own prone to consuming "the other death" is conducive to the rise and expansion of terrorism. While we are disgusting (or shocked) for what recently happened in Paris, Nice and other important capitals, we do not let stopping to watch!

ORGANIZATION OF THE BOOK

The present book was oriented to students, scholars, practitioners concerned by the fields of cyber-security in a moment where information is vital for strategic success decision making process. I will explain why cyber-terrorism cannot be dissociated from terrorism in spite of their differences. For those readers who want to open the pages of this book, I should stress, it is intended to re-discover new horizons of research for the decades to come.

As the previous argument given, it is noteworthy that this book is organized in three sections, which deal with different aspects of cyber-terrorism, risk perception, democracy and civil discontents. Over recent decades, specialists focused on the global nature of capitalism and its derived risks. The fact is that in a world where travels, news and information go faster than other periods, the effects of terrorism are much worse in urban cities than other rural zones.

Chapter 1 provides with a fresh view to implement in models that are oriented to prevent web parameter tampering attacks. This model checks parameters with stored database server for validation, at the time it validates resulted outcomes.

Chapter 2 thematises on the possibility to monitor cyber-attacks and risks by means of the implementation of a cyber-security centre. Not only cyber-security represents a chief factor to enhance the well-being of nations, but also is vital to struggle against terrorism. While a centre is created two external requirements should be met, authorative and normative, which are adjoined to internal quests as marketing organization and sales. All operational consideration to implement an efficient centre for control and surveillance in cyber-spaces is adamantly discussed in this recommending chapter.

Chapter 3 reviews the necessary big data to predict social and individual behaviour in terrorism hot-spot or conflict-driven zones. By the exploration of different key variables as age, gender composition, nationality, origin, social status, this model hints interesting results which would help to boost security in zones whipped by "the scourge of terrorism".

Chapter 4 brings into question to what extent existing infrastructure plays a leading role in the configuration of security. Besides focusing on financial and electric power as potentially risks for next cyber-attacks, he holds the thesis that transport infrastructure should be analysed alternating quantitative and qualitative methods instead. A lot of cyber-attacks to transportation sectors have taken place over the recent years and of course this offers a fertile ground to understand the intersection of terrorism and mobility.

Chapter 5 introduces a short debate respecting to how protestant Reform developed a climate of distrust which is conducive not only to terrorism but risk perception. Theoretically speaking while technology would be used to make of this world a safer place, the needs to anticipate future events a discourse which is expressed under the doctrine of precautionary principle, to avoid threats and risks.

Chapter 6 calls the attention to the atmosphere of distrust Governments and their agencies are facing today. Part of this derives from the fact states uses obsolete tactics, which often defy on the democratic institutions, but what is more than important to discuss is to what extent, digital technologies open the doors for new opportunities for public bodies. This chapter is based on the study case of No 10 Downing Street e-petitions Initiative that engage lay-citizens to take an active role in the process of security within established democratic institutions.

Chapter 7 triggers a hot-debate in the last chapter of the second section, examining the Matrix Saga as the precondition to understand how the meaning of reality can be politically tergiversated to discipline the citizenry. Matrix represents more than a film, it exhibits an old concern from philosophers to harmonize the counter-and-balance powers of democracy with the need of security.

Chapter 8 echoes the original concerns of NSC (National Security Council) in view of the rise of computers, software and technologies which can be disposed towards criminal acts. Per these specialists, cyber terrorism has the same or worse disastrous effects than classic terrorism. This chapter provides a conceptual discussion on three different arguments respecting to cyber-terrorism, which merits to be included in this project.

Chapter 9 discusses the problems of democracy to articulate individual civil rights to protest, with civil chaos. This work focuses on the rise of civil disobedience as a political factor to produce political instability, which in some condition can be functional to some terrorist cells. Dialoguing with Rawls and other philosophers about the nature of state, Crosston devotes considerable resources to find a valid solution to the dichotomy between surveillance vs. democracy. The same applies for the use of Twitter that proved to be a valid instrument for channelling social discontent in developed and underdeveloped nations.

Chapter 10 sheds light on the problem of social discontent and digital technologies. The culture of protest found in twitter a good ally, since the mobilization of masses turned easier in our times, in view of the rise of informational technology or ICT than the society of our parents. Though at a preliminary stage, governments should control social media in order to protect their citizens, providing health services or medical care in the protest, it is safe to say that these protests evinces serious political crises in view of the public discredit politicians and leaders suffer.

Chapter 11 centres on what specialists dubbed as "the irony of democracy". The dilemma says that social order is convened by means of the introduction of lesser-evil doctrine. As it was originally discussed by Michael Ignatieff, terrorism and "the war on terror" undermined the democratic institutions as well as the autonomies of branches and agencies. Sociologically speaking, the roots of our current atmosphere of terror come from our incapacity to understand its connection with rationality. Indeed, beyond the polemics terrorism has woken up in recent decades, a much deeper discussion is necessary to understand that terrorism and economy are inevitably entwined. Here, I toy with the idea that in the society of "terror", the inflation of risk in charge of mass media not only contributes to the formation of an omnipresent state but also engenders a neo-individualism (universalism) which leads very well to substantial changes in pre-existing jurisprudence. The paradox lies in the fact in the name of human right violations; the doctrine of lesser evil is eroding the basis of Athenian democracy.

Hopefully this guide-book helps readers to understand not only the problem of terrorism, but also the role of technology as a double-edged sword to protect citizens. I would thanks IGI global for the opportunity to publish this project as well as all well-reputed authors and contributors who have done of this a masterful work, highly recommended by scholars, policy makers and students concerned by topics of this calibre.

Maximiliano E. Korstanje
University of Palermo, Argentina & University of Leeds, UK
Buenos Aires, Oct 19 of 2016

REFERENCES

Eid, M. (Ed.). (2014). *Exchanging terrorism oxygen for media airwaves: The age of terroredia*. Hershey, PA: IGI Global. doi:10.4018/978-1-4666-5776-2

Foucault, M. (2003). *Society must be defended: Lectures at the Collège de France* (D. Macey, Trans.). New York: Picador.

Howie, L. (2011). *Terror on the screen: Witnesses and the re-animation of 9/11 as image-event, popular culture*. Washington, DC: New Academia Publishing, LLC.

Ignatieff, M. (2013). *The lesser evil: Political ethics in an age of terror*. Princeton, NJ: Princeton University Press.

Korstanje, M. E. (2015). *A difficult world, examining the roots of capitalism*. New York: Nova Science Pubs.

Korstanje, M. E. (2016a). *The rise of Thana capitalism and tourism*. Abingdon, UK: Routledge.

Korstanje, M. E. (2016b). *Terrorism in the global village*. New York: Nova Science Pubs.

McCauley, C., & Moskalenko, S. (2008). Mechanisms of political radicalization: Pathways toward terrorism. *Terrorism and Political Violence*, *20*(3), 415–433. doi:10.1080/09546550802073367

Moskalenko, S., & McCauley, C. (2011). The psychology of lone-wolf terrorism. *Counselling Psychology Quarterly*, *24*(2), 115–126. doi:10.1080/09515070.2011.581835

Naisbitt, J. (1984). *Megatrends: Ten new directions transforming our lives (No. 04; HN59. 2, N3.)*. New York: Warner Books.

Piazza, J. A. (2008). Incubators of terror: Do failed and failing states promote transnational terrorism? *International Studies Quarterly*, *52*(3), 469–488. doi:10.1111/j.1468-2478.2008.00511.x

Verton, D., & Brownlow, J. (2003). *Black ice: The invisible threat of cyber-terrorism*. Osborne.

Section 1
Cyber Security in Global Times

Chapter 1
A Simple Solution to Prevent Parameter Tampering in Web Applications

Oğuzhan Menemencioğlu
Karabük University, Turkey

İlhami Muharrem Orak
Karabük University, Turkey

ABSTRACT

Business over the internet such as banking and several online services are growing rapidly. Similarly, social media web portals are also getting more and more involved in our daily life. Since these applications are popular and consist of personal and valuable data, they attract malicious attacks to their vulnerable points. The weakness can also be faced in all businesses and institutions that do not care the necessary security steps. The web parameter tampering is one of the major attacks which is based on the modification of parameters. In order to prevent the parameter tampering, a novel and simple mechanism is implemented by verifying the validity. The mechanism is based on a deterministic finite state machine. Beside this static method, the system also has run time validation which leads for the usage of hybrid analysis approach. As an evaluation, performance assessment of the algorithm is done for real time attacks targeting a web site.

DOI: 10.4018/978-1-5225-1938-6.ch001

INTRODUCTION

Internet becomes essential for connecting companies, service providers and users online to each other. In spite of their wide usage, the web applications have some vulnerabilities. Scripting languages such as PHP, communicate with database using low-level queries as strings API. These languages represent data and code with the strings instead of more sophisticated APIs. Therefore application code can include database queries which takes unchecked user inputs (Wassermann & Su, 2007). Although the most common vulnerability is related to database implementation, there are also some other weak points of the web applications such as cross-site scripting. It will be useful to understand the vulnerabilities and to introduce reliable and effective solution for preventing the attacks.

There are different approaches to detect and prevent tampering attacks. For Structured Query Language Injection Attacks (SQLIAs), a taxonomy was provided by Chung et al. (Chung, Wu, Chen, & Chang, 2012). Figure 1 illustrates their taxonomy. Details of taxonomy and concerned concepts are provided in below. This taxonomy can be considered as general classification of all types of vulnerabilities.

First approach stated in the taxonomy is static analysis. Zhang et al. proposed a method which can be classified as static analysis, to eliminate integrity problems by trying to detect absence of integrity constraint enforcement (Zhang et al., 2011).

Figure 1. Taxonomy of approaches

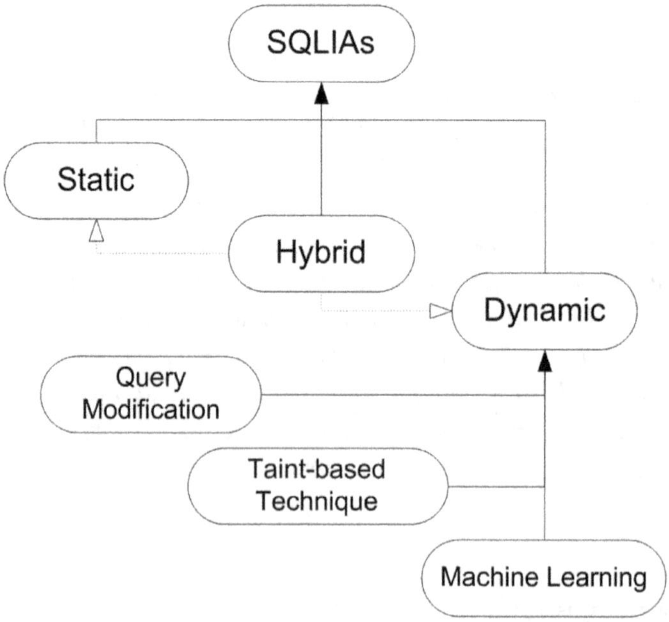

Dynamic analysis is another widely proposed approach. Ogheneovo and Asagba used a dynamic method comparing real time user queries with syntactic structure of the queries defined by parse tree (query modification) (Ogheneovo & Asagba, 2013). Chan et al. introduced a dynamic method using machine learning. They discovered associative patterns by using fuzzy logic. They generated and pruned the rules by using Apriori algorithm for validating input values, input field lengths, and SOAP size (Chan, Lee, & Heng, 2013, 2014). On the one hand, Chung et al. proposed a method that checks the queries whether they are legitimate. If queries are not legitimated, then they are processed with vulnerability detectors. They integrated their algorithm with the well-known three vulnerability detectors instead of developing a new detector (Chung et al., 2012). Jang and Choi used regular expression and size of user query result to control inputs (Jang & Choi, 2014). Kim and Lee used support vector machines to classify SQLIAs (Kim & Lee, 2014).

On the other hand, in some studies hybrid methods were proposed by using static analysis and providing dynamic approach with runtime detector. In this respect, Lee et al. decreased the complexity of the algorithm from $O\left(n^3\right)$ to $O\left(n\right)$ by comparing static SQL queries with dynamically generated queries after removing the attribute values (Lee, Jeong, Yeo, & Moon, 2012). The AMNESIA method proposed by Halfond and Orso was highly referred in other studies especially for benchmarking. They identified hotspots and built a model by using Non-Deterministic Finite-State Machine (NDFSM). They also check user queries dynamically to confirm the static model (Halfond & Orso, 2005). Balasundaram and Ramaraj controlled query infrastructures and validated the queries by checking length of queries and restricting the queries such as drop, delete, etc. (Balasundaram & Ramaraj, 2012). Muthuprasanna et al. used NDFSM for static analysis and runtime validator for comparing queries (Muthuprasanna, Ke, & Kothari, 2006).

BACKGROUND

Web applications may include some vulnerabilities and the definitions of these type vulnerabilities may vary. This section defines the vulnerability types and highlights the types consisting parameter tampering. The details and examples of the vulnerabilities consisting parameter tampering are provided. Static, dynamic, and hybrid analysis approaches are mentioned and performance approaches of reviewed papers are analysed.

Types of Vulnerabilities

Web parameter tampering, also called form-field, is not well represented due to classification difficulty and could be categorized as an authentication error (Christey & Martin, n.d.). However, in our study, definitions provided by The Open Web Application Security Project (OWASP) are taken into consideration. OWASP released Top 10 list for attacks as (OWASP, n.d.):

1. Injection,
2. Broken Authentication and Session Management,
3. Cross-Site Scripting (XSS),
4. Insecure Direct Object References,
5. Security Incorrect Configuration,
6. Sensitive Data Exposure,
7. Missing Function Level Access Control,
8. Cross-Site Request Forgery (CSRF),
9. Using Components with Known Vulnerabilities, and
10. Un-validated Redirects and Forwards.

Parameter tampering is mainly involved in “Injection”, “Insecure Direct Object References”, “Missing Function Level Access Control” and “Un-validated Redirects and Forwards”. According to OWASPs’ top 10 list, they are in 1, 4, 7 and 10th order respectively.

- **Injection:** The attack attempts to exploit the vulnerabilities between backend server and database server. Since SQL commands are separated by semicolon, attackers can use this structure for the attacks such as “*; drop table {tablename};*”. This situation not only occurs in string parameters but also in some other parameters (Gollmann, 2008). Attackers can also taint the parameter value in browser to send “*‘ or ‘1’=’1*”. For example:

```
http://muh.karabuk.edu.tr/?id=' or '1'='1
```

- **Insecure Direct Object References:** Some authorized system users sometimes can attack the system by trying to change parameter values to access objects which they are not authorized for. Applications do not always verify user authentication for each object. Attackers tamper parameters to access to the intended customer’s account or any user’s account. For example:

```
http://muh.karabuk.edu.tr/ac/?act=ntacct
```

- **Missing Function Level Access Control:** An authorized system user may try to change the URL or a parameter to access a privileged function. Incorrect configuration or missing proper code checks cause this vulnerability. Detecting the imperfection is easy but identifying specific URLs or functions is very difficult.
- **Un-validated Redirects and Forwards:** Applications usually redirect users to same or some other pages by passing parameters. Attackers use un-validated parameters to violate the system. Detection of redirecting attacks is easier than that of forwarding attacks (OWASP, n.d.).

Analysis of Approaches

As mentioned before, static, dynamic or hybrid approach is used for vulnerability analysis (Chung et al., 2012; Lee et al., 2012; Natarajan & Subramani, 2012; NIST, n.d.). Static approach analyses the code without execution (Natarajan & Subramani, 2012). It focuses on program codes as text. It requires rewriting web applications (Lee et al., 2012). There are some forms of static analysis e.g. software metrics and reverse engineering. In analysing the code statically several techniques can be used such as data flow analysis, model checking and abstract interpretation (Natarajan & Subramani, 2012). Commonly JSA (Java String Analysis) is used to perform static analysis (Lee et al., 2012; Muthuprasanna et al., 2006). Instead of manual tests of web applications, dynamic analysis analyses the vulnerabilities during execution time (Natarajan & Subramani, 2012). It focuses on instructions in run time. It has two types: dynamic taint analysis and forward symbolic execution. Dynamic taint analysis observes computations affected by user input. Forward symbolic execution uses a logical formula (Schwartz, Avgerinos, & Brumley, 2010). Dynamic analysis checks web server responses for each input. It does not need modification of web application. On the other hand hybrid method simultaneously analyses web pages and generates SQL queries to test (Lee et al., 2012).

Performance Approaches of the Algorithms

In any kind of preventing method, besides success rate, time performance is also important for its preference. In this respect, when the methods provided in literature are analyzed, it can be seen that in some studies, method is provided but evaluation of the real time performance or/and accuracy is not given. However, there are some others which provided both of them. Lee et al. calculated complexity of their algorithm instead of real time evaluation (Lee et al., 2012). Ogheneovo and Asagba compared their algorithm with public and open benchmark of AMNESIA for accuracy and performance assessment. They also provided a real time evaluation results but they did not detail their evaluation method. By considering the technique used in their algorithms, it can be predicted that high overhead time is due to parse tree (Ogheneovo & Asagba, 2013). Chan et al. provided accuracy assessment results. They also stored request and response times of web service for real time evaluation. Although their algorithm used SOAP size, it responds to detection and prevention process in 1 second regardless of SOAP size (Chan et al., 2014). Chung et al. discussed their algorithm complexity and provided real time total execution time (Chung et al., 2012). Jang and Choi provided accuracy and time complexity assessment for their algorithm. They tested their algorithm on Amnesia system for benchmarking. They also provided time measurements for real time execution (Jang & Choi, 2014). Kim and Lee provided accuracy assessment and real time execution results for detector and conversion of a query to n-dimensional feature vector (Kim & Lee, 2014). Halfond and Orso conducted complexity analysis and give the average real time execution results (Halfond & Orso, 2005). Balasundaram and Ramaraj provided only real time execution time of algorithm (Balasundaram & Ramaraj, 2012). Muthuprasanna et al. provided time evaluation of offline process for static analysis characteristic. They also provided evaluation of run time process (Muthuprasanna et al., 2006).

MATERIALS AND METHODS

In this chapter a mechanism which uses Deterministic Finite State Machine (DFSM) to validate parameters safety, is proposed. The states in the mechanism are firstly identified. On the other hand, parameter checking is also performed on runtime.

In this section, the preliminary definitions used by the mechanism are provided. The architecture of new mechanism is also described. Then the innovation performed by our approach is provided.

Preliminary

Parameters used in web applications usually occurred in one form of the three common types: numeric, non-numeric and user id parameters. Hence, parameters are classified in three types to control them. In order to prevent or detect tainted parameters, a DFSM is defined for each parameter type. Numeric parameter control is provided by using $DFSM_1$ that accepts only numbers. Non-numeric parameters are controlled by using $DFSM_2$ that accepts only valid strings. On the other hand, DFSM_3 accepts only numbers which have valid user ids.

Definition 1: $DFSM_1$ is a deterministic finite automaton (DFA) and defined as $(\Sigma, S, s_0, \delta_1, F)$, where $\Sigma = \{0,...,9\}$, $S = \{0,1,2\}$, $s_0 = 0$, $\delta_1 : S \times \Sigma \rightarrow S$, $F = \{1\}$, Σ is a finite, non-empty set of symbols (input alphabet); S is finite, non-empty set of states; s_0 is an element of S (initial state); δ_1 is state-transition function and F is set of final states (a subset of S).

Figure 2 (a) illustrates the $DFSM_1$ transition diagram, which defines δ_1.

Definition2: $DFSM_2$ isdefinedas $(\Sigma, S, s_0, \delta_2, F)$, where $\Sigma = \{a,...,z,0,...,9,!..@\}$, $S = \{0,1,2\}$, $s_0 = 0$, $\delta_2 : S \times \Sigma \rightarrow S$, $F = \{1\}$ and $I = \{i_0, i_1, i_2, i_3\}$.

w, x, y, z are user inputs which are unique symbols of alphabet and $w, x, y, z \in \Sigma$, $f_1 : \Sigma \rightarrow D$ and $g_1 : \Sigma \rightarrow \{0,1\}$, g_1 is a function defined as (1) and used in I.

Figure 2. (a) $DFSM_1$ transition diagram, (b) $DFSM_2$ transition diagram, (c) $DFSM_3$ transition diagram

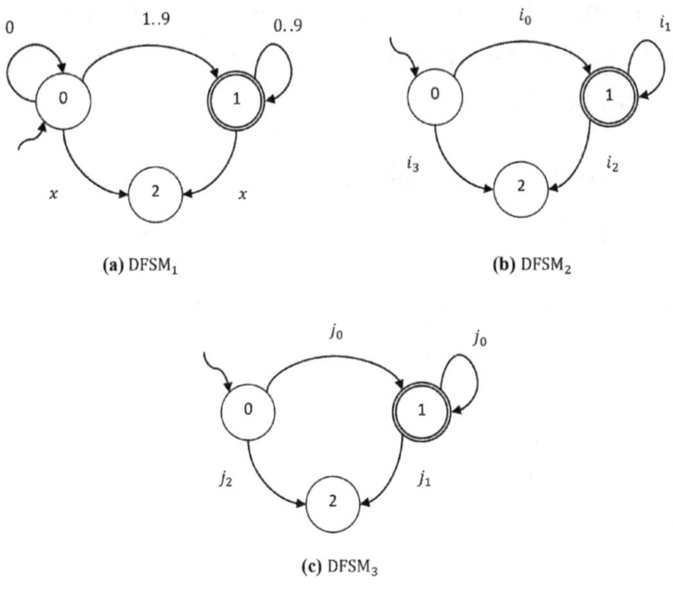

$$g_1(X) = \begin{cases} 1 & if \quad \exists d_w \,|f(w) = d_w, \\ 1 & if \quad \exists d_x \,|f(x) = d_x, \\ 0 & if \quad \nexists d_y \,|f(y) = d_y, \\ 0 & if \quad \nexists d_z \,|f(z) = d_z, \end{cases} \tag{1}$$

I is defined as (2).

$$I = \begin{cases} i_0 & if \quad g_1(w) = 1,\ \delta_2(s_0, w) = s_1, \\ i_1 & if \quad g_1(x) = 1,\ \delta_2(s_1, x) = s_1, \\ i_2 & if \quad g_1(y) = 0,\ \delta_2(s_1, y) = s_2, \\ i_3 & if \quad g_1(z) = 0,\ \delta_2(s_0, z) = s_2, \end{cases} \tag{2}$$

D is a set of parameters defined on application development process by software developers and stored in a database. f_1 function checks if any w, x, y, z parameter exists in the database. In the other words, f_1 function executes $\sigma_{Value} = w^{(PARAMETER)}$ operation. Therefore, f_1 function is a one-to-one function.

g_1 function returns 1 if the parameter is defined and valid. Otherwise it returns to 0 through f_1 function. Figure 2(b) illustrates the $DFSM_2$ transition diagram. In other words, transition diagram defines δ_2.

Definition 3: $DFSM_3$ is defined as $(\Sigma, S, s_0, \delta_3, F)$, where $\Sigma = \{0, \ldots, 9\}$, $S = \{0,1,2\}$, $s_0 = 0$, $\delta_3 : S \times \Sigma \to S$, $F = \{1\}$ and $J = \{j_0, j_1, j_2\}$.

w, x and y are user inputs and unique symbols of alphabet. $w, x, y \in \Sigma$, $f_2 : \Sigma \to D$ and $g_2 : \Sigma \to \{0,1\}$, g_2 is a function defined as (3) and used in J.

$$g_2(Y) = \begin{cases} 1 & if \quad \exists d_w \,|f(w) = d_w, \\ 0 & if \quad \nexists d_x \,|f(x) = d_x, \\ 0 & if \quad \nexists d_y \,|f(y) = d_y, \end{cases} \tag{3}$$

J is defined as (4).

$$J = \begin{cases} j_0 & if \quad g_2\left(w\right) = 1,\ \delta_3\left(s_0, w\right) = s_1, \\ j_1 & if \quad g_2\left(x\right) = 1,\ \delta_3\left(s_1, x\right) = s_1, \\ j_2 & if \quad g_2\left(y\right) = 0,\ \delta_3\left(s_1, y\right) = s_2, \end{cases} \tag{4}$$

D is a set of all User ID's stored in database. f_2 function checks if any "user ID" exists in the database. In other words, f_2 function executes $\sigma_{Id} = w^{(PARAMETER)}$ operation. Therefore f_2 function is also a one-to-one function.

g_2 function returns 1 if any user ID is defined and valid or returns 0 otherwise through f_2 function. Figure 2(c) illustrates the $DFSM_3$ transition diagram which defines δ_3.

Validation process is defined by Def.1, Def. 2 and Def. 3. User ID, numeric and non-numeric parameters are controlled by the mechanism.

Proposed System Architecture

Figure 3 illustrates the main flow chart of the detection mechanism. Each parameter should be checked by mechanism before execution. Un-validated parameters and attack attempts are detected and stored by the mechanism. The requests having legitimate parameters are redirected to targeted pages. State machine definitions describe the steps of the validation.

Figure 4 illustrates the novel web based architecture. In proposed design database not only stores data as usual but also stores legitimate parameters. Application server checks validation of parameters as well as implementing business logic as middle layer.

New architecture makes a validation check for requests instead of directly performing inquiry. If the request includes un-validated parameters, the mechanism stops the inquiry process. If parameters are legitimated, then mechanism allows the inquiry to be processed.

Implementation

In order to evaluate the new approach, a prototype, "Academic Curriculum Vitae Based Faculty Information System" is developed on Karabük University Engineering Faculty web site (Menemencioğlu, Sonuç, Karaş, & Orak, 2013). This application provides a platform for self-accessing, managing, and manipulating personal data and publishing it for web visitors.

Figure 3. Mechanism flow chart

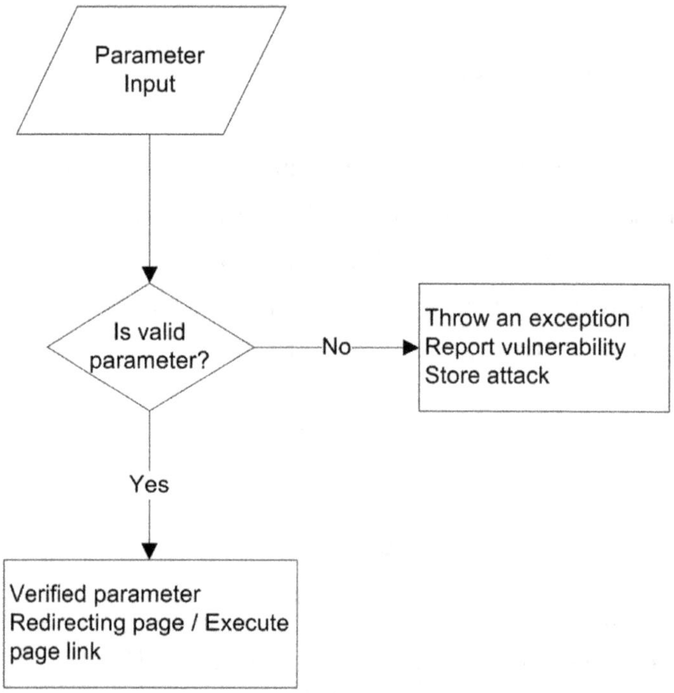

Figure 4. Architecture of supposed method

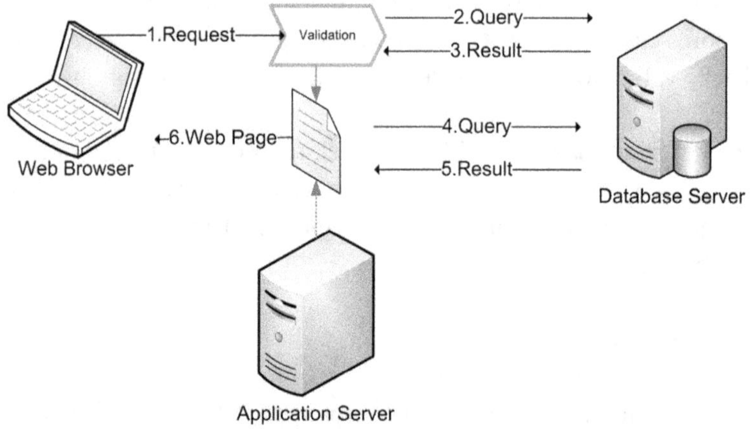

Application server implements business logic and validation process. After the request of client, first step is validation. The validation mechanism checks parameters as defined in previous sections. Some of the parameters are controlled by mechanism given in Def. 1 while some others are controlled by Def. 2 and 3. If parameters are valid, then the requested web page is served by application server.

Whole mechanism is implemented with PHP script language. Parameters are processed by the mechanism. The proposed mechanism checks the parameters with the ones stored in database server for validation, and evaluates the results. In case of an attack, it stores IP address of the attacker, and declines the requested operation and warns the user. Otherwise, business logic continues on the process.

The mechanism is performed on a Microsoft Windows Server 2003 Enterprise running on a system with Dual Core AMD Opteron CPU (2.39 GHz) and 8 GB memory. The server is not a dedicated application server. Since the application server also runs as database server, it has high overhead. MySQL is used as database in this system.

SOLUTIONS AND RECOMMENDATIONS

In this section, the results of proposed technique are analyzed and the evaluation of the approach is discussed by comparing with the other methods in terms of execution time.

Results of Proposed Method

This chapter proposes a novel idea which is very simple and easy to implement. For detecting and preventing web parameter tampering attacks, DFSM and database are used. When a user posts a web parameter or tampers a parameter, the mechanism handles it whether the parameter is valid or not. The details of the results are mentioned below. This mechanism prevents four types of risks which are "Injection", "Insecure Direct Object References", "Missing Function Level Access Control", "Un-validated Redirects and Forwards".

The mechanism that is proposed, achieves successful results in preventing parameter tampering attacks. The process of parameter checking in the database server can be done in $\theta(1)$ time. Query time depends on workload of server. If parameter tampering is detected, then there will be time cost for storing attack too. Even that time complexity is still $\theta(1)$.

Results of Real World Attacks

The data set includes three months' data. During this period, 54,098 attacks are detected. At the same period, 825,930 sessions are recorded. Figure 5(a) shows the distribution of vulnerabilities of total sessions.

Figure 5(b) shows the distribution of the parameter types. It can be seen that user-id parameter vulnerability is not encountered.

Results of Performance

One of the most important performance metrics is accuracy. Recent detection and prevention methods have reached very low error rates (lower than 0.1%) (Chung et al., 2012). Another performance metrics are execution time (Chung et al., 2012; Jang & Choi, 2014), and cost of implementation (Jang & Choi, 2014). This study focuses on an algorithm which has low execution time to improve performance. The performance metrics and evaluation methods of different researches are analyzed. Table 1 provides the performance metrics of researches which are used for evaluation. Table 1 also includes researches' methods and classification in analysis approaches.

In order to evaluate the performance, a runtime analysis is performed. When numeric and string parameter types are considered together, proposed detection mechanism responses in average 1.2 millisecond (ms). For numeric type of parameters, the detection mechanism response in average 1.1 ms, but for string types of parameters average response time is 1.4 ms. A chauvenet test is implemented on raw data to obtain more reliable results. Overview of the results is shown in Table 2.

In string types of parameters, some other overheads are measured besides the execution time of algorithm to make better evaluation. These are detection of time including insertion, execution time of queries having selection and insertion, and page load time. All these results are gathered in the same time of execution. Table

Figure 5. (a) Session and attack ratio, (b) distribution of parameter types

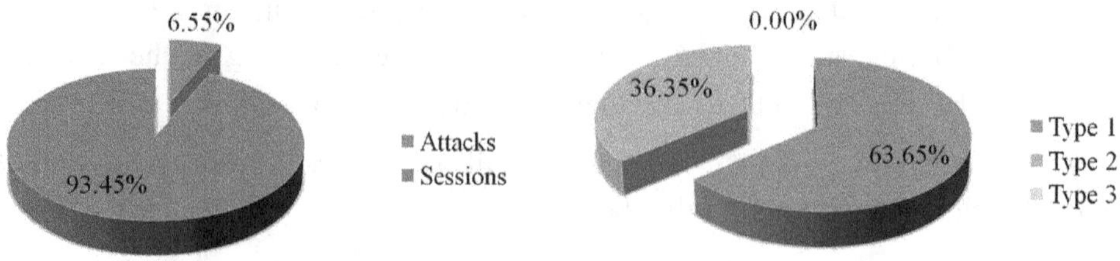

(a) Sessions & Attacks

(b) Types

Table 1. Performance metrics, methods, and classification

	Real Time Detection	Real Time Evaluation	Accuracy	Code Adjustment	Method	Class
Zhang et al. [3]	○	○	●	◉		Static
Lee et al. [9]	●	○	●	●	Structural check	Hybrid
Ogheneovo & Asagbas [4]	●	◉	●	○	Parse tree	Dynamic
Chan et al. [6]	●	◉	●	●	Fuzzy logic, apriori	Dynamic
Chung et al. [2]	●	●	○	○	Hot queries	Dynamic
Jang & Choi [7]	●	◉	◉	●	Regular expression	Dynamic
Kim & Lee [8]	◉	◉	●	●	SVM and feature vector	Dynamic
Balasundaram & Ramaraj [11]	●	●	○	●	Structural check	Hybrid
Halfond & Orso [10]	●	●	●	○	NDFSM	Hybrid
Muthuprasanna et al. [12]	●	◉	◉	○	NDFSM	Hybrid
Menemencioğlu &Orak	●	●	○	◉	DFSM	Hybrid

Symbols: ●: supported - required, ◉: partially supported – required, ○: not supported – required.

Table 2. Measurements after chauvenet test

	Detect	Detect + Insert	Select	Insert	Pageload
Avg	0.0014	0.0188	0.0002	0.0174	0.1397
Std Dev	0.0002	0.0024	0.0000	0.0024	0.2047
Min	0.0011	0.0137	0.0002	0.0123	0.0152
Max	0.0021	0.0274	0.0002	0.0259	0.9144
Max-Min	0.0010	0.0137	0.0001	0.0136	0.8992
Max/Min	1.9403	1.9997	1.4069	2.1032	60.0452

3 provides the normalization parameters. In addition, Figure 6 provides correlation of different processes and the impact of the processes to each other.

Since the simplicity and the effectiveness of the algorithm, its execution time is very short. In the exaction of the algorithm, application server, database server, and network are the parameters for the performance. Since the proposed mechanism runs on non-dedicated server, one should expect slower performance.

Figure 6. shows clearly that our algorithm is affected by the overheads of the server. Detection and "select" query peaks overlap with each other. It is expected result because detection includes "select" query as a part. "Insert" and detection peaks do not overlap completely with each other but almost entire "insert" query peaks coincide with "select" query peaks. From the Figure 6 it can be seen that the server workload is not constant. The peaks show heavy workload on the server

Table 3. Normalization values of properties

	Normalization Parameter
Detect	100
Detect + Insert	10
Select	500
Insert	10
Pageload	0.5

Figure 6. Graph of random selected data from results

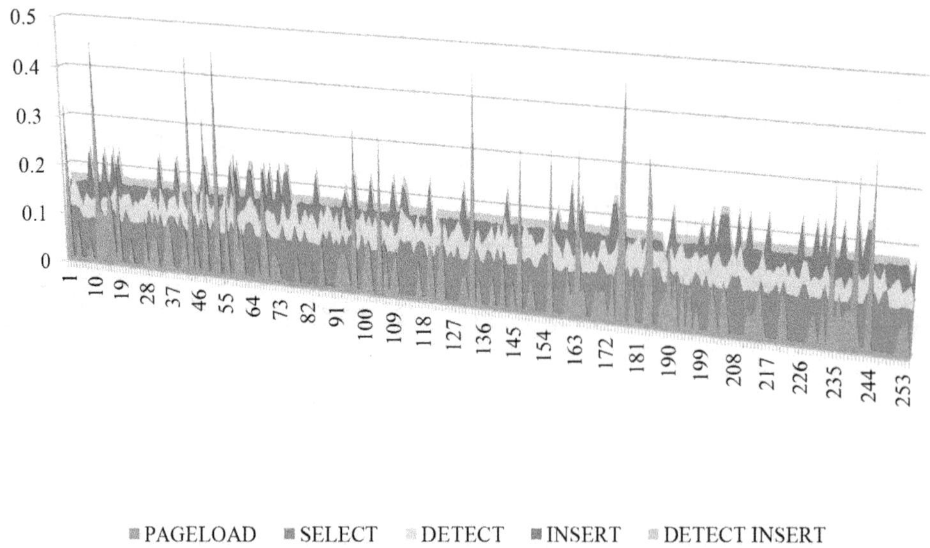

which results in delay in response time. In addition, the page load peak and the other peaks related to database operations do not overleap exactly. Therefore, being non-dedicated server may not directly have impact on overheads. It can be caused by the quantity of system sources reserved for services.

Table 4 provides the brief properties and the results of current methods for comparing with our method.

FUTURE RESEARCH DIRECTIONS

In future work, stored attacks can be transformed to ontologies. Measuring similarity between ontologies and semantic density analysis etc. can be calculated. Ontologies can be analyzed for inferring implicit knowledge about vulnerabilities by using semantic web infrastructure.

Table 4. Comparison of performance

	Year	Real Time (~ or ≥) (ms)	Method	System
Ogheneovo & Asagbas [4]	2013	3500	Average query response time	JAVA, MySQL
Chan et al. [6]	2014	1000	SOAP size and request response time	Intel Pentium Dual CPU E2160 1.80 GHz & 1.79 GHz, 0.99 GB RAM, Microsoft Windows Server 2003
Chung et al. [2]	2012	750	Execution time	Intel Core 2 CPU 2.4 GHz, 4 GB RAM, Windows Vista, Java
Jang & Choi [7]	2014	720	Average execution time	Intel Core i5 2.40 GHz, 8.00 GB RAM, Ubuntu V 9.0, Java
Kim & Lee [8]	2014	505.5	Average of classifier, vector converter, sequence converter, vice versa converter time, and detection time	-
Balasundaram & Ramaraj [11]	2012	30	Execution time	.NET
Halfond & Orso [10]	2005	25	Differences of request response time	-
Muthuprasanna et al. [12]	2006	4.3	Execution time	Dual Core
Menemencioğlu &Orak	2016	1.4	Execution time	AMD Dual Core Opteron 2.39 GHz, 8 GB RAM, Microsoft Windows Server 2003, PHP, MySQL

Symbols: -: no evaluation.

CONCLUSION

Parameter tampering is very important vulnerability for web applications. A mechanism for preventing parameter tampering is proposed based on the DFSM. This mechanism attempts to prevent four most common risk types "Injection", "Insecure Direct Object References", "Invalidated Redirects" and "Forwards and Missing Function Level Access Control" which involve parameter tampering.

The proposed mechanism uses hybrid analysis approach. It is very simple and effective in terms of architecture and implementation. The cost of the implementation and also the overhead of the algorithm are very low. The results are very successful when compared with other methods. The performance can be increased by using standalone server instead of shared environment.

In this study, the detection system is implemented on a faculty information system. Although it does not keep very valuable data, the attack ratio is 7 percent of all sessions. It is obvious that more important systems such as online banking and online services will expose to more attacks. Therefore, this kind of preventing mechanism will be critically necessary for each web applications.

REFERENCES

Balasundaram, I., & Ramaraj, E. (2012). An efficient technique for detection and prevention of SQL injection attack using ASCII based string matching. *Procedia Engineering, 30*(2011), 183–190. http://doi.org/10.1016/j.proeng.2012.01.850

Chan, G. Y., Lee, C. S., & Heng, S. H. (2013). Discovering fuzzy association rule patterns and increasing sensitivity analysis of XML-related attacks. *Journal of Network and Computer Applications, 36*(2), 829–842. doi:10.1016/j.jnca.2012.11.006

Chan, G. Y., Lee, C. S., & Heng, S. H. (2014). Defending against XML-related attacks in e-commerce applications with predictive fuzzy associative rules. *Applied Soft Computing, 24*, 142–157. doi:10.1016/j.asoc.2014.06.053

Christey, S., & Martin, R. A. (n.d.). *CWE - Vulnerability Type Distributions in CVE.* Retrieved April 17, 2015, from http://cwe.mitre.org/documents/vuln-trends/index.html

Chung, Y., Wu, M., Chen, Y., & Chang, W. (2012). *A Hot Query Bank approach to improve detection performance against SQL injection attacks. In Computers & Security*. Elsevier Ltd. doi:10.1016/j.cose.2011.11.007

Gollmann, D. (2008). Securing Web applications. *Information Security Technical Report, 13*(1), 1–9. doi:10.1016/j.istr.2008.02.002

Halfond, W. G. J., & Orso, A. (2005). AMNESIA: Analysis and Monitoring for NEutralizing SQL-injection Attacks. In *Proceedings of the 20th IEEE/ACM International Conference on Automated Software Engineering* (pp. 174–183). http://doi.org/ doi:10.1145/1101908.1101935

Jang, Y.-S., & Choi, J.-Y. (2014). Detecting SQL injection attacks using query result size. *Computers & Security, 44*, 104–118. doi:10.1016/j.cose.2014.04.007

Kim, M. Y., & Lee, D. H. (2014). Data-mining based SQL injection attack detection using internal query trees. *Expert Systems with Applications, 41*(11), 5416–5430. doi:10.1016/j.eswa.2014.02.041

Lee, I., Jeong, S., Yeo, S., & Moon, J. (2012). A novel method for SQL injection attack detection based on removing SQL query attribute values. *Mathematical and Computer Modelling, 55*(1-2), 58–68. doi:10.1016/j.mcm.2011.01.050

Menemencioğlu, O., Sonuç, E., Karaş, İ. R., & Orak, İ. M. (2013). Academic Curriculum Vitae based Faculty Information System. In XV (pp. 1123–1127). Antalya, Turkey: Akademik Bilişim Conference Proceedings. Retrieved from http://ab.org.tr/ab13/kitap/menemencioglu_sonuc_AB13.pdf

Muthuprasanna, M., Ke, W., & Kothari, S. (2006). Eliminating SQL injection attacks - A transparent defense mechanism. *Proceedings of the Eighth IEEE International Symposium on Web Site Evolution, WSE 2006*, (pp. 22–30). http://doi.org/ doi:10.1109/WSE.2006.9

Natarajan, K., & Subramani, S. (2012). Generation of Sql-injection Free Secure Algorithm to Detect and Prevent Sql-Injection Attacks. *Procedia Technology, 4*, 790–796. doi:10.1016/j.protcy.2012.05.129

NIST. (n.d.). *Web Application Vulnerability Scanners*. Retrieved June 11, 2013, from http://samate.nist.gov/index.php/Web_Application_Vulnerability_Scanners.html

Ogheneovo, E., & Asagba, P. (2013). A Parse Tree Model for Analyzing And Detecting SQL Injection Vulnerabilities. *West African Journal of Industrial and Academic Research, 6*(1), 33–49. Retrieved from http://www.ajol.info/index.php/wajiar/article/view/87436

OWASP. (n.d.). *Category:OWASP Top Ten Project - OWASP*. Retrieved April 17, 2015, from https://www.owasp.org/index.php/Top_Ten

Schwartz, E. J., Avgerinos, T., & Brumley, D. (2010). All You Ever Wanted to Know About Dynamic Taint Analysis Forward Symbolic Execution (but might have been afraid to ask) A Few Things You Need to Know About Dynamic Taint Analysis Forward Symbolic Execution (but might have been afraid to ask) The Root. *Policy*, 1–5.

Wassermann, G., & Su, Z. (2007). Sound and precise analysis of web applications for injection vulnerabilities. *ACM SIGPLAN Notices*, *42*(6), 32. doi:10.1145/1273442.1250739

Zhang, H., Kuan Tan, H. B., Zhang, L., Lin, X., Wang, X., Zhang, C., & Mei, H. (2011). Checking enforcement of integrity constraints in database applications based on code patterns. *Journal of Systems and Software*, *84*(12), 2253–2264. doi:10.1016/j.jss.2011.06.044

ADDITIONAL READING

Alpuente, M., Ballis, D., & Romero, D. (2014). A rewriting logic approach to the formal specification and verification of web applications. *Science of Computer Programming*, *81*(246016), 79–107. doi:10.1016/j.scico.2013.07.014

Avancini, A., & Ceccato, M. (2013). Comparison and integration of genetic algorithms and dynamic symbolic execution for security testing of cross-site scripting vulnerabilities. *Information and Software Technology*, *55*(12), 2209–2222. doi:10.1016/j.infsof.2013.08.001

Awoleye, O. M., Ojuloge, B., & Ilori, M. O. (2014). Web application vulnerability assessment and policy direction towards a secure smart government. *Government Information Quarterly*, *31*(S1), S118–S125. doi:10.1016/j.giq.2014.01.012

Chen, C. M., Huang, J. J., & Ou, Y. H. (2015). Efficient suspicious URL filtering based on reputation. *Journal of Information Security and Applications*, *20*, 26–36. doi:10.1016/j.jisa.2014.10.005

Dafydd, S., & Marcus, P. (2011). The Web Application Hacker's Handbook: Finding and Exploiting Security Flaws 2nd Edition (Second Edi). Indianapolis, Indiana: John Wiley & Sons. Inc.

Deepa, G., & Thilagam, P. S. (2016). Securing web applications from injection and logic vulnerabilities: Approaches and challenges. *Information and Software Technology*, *74*, 160–180. doi:10.1016/j.infsof.2016.02.005

Hydara, I., Sultan, A. B. M., Zulzalil, H., & Admodisastro, N. (2015). Current state of research on cross-site scripting (XSS) - A systematic literature review. *Information and Software Technology*, *58*, 170–186. doi:10.1016/j.infsof.2014.07.010

Johns, M. (2014). Script-templates for the content security policy. *Journal of Information Security and Applications*, *19*(3), 209–223. doi:10.1016/j.jisa.2014.03.007

Li, Y.-F., Das, P. K., & Dowe, D. L. (2014). Two decades of Web application testing—A survey of recent advances. *Information Systems*, *43*, 20–54. doi:10.1016/j.is.2014.02.001

Luo, J. Z., & Yu, S. Z. (2013). Position-based automatic reverse engineering of network protocols. *Journal of Network and Computer Applications*, *36*(3), 1070–1077. doi:10.1016/j.jnca.2013.01.013

Maerien, J., Michiels, S., Hughes, D., Huygens, C., & Joosen, W. (2015). SecLooCI: A comprehensive security middleware architecture for shared wireless sensor networks. *Ad Hoc Networks, 25, Part A*, 141–169.

Pinzón, C. I., De Paz, J. F., Herrero, Á., Corchado, E., Bajo, J., & Corchado, J. M. (2013). IdMAS-SQL: Intrusion Detection Based on MAS to Detect and Block SQL injection through data mining. *Information Sciences*, *231*, 15–31. doi:10.1016/j.ins.2011.06.020

Prokhorenko, V., Choo, K. K. R., & Ashman, H. (2016a). Context-oriented web application protection model. *Applied Mathematics and Computation*, *285*, 59–78. doi:10.1016/j.amc.2016.03.026

Prokhorenko, V., Choo, K. K. R., & Ashman, H. (2016b). Web application protection techniques: A taxonomy. *Journal of Network and Computer Applications*, *60*, 95–112. doi:10.1016/j.jnca.2015.11.017

Salas, M. I. P., & Martins, E. (2014). Security testing methodology for vulnerabilities detection of XSS in web services and WS-security. *Electronic Notes in Theoretical Computer Science*, *302*, 133–154. doi:10.1016/j.entcs.2014.01.024

Salas-Zárate, M. D. P., Alor-Hernández, G., Valencia-García, R., Rodríguez-Mazahua, L., Rodríguez-González, A., & López Cuadrado, J. L. (2015). Analyzing best practices on Web development frameworks: The lift approach. *Science of Computer Programming*, *102*, 1–19. doi:10.1016/j.scico.2014.12.004

Shahbaz, M., McMinn, P., & Stevenson, M. (2015). Automatic generation of valid and invalid test data for string validation routines using web searches and regular expressions. *Science of Computer Programming*, *97*(P4), 405–425. doi:10.1016/j.scico.2014.04.008

Shar, L. K., & Tan, H. B. K. (2013). Predicting SQL injection and cross site scripting vulnerabilities through mining input sanitization patterns. *Information and Software Technology*, *55*(10), 1767–1780. doi:10.1016/j.infsof.2013.04.002

Skrupsky, N., Bisht, P., Hinrichs, T. L., Venkatakrishnan, V. N., & Zuck, L. D. (2013). TamperProof: a server-agnostic defense for parameter tampering attacks on web applications. *Codaspy*, 129–140. http://doi.org/10.1145/2435349.2435365

Uzunov, A. V. (2016). A survey of security solutions for distributed publish/subscribe systems. *Computers & Security*, *61*, 94–129. doi:10.1016/j.cose.2016.04.008

Zhu, J., Xie, J., Lipford, H. R., & Chu, B. (2014). Supporting secure programming in web applications through interactive static analysis. *Journal of Advanced Research*, *5*(4), 449–462. doi:10.1016/j.jare.2013.11.006 PMID:25685513

KEY TERMS AND DEFINITIONS

Dynamic-Analysis: The vulnerabilities analyze is processed during execution time by checking web server responses without needing modification of application codes.

Hybrid-Analysis: The vulnerabilities analyze of the program codes and testing the generated SQL queries are processed simultaneously in hybrid analysis.

Injection: In any type of parameters, especially string type of parameters, attackers attempt to insert malicious commands after semicolon such as "*; drop table {tablename};*".

Insecure Direct Object References: An attack type of unauthorized account or object access by attempted some authorized system users.

Missing Function Level Access Control: A privileged function based attack which is attempted by some authorized system users and by changing URL or parameter.

Static-Analysis: The vulnerabilities analyze is processed without execution by checking program codes as text. The analyze requires rewriting application codes.

Un-Validated Redirects and Forwards: Attackers violate the applications by un-validated parameters when passing parameters to redirect any pages.

Vulnerability: Vulnerability that weak points of applications which attract malicious attacks on, lead harmful destructions.

Chapter 2
Cyber Security Centres for Threat Detection and Mitigation

Marthie Grobler
Council for Scientific and Industrial Research, University of Johannesburg, South Africa

Pierre Jacobs
Council for Scientific and Industrial Research, University of Johannesburg, South Africa

Brett van Niekerk
University of KwaZulu-Natal, South Africa

ABSTRACT

With the continuing evolution of cyber threats, it is only a matter of time before an organisation will suffer a major breach or there is an incident of national significance. This necessitates monitoring to detect possible incidents and mechanisms to respond and recover from breaches. This chapter provides an overview of structures to aid in threat detection and incident recovery. Security Operation Centres (SOCs), Computer Security Incident Response Teams (CSIRTs), and Security Intelligence Centres (SICs) will be covered, and the differences, benefits and limitations will be discussed. Guidance for the implementation of these security capabilities within organisations will be provided.

DOI: 10.4018/978-1-5225-1938-6.ch002

INTRODUCTION

Threats within the cyber domain are becoming more prominent. It is therefore imminent that threats need to be mitigated as far as possible. This chapter investigates how monitoring and incident response can be applied to assist cyber security centres to detect and mitigate threats. For the purpose of this publication, cyber security centres include the structures known as Security Operation Centres (SOCs), Computer Security Incident Response Teams (CSIRTs), and Security Intelligence Centres (SICs). This chapter indentifies the need for both monitoring and incident response, and the benefits that these actions bring about for organisations making use of the cyber security centre's services. The chapter further looks at the benefits and challenges faced by cyber security centres in threat mitigation, and provides some guidance in terms of setting up these centres. The aim of the chapter is to assist cyber security centres in improving their threat detection and mitigation capabilities. The chapter investigates the need for and benefits of monitoring and incident response, and looks into the structures required to facilitate monitoring and threat detection, structures to facilitate incident handling, and structures to facilitate advanced threat detection and mitigation. The chapter concludes by providing guidance in terms of cyber security threat detection and mitigation.

THE NEED FOR MONITORING AND INCIDENT RESPONSE

Within the ever changing security environment, it is very important to adhere to constant monitoring. Not only is the security environment continuously changing, but new challenges arise daily in terms of new vulnerabilities and new attack types. Security attacks are now more likely to be targeted, purposeful and organised, posing a much more directed threat for an organisation. As a result of the increased level of connectivity between systems, sensitive data is faced with security, integrity and credibility issues (Gandotra, Singhal & Bedi, 2009).

In the view of the changing threat landscape, it is imperative that more attention be paid to system monitoring (Gandotra, Singhal & Bedi, 2009). In order to manage these aspects, all organisational systems should be monitored to such an extent that any incidents can be managed and appropriately responded to on a timeous basis. Thus, threat mitigation within an organisation needs to be planned and executed in an ordered way in order to enhance opportunities and reduce threats to the sensitive data within the organisation. As part of a holistic cyber security strategy, authorities will deploy security controls in supporting the improvement of their entity's cyber security posture. The SysAdmin, Audit, Network, Security institute (SANS) groups these controls into technical security controls, administrative security controls and physical security controls (Northcutt, 2009), while the National Institute of Standards and Technology (NIST, 2014) uses a grouping of Know, Prevent, Detect, Respond and Recover security controls.

In order to keep concepts simple, SANS taxonomy of technical, administrative and physical security controls is used. A technical security control could be anything from a firewall or an Intrusion Prevention System (IPS), to end-point protection in the form of anti-virus or anti-malware software. These technical controls need to be monitored to ensure that they work as intended, and to detect possible attacks and anomalies as they happen. Monitoring is expressed as a requirement by various authoritative documents (laws, acts, treaties and regulations) and normative documents (standards, frameworks, policies and best practices). Monitoring is typically done from a SOC or a SIC, and the primary technology used is the Security Incident and Event Monitoring (SIEM) tool (Zimmerman, 2014).

The process is simplified to comprise four steps, as shown in Figure 1 (MITRE, 2016).

In the threat mitigation and management process, the threats are firstly identified by means of monitoring events. All identified events are assessed according to probability and consequence. Consequences may include cost, schedule, technical performance and capability, amongst others. If the event or combination of events is elevated to an incident, the threat is identified. Secondly, the impact of threats is assessed. Thirdly, an analysis is conducted of the threat prioritisation. All threat events assessed as medium or high criticality is monitored and, if required, moved to the incident response queue. Low critical threats may be further monitored.

Figure 1. Threat mitigation and management
Adapted from MITRE, 2016.

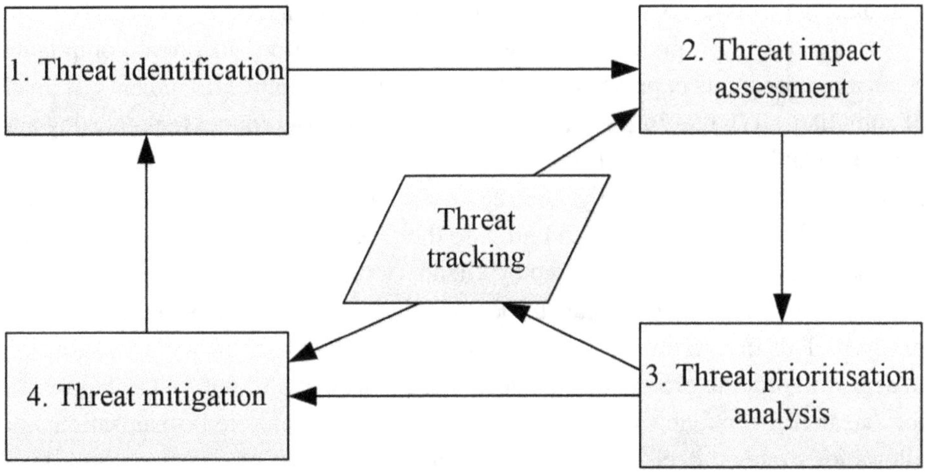

Lastly, the threat is mitigated. During this step, incident response is actioned. Throughout the threat mitigation and management process, monitoring needs to be conducted to ensure that the status of the threats is known to the organisation (MITRE, 2016). As such, the identification of threats, as an output of the monitoring process, is vital to the analysis of potential risks and identification of remedial measures needed to be addressed by an organisation (Gandotra, Singhal & Bedi, 2009).

When determining a strategy for cyber security monitoring of assets, organisations need people, processes and tools/technology to support the effort. Monitoring on its own is of little value unless the information collected is analysed, and actionable intelligence is derived from it. Another aspect to consider, is legal and regulatory requirements which might influence the type of information which can be collected – in some instances, permission is needed from information owners before their information can be collected and monitored. In other instances, the law will prescribe that certain information types (such as financial data) must be monitored. These requirements will differ from country to country, and together with organisational governance, authoritative and normative requirements will form the basis on which the strategy is built.

THE BENEFITS OF MONITORING AND INCIDENT RESPONSE

Having a robust monitoring programme in place will ensure that attacks, anomalies and system abuses are detected and reacted on in a timely fashion. It has to be kept in mind that a monitoring programme does not prevent attacks, but will assist the cyber security centre in identifying attacks and allowing timeous mitigation actions or controls to be deployed. Monitoring should not be seen as a preventative measure, but rather an activity to assist in the early detection of threats.

A monitoring programme also allows for organisations to prove compliance against requirements expressed in authoritative and normative documents. If proof of compliance is a requirement from the monitoring programme, the reporting and storage capabilities of the toolset should be considered, as well as its ability to collect data from different sources. The toolset should also allow for compliance checking against various requirements, and provide the flexibility to customise compliance parameters and reports, as required by clients.

Cyber security monitoring can further assist in the classification and categorisation of assets. This in turn allows organisations to determine suitable technical controls to deploy in order to protect those assets. The events being monitored also provides for a near real-time view of what is happening in the monitored organisation, and allows for strategic decision making and situational awareness. A very high-level monitoring framework is provided in Figure 2. This framework should only be used as a reference, and is not intended to replace any of the available monitoring frameworks – such as "CAESARS Framework Extension: An Enterprise Continuous Monitoring Technical Reference Model", "Continuous Monitoring Reference Model Workflow, Subsystem, and Interface Specifications" and "Applying the Continuous Monitoring Technical Reference Model to the Asset, Configuration and Vulnerability Management Domains" by NIST (2012).

The use of technical controls and their efficacy can be measured through the monitoring programme. In measuring the effectiveness of technical controls, measures of effectiveness (MoEs) and measures of performance (MoPs) should be determined beforehand. This will aid in deciding what types of events to monitor. The high-level framework has a placeholder for requirements expressed by organisational and national authoritative and normative documents, followed by a placeholder to identify the monitoring requirements or functions which should be offered by the monitoring capability. These would be influenced by the requirements expressed in the authoritative and normative documents. The functional requirements

Figure 2. High-level monitoring framework

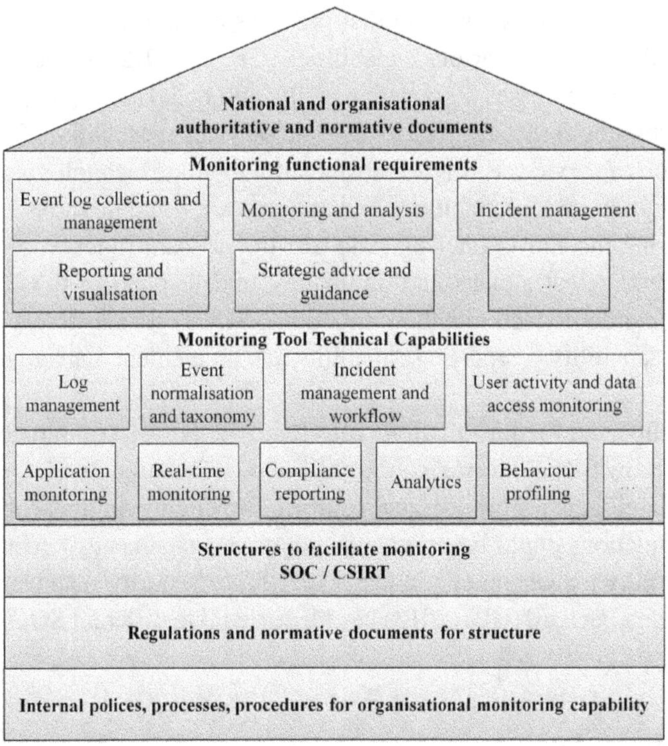

will serve as input into the determination of the monitoring tool's technical requirements. During the determination of the technical requirements, MoEs and MoPs should also be determined. A decision on the structure which will be used to facilitate the monitoring capability needs to be made, followed by the regulations and normative requirements applicable to the structure. These requirements will serve as input into the development of structure specific internal policies, processes and technology specific procedures.

The next sections will look at the respective structures focusing on the facilitation of monitoring and threat detection, incident handing, and advanced threat detection and mitigation.

STRUCTURES TO FACILITATE MONITORING AND THREAT DETECTION: SOCs

SOCs are the most commonly used structure to assist in the facilitation of cyber security monitoring and threat detection. Organisations typically build a SOC to facilitate the monitoring and incident management capability requirement. This section will provide information on SOCs, the approach to building SOCs, and provide a SOC framework.

A SOC can be seen as a system that provides capabilities. A service provides functions and consists of various capabilities (Graves, 2012), while a capability is a combination of processes, knowledge, skills and behaviours (people), tools and systems (technology) (Strategy&, 2012). The services thus realises functions. The determination of services is important since it will guide technology selection and process definition. SOC capabilities and services are listed in Table 1.

Considering the above, one can state that the primary function of a SOC is to monitor assets, detect attacks and anomalies, and respond to those attacks and anomalies. In order to facilitate these functions, SOCs provides threat intelligence, a workflow capability, a reporting capability and an incident management capability.

As a result of all the information collected by a SOC, a compliance capability can be offered by SOCs, leveraging the collected information. A SOC can also provide threat intelligence from external sources. Depending on the scope of a SOC, additional functions could be provided, such as management of security devices and vulnerability management – in fact, everything "security" can be offered from a SOC. Hewlett Packard (HP) (2013) has identified five different SOC generations. These are:

Table 1. SOC capabilities services

SOC Capabilities	SOC Services
Event log collection and management	Counter intelligence
Real-time monitoring and analysis ability	Surveillance
Incident management	Integrated threat intelligence.
Reporting and visualisation	Incident response, incident management
Providing strategic advice and guidance or threat intelligence	Security knowledge management and security pre-warning
Forensic and investigative functionality	Aggregation and analysis of intelligence data
Vulnerability management	Correlation of content intelligence data
Business impact assessment	Analytical intelligence capabilities
Change management	Workflow automation
Network and security device management	24x7 monitoring
Security awareness training	Forensic analysis
	In-house research
	Reporting
	Vulnerability management
	Risk management
	Asset management and criticality rating

Jacobs, 2015.

- First-generation SOCs (1975 – 1995) were ad hoc, understaffed and utilised emerging technologies. Their responsibility was mainly to monitor and manage security technical controls.
- Second-generation SOCs (1996 – 2001) emerged in a period characterised by malware outbreaks. This era saw the beginning of vulnerability tracking and formalised system patching. During this period security monitoring and management services were commercialised, constituting the advent of Managed Security Service Providers (MSSPs). The big focus during this period was on intrusion detection systems. The concept of security information event monitoring (SIEM) was introduced towards the end of this period.
- Third-generation SOCs (2002 – 2006) were marked by the organisation and expansion of cyber crime. Malware changed from worms to targeted attacks. Crisis and incident management were formalised by Computer Incident Response Teams (CIRTs).
- Fourth-generation SOCs (2007 – 2012) could be recognised by the publicity of a politically motivated threat landscape, which initiated the advent of hacktivism. Organisations began to realise that intrusions will happen, re-

gardless of the technical and other preventative technologies that are in place. The focus during this era has shifted from detection and prevention to exfiltration, detection and containment.

- Fifth-generation SOCs or SICs (2013 –) are still evolving. Most security products provide end-point solutions using signatures to detect malware. However, the threat comes from human adversaries. This SOC recognises the changes in the threat landscape and approaches the challenge holistically by providing counter-intelligence, surveillance, criminal psychology and analytical thinking training to augment the deployed technology. These SOCs also realise that security programmes need to be active, engaged and intelligent. This will be discussed below as a Security Intelligence Centre.

AN APPROACH TO BUILDING SOCs

When considering scope and services on offer, it is important to note that services offered exclusively by SOCs are real time monitoring and response, as well as the management of security controls such as firewalls and IPSs. SOC services are delivered using two distinct models:

- **Internal SOCs or SOCs Serving the Internal Security Requirements of an Organisation:** This model serves a single (internal) customer. The deployment could be centralised, or in the case of global organisations, distributed.
- **MSSPs Providing SOC Services-as-a-Service to Organisations:** This model serves different customers form different industries, and even different countries (Rothke, 2012; Gartner, 2013).

Some considerations to keep in mind: MSSP services will typically be cheaper, more objective and dedicated and experienced resources may be available, providing a 24x7 monitoring and incident response capability. SOCs can be situated locally, or be geographically and globally dispersed (KPMG, 2012). There are advantages and disadvantages for each approach, specifically in terms of cost and skills.

SOC Framework

SOCs need to put security measures in place in order to prevent security incidents and foster trust with between itself and its stakeholders. To this effect, a SOC should ensure that it employs the highest level of ICT security. Identifying applicable frameworks and standards to guide and steer the building of SOCs could be a daunting task. It is not sufficient to rely on a single source framework, since this will not cover the full spectrum required by SOC functionality. For the purpose of developing a framework, one has to view a SOC as a business entity on its own, consisting not only of ICT but covering all business domains. These include financial, sales and marketing, legal, logistical and human resource aspects.

It is therefore recommended to apply a number of applicable frameworks and standards to guide and steer the building of SOCs:

- ITIL focuses mainly on operational ICT aspects, but not the rest of the business domains (Ball, 2006).
- CoBIT 5 focuses on Governance of Enterprise IT and lists seven information criteria that should be addressed: confidentiality, integrity, availability, effectiveness, efficiency, reliability and compliance (ISACA, 2012).
- ISO/IEC 27001:2013 and ISO/IEC 27002:2013 is a standard covering information security, and is a robust and internationally accepted standard (ISO/IEC, 2009). Certifying against a standard is a double edged sword, in that clients can be assured that the SOC has ICT security measures and control in place, and negates the need for clients to conduct their own security tests in the case of a MSSP.
- The Business Process Framework (enhanced Telecommunications Map or eTOM) from the Telecommunications Management Forum's New Generation Operations Systems and Software (NGOSS) initiative covers the business aspects lacking in ITIL, CoBIT and ISO/IEC 27001:2013 and ISO/IEC 27002:2013 (Jiejin, 2009).

To address these aspects, a combination of these frameworks and standards should be applied. The eTOM framework, augmented by ITIL, CoBIT and ISO/IEC 27001:2013 and ISO/IEC 27002:2013 allows for a robust, scalable framework covering all aspects of business – not only ICT. These frameworks, standards and best practices are contextualised in Figure 3.

Challenges

There are a number of challenges that a SOC could face:

- There is a lack of consensus on basic definitions and terminology, for example when something can be classified as an event, incident or threat.
- There is confusion on the services and functions between SOCs and CSIRTs, despite several authoritative documents outlining the differences.
- There is a lack of management and financial commitment.

Figure 3. Contextualisation of standards, frameworks, and best practices
Source: Van Niekerk & Jacobs, 2015.

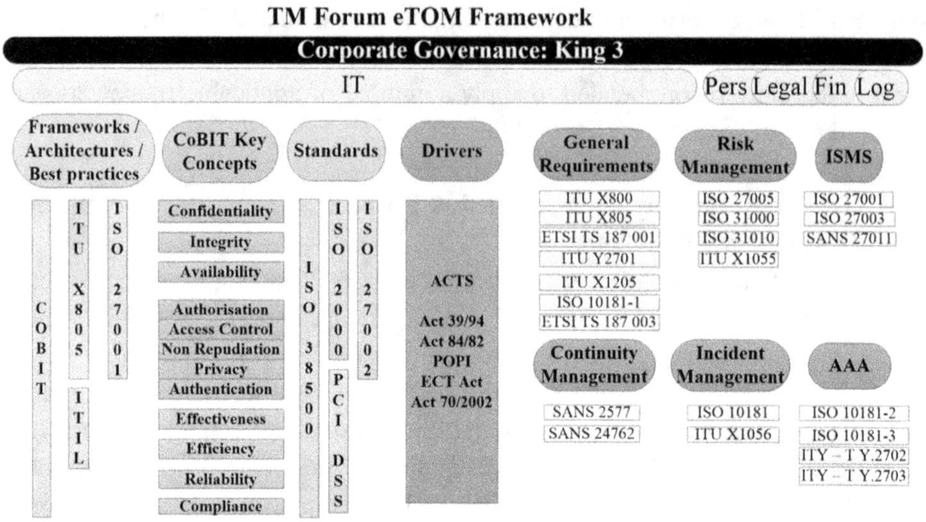

- Facilities are not always adequate and do not match the operational requirements.
- There is a lack of skilled and experienced staff.
- Selection of technology is done based on sales talks and not a scientific selection process.
- Disconnect between SOC and business processes.
- Value proposition not realised due to inadequate reporting.

STRUCTURES TO FACILITATE INCIDENT HANDLING: CSIRTs

CSIRTs are the most commonly used structure to assist in the facilitation of cyber security incident handling. This section will provide information on CSIRTs, as well as the integration between CSIRTs and SOCs.

One of the primary objectives to setting up a CSIRT is deciding what services it will provide. Although there is no pre-defined set of services that a CSIRT should offer, it is recommended that some core services be chosen to begin building a public profile (ENISA, 2006). According to the European Network and Information Security Agency (ENISA) guidelines, there are a number of services and functions that a CSIRT can deliver (ENISA, 2009). It is not common for any CSIRT to provide all the services and functions. Therefore, the selection of the service package is a crucial decision in prioritising which services and functions to provide to the constituents.

Some confusion could exist as to whether a SOC or CSIRT is the applicable structure. Organisations need to carefully assess what their requirements are. The "Handbook for Computer Security Incident Response Teams (CSIRTs)" (West-Brown, Stikvoort, Kossakowski, Killcrece and Ruefle, 2003) lists the typical services offered by a CSIRT. Monitoring of ICT assets, infrastructure and technical security controls is not a service common to a CSIRT. That is not to say a CSIRT cannot perform such a function. In fact, one of the services offered by CSIRTs is intrusion detection services, but there is a caveat that this service is very difficult to offer and is often outsourced. West-Brown, Stikvoort, Kossakowski, Killcrece and Ruefle (2007) categorise the CSIRT services as proactive, reactive and security quality management services. The categories and their services are displayed in Table 2.

Reactive services are triggered by an event or request, such as a report of a compromised host, wide-spreading malicious code, software vulnerability, or something that was identified by an intrusion detection or logging system. Reactive services are the core component of CSIRT work. This class of services aims at handling incidents and mitigating cyber security threats. These services are designed to respond to requests for assistance, reports of incidents from the CSIRT constituency, and any threats or attacks against CSIRT systems (ENISA, 2006).

Proactive services provide assistance and information to help prepare, protect and secure constituent systems in anticipation of attacks, problems, or events. Performance of these services will directly reduce the number of incidents in the future. This class of services aims to reduce the number of cyber security threats and incident through the implementation of preventive measures, such as awareness building and training (ENISA, 2006).

Security quality management services augment existing and well-established services that are independent of incident handling and traditionally performed by other areas of an organisation such as the IT, audit or training departments. If the CSIRT performs or assists with these services, the CSIRT's point of view and expertise can provide insight to help improve the overall security of the organisation and identify risks, threats and system weaknesses. These services are generally proactive but contribute indirectly to reducing the number of incidents. This class of services are designed to improve the overall security through consultancy and educational measures (ENISA, 2006).

Other factors that have an impact on the service selection include the mission statement and the available resources. As such, the recommendation of the selection of CSIRT services is based on two attributes, namely priority and feasibility of the service. Priority is based on the mission statement and the objectives of the CSIRT, whilst the feasibility scale looks at how achievable the services are. The criteria for feasibility primarily look at the resources required and the maturity of the CSIRT.

Table 2. CSIRT services

Reactive Services	Proactive Services	Service Quality Management Services
Alerts and warnings	Announcements	Risk analysis
	Incident Handling	
• Incident analysis • Incident response on-site • Incident response support • Incident response coordination	Technology watch	Business continuity and disaster recovery planning
	Vulnerability Handling	
• Vulnerability analysis • Vulnerability response • Artefact response coordination	Security audit or assessments	Security consulting
	Artefact Handling	
• Artefact analysis • Artefact response • Artefact response coordination	Configuration and maintenance of security tools, applications and infrastructures	Awareness building
	Development of security tools	Education and training
	Intrusion detection services	Product evaluation and certification
	Security related information dissemination	

SOCs offer similar services, but with the core focus of SOCs being on monitoring of ICT assets, infrastructure, and security technical controls.

An Approach to Building CSIRTs

CSIRTs can serve a single organisation or multiple different organisations from different industries. CSIRTs can also be established to provide a service at both a national and public level. ENISA mentions four different ways in which a CSIRT could be structured (ENISA, 2015). These are:

- As an independent organisation with its own staff (equivalent to a MSSP).
- As part of an existing organisation (equivalent to an internal or local SOC).
- Campus, or distributed model (equivalent to a distributed, or global SOC).
- Voluntary model where people or groups with a shared interest get together and provide advice and support to each other.

CSIRT Framework

It is important to understand the basic framework of a CSIRT before implementing any of the recommended services or functions. The framework establishes a few fundamental questions such as "What does the CSIRT do?", "For whom?", "In what setting?" and "In collaboration with whom?". These questions appear trivial, but without a clear understanding of the answers, the CSIRT might struggle to respond correctly to incidents. The aim of the framework is to identify the following:

- Mandate and mission of the CSIRT.
- Constituency of the CSIRT.
- Relation to other CSIRT teams (Carnegie Mellon University SEI - CERT Coordination Centre, 2003).

Figure 4 shows how the CSIRT's mission statement can be used to derive the relevant policies, quality and services of the CSIRT, and how these relate directly amongst each other and the resultant CSIRT procedures.

Figure 4. CSIRT quality and service framework
Source: Carnegie Mellon University SEI - CERT Coordination Centre, 2003.

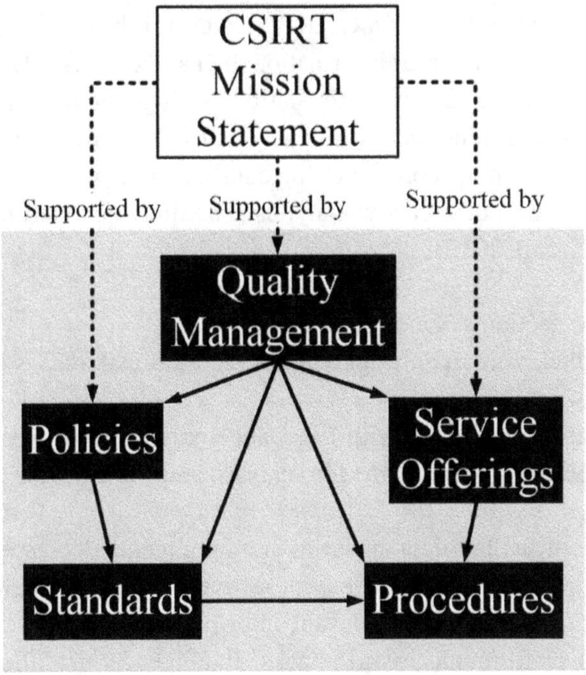

Challenges

There are a number of challenges that a CSIRT could face:

- There is a lack of consensus on basic definitions and terminology, for example when can something be classified as an event, incident or threat.
- There is confusion on the services and functions between SOCs and CSIRTs, despite several authoritative documents outlining the differences.
- There is a lack of management and financial commitment.
- Facilities are not always adequate and do not match the operational requirements.
- Disconnect between CSIRT and business processes.
- CSIRT services selection not aligned with business requirements.
- Difficult to become integrated into the international CSIRT community and building the resulting international network of trust.

STRUCTURES TO FACILITATE ADVANCED THREAT DETECTION AND MITIGATION: SICs

SICs are the most commonly used structure to facilitate advanced threat detection and mitigation. A SIC exhibits the monitoring capability of a SOC and some of the incident response capabilities of a CSIRT. However, it has a deeper ability for the analysis of external and internal information than a traditional SOC. This often results in SICs being termed as a next-generation, Fifth-generation or intelligence-led SOC, or even a cyber defence centre (BAE Systems, 2015; RSA, 2013; RSA, 2014). SICs are often based on the concept of big data analytics, which can be considered as the third-generation of security analytics. The three generations are (CSA Big Data Working Group, 2013):

- Intrusion detection systems (IDS).
- SIEMs, which were mainly aggregators (a traditional SOC is built around this technology).
- Second-generation SIEM with Big Data Analytics, which can provide long-term correlations, reduce time for analysis, and work with more diverse data.

The purpose of the big data analytics and intelligence-led aspects are to security operations from a reactive mode to proactively detect and identify threats and compromises quicker, and feed relevant information to response teams to more effectively recover from successful attacks. Intelligence also allows for security devices to recognise new threats attempting attacks and therefore improves the prevention of breaches.

Levels of Intelligence

The intelligence consumed and produced by a SIC can be categorised into four levels depending on the audience and application to a time frame. These are illustrated in Table 3.

Strategic and operational intelligence tends to be more in human readable format and released by vendors or government sources as whitepapers, advisories and alerts. Technical intelligence provides indicators of compromise which can be used by security devices through automated feeds, and therefore would be associated more with subscription-based cyber threat intelligence feeds. Tactical intelligence may be a mixture, and includes threat tactics, techniques and procedures.

Sources of Data and Intelligence

Internal sources of data for a SIC are similar to those that a traditional SOC would aggregate and use for monitoring, and may include (IBM, 2013; Marty, 2009):

- Network and security devices and applications.
- Servers and endpoint machine system logs.
- Application logs.
- Packet capture and inspection.
- Users and application behaviour.

Table 3. Levels of intelligence

	Strategic	Operational	Tactical	Technical
Audience	• Board • Executives • Government	• Defenders • Senior security management • Analysts	• Analysts • Senior security management • Architects • System admins	• Analysts • Incident response • Security devices
Time-frame	Long-term	Short-term	Long-term	Immediate
Scope	General	• Industry sector • Community	Organisation	Organisation
Focus	• Political • Social • Economic • Behavioural	• Adversary campaigns • Specific incoming attacks	Attacker tactics, techniques and procedures	Indicators of compromise

Adapted from Chismon and Ruks (2015) and INSA Cyber Intelligence Task Force (2014).

- Configuration, vulnerability and compliance data.

External sources of data can be considered as an aspect of the intelligence that provides security intelligence with the proactive detection and identification capability. These sources may include (Chismon & Ruks, 2015; INSA Cyber Intelligence Task Force, 2015):

- Commercial (subscription of free) cyber threat intelligence feeds.
- Community shared threat and vulnerability information.
- Vendor supplied threat and vulnerability information.
- Social media and paste sites (e.g. PasteBin).
- Alerts from government intelligence and national CSIRTs.

Often the feeds of signatures and indicators of compromise are confused with intelligence. This is just one aspect, which enables the production of intelligence. The feeds by themselves are just data. By conducting the analysis with the internal data intelligence that is relevant to the organisation is derived, an actionable in that the response teams are able to recover from any incidents (Dalziel, 2015; Symantec, 2015).

Data Analysis and Application

Two broad sets of analysis and detection can be done within a SIC: correlation and anomaly detection. Correlation focuses on logs, events, flows, IP reputation, indicators of compromise and signatures to detect and identify threats. This is mostly effective against known threats due to the signatures and indicators of compromise. Creating baselines against normal behaviour will allow for anomalies to be detected, which could be an indicator of both known and previously unknown threats as well as insider threats (IBM, 2013). It is important to plan for a situation where previously unknown or emerging attacks could be targeting the organisation (Symantec, 2015), as these may require more in-depth analysis than simply correlating against known signatures. Retrospective analysis can be performed on archived data to investigate any previous events that could indicate prior compromise. The retrospective analysis is particularly important when building a timeline of a possible prolonged campaign from more recent security events (Seward, 2016; Li & Clark, 2015).

As with a SOC, the data and analysis results need to be presented to the relevant stakeholders. Visualisation and tailoring the reports to the various audiences is important. Ensure that the reports and visualisations are meaningful to the audience. Marty (2009) and Jacobs and Rudis (2014) provide good overviews and explanations of analysis and visualisation for information security. When designing reports, it is important not to forget the executives as an audience (Symantec, 2015), as the intelligence may aid in strategic risk-based decisions, and they provide the budget for projects and operations such as a security intelligence centre.

In order to proactively detect, identify, and respond to threats, analysts in a SIC can embark on threat hunting. This is particularly useful for finding stealthy advanced persistent threats that evade traditional security devices and monitoring (ITWeb, 2015; Sqrrl, 2016). Threat hunting, like traditional intelligence analysis, is based on hypotheses, which are then used to direct searches for anomalous behaviour or patterns (ITWeb, 2015; Sqrrl, 2016).

The analysis of the data sources is key in a SIC. The multiple internal and external sources can provide greater context and insight into what a traditional SOC would only be able to monitor. The intelligence provides for improved threat detections, more capable responses, and more accurate risk-based decision making. It also allows for proactive threat hunting, thereby mitigating the impact breaches could have on the organisation.

An Approach to Building SICs

A SIC can be built from scratch, or by 'upgrading' a traditional SOC. As such, this section will focus on the development of a cyber threat intelligence capability, which will integrate into the SOC.

Chismon and Ruks (2015) and Symantec (2015) provide a number of 'quick wins' to start implementing threat intelligence within an organisation. These include:

- Identify informal threat intelligence attempts within the organisation and provide support and direction to allow them to grow and align with the business. Help integrate them with the business requirements and formalise planning to maximise the impact.
- Determine how senior management perceive cyber threats. Create a list of competitors, nations, and other groups who may benefit from attacking your organisation, or have animosity towards it. Conduct open source investigations to assess the relevance of any threats that they are concerned about.
- Establish relationships with operations, finance, legal, and governance, risk and compliance. With them, assess the success factors and risks to the business.
- Interact with the immediate community and industry sector to identify threats that may not have been identified.
- Investigate the public perception of the organisation for the period immediately preceding any significant attack attempts.

- Develop a contact list for anyone who needs to be notified of an impending attack warning.
- Subscribe to email and RSS alerts of vendor-produced reports and white-papers related to threats, particularly for emerging threats. Extract the relevant information and introduce this into any architecture or system implementations or upgrades. Where relevant, update any security devices with relevant signatures.
- Subscribe to free feeds with indicators of compromise or signatures that can be used to update security devices.
- Participate in forums and with trusted partners and clients in your community to develop a joint view of threats. Use this to initiate informal information sharing and warnings.

To identify intelligence providers, the scope of the offering needs to be assessed. Many providers only offer a limited set of indicators, which may not be relevant to the organisation. An ideal offering would include not only the indicators, but strategic and operational contexts, malware analysis, and related indicators in the event of an advanced persistent threat. Ideally, there should be the capability to customise the alerts and feeds to your needs, and prioritise them related to the relevance to your organisation (SearchSecurity, 2016).

Challenges

There are a number of challenges that a SIC could face, including:

- SICs are expensive to set up, consequently there are limited implementations.
- Large quantities of data, which traditional SIEM solutions are unable to correlate in near-real time (CSA Big Data Working Group, 2013; Sensage, 2014; Seward, 2016).
- Intelligence sources are not relevant to the specific environment (Dalziel, 2015; Symantec, 2015).
- Relevant information needs to be filtered out from the noise to provide actionable intelligence for a variety of audiences, from executives to incident responders (Gallego, 2015; Seward, 2016).
- There is a lack of established or defined processes for intelligence.
- Executives are not considered as a key stakeholder or intelligence consumer.
- There is a reliance purely on raw threat data from feeds.
- There is a focus on common attacks and not planning for emerging threats (Symantec, 2015).

GUIDANCE IN TERMS OF CYBER SECURITY THREAT DETECTION AND MITIGATION

This chapter looks at structures required to facilitate monitoring and threat detection, structures to facilitate incident handling, and structures to facilitate advanced threat detection and mitigation. This section will provide guidance in terms of the evolution of cyber security threat detection and mitigation, by integrating the best of SOCs, CSIRTs and SICs.

Reflecting on the CSIRT services and the services offered by SOCs, one can see that there is a great overlap in terms of services offered. When comparing SOC and CSIRT services with one another, the service overlap is clearly visible, refer to Table 4.

Table 4. Comparison of CSIRT and SOC services

CSIRT Unique	Common	SOC Unique
Security audits or assessments against standards	Configuration and maintenance of tools, applications and infrastructure	Insider threat case support and investigation
-	Incident handling	Reporting and trending
	Vulnerability handling	Border protection device observations and measurements (O&M)
	Artefact handling	SOC infrastructure O&M
	Forensics handling	Sensor tuning and maintenance
	Announcements	Custom signature creation
	Security related information dissemination	Real-time monitoring
	Business continuity and disaster recovery planning for constituents	Vulnerability scanning and assessment
	Development of security tools	Network mapping
	Intrusion detection services	Penetration testing
	Technology watch	Situational awareness
	Risk analysis	Media relations
	Alerts and warnings	-
	Security consulting	
	Awareness building	
	Education and training	
	Product evaluation and certification	

ENISA, 2015; Jacobs, 2015; SEI, 2015.

Building on this information, the next sections will provide guidance and possible integration of these respective structures to develop a cyber security centre framework.

Cyber Security Centre Framework

Most organisations would already have identified frameworks, standards and best practices to guide and steer their operations. It would however be impractical to identify a standard framework or best practice for each and every aspect and function of these cyber security centres. Paradia (2012) identified process standards, frameworks and methodologies to be used when building service operation centres. Although not cyber security centre specific, one could argue that these are also applicable, since they will provide a security service in essence. Table 5 lists the frameworks and standards as identified by Paradia (2012).

People, Processes, and Tools Framework Specific to Cyber Security Centres

There are a variety of frameworks which could be considered when building these cyber security centres. Over and above the fact that it comprises people, processes and tools, the capability as a whole needs to be managed, governance requirements must be catered for, and quality and business process management and project management need to be applied.

In the planning, building and running of cyber security centres, it is helpful to view it as a system and the SIEM as a "system of systems". In doing so, systems engineering principles could be used in the building of the cyber security centre. Following this approach allows for the determination of authoritative and organisational normative and governance requirements. The organisational business requirements and business drivers also need to be clarified, since these would serve as input when the cyber security centre functional specifications are being developed. Cyber security centre services can either be offered as an in-house service, or outsourced to MSSPs. This is a business decision that needs careful consideration.

When building a cyber security centre, one has to keep in mind that there are external requirements that needs to be satisfied (authoritative and normative), as well as internal requirements. The internal requirements are centre specific, and include aspects such as internal security, marketing to the organisation, and sales if the centre serves as a MSSP. Furthermore, there are operational considerations, such as incident management and monitoring processes, as well as centre specific policies processes and technology specific procedures. As an example, SOCs could choose to conform to ISO/IEC 27001:2005 to satisfy security requirements, and ITIL

Table 5. Cyber security centre processes, frameworks, and methodologies

Category/Discipline	Type	Framework/Standard/Methodology
Quality management and business process management	Frameworks that focus on quality standards, applied to specific IT domains	TQM (Total Quality Management)
		ISO 9000
		ISO/IEC 2000
		TOGAF
		ISO/IEC 27001
Quality improvement	Frameworks that focus on assessment and improvement of processes, performance or other, not focussing on how-to aspects of operating the IT	CMMI (Capability Maturity Model Integration)
		Six Sigma
		IT Balanced Scorecard
IT governance	Frameworks that focus on how to organise the IT function in terms of responsibilities, control and organisation	ISO/IEC 38500
		COBIT
		M_O_R (Management of Risk)
Information management	Frameworks that focus on how to perform and organise certain aspects of information management such as procurement, service management and requirements	ITIL
		eTOM
Project management	Frameworks that focus on project, programme, and portfolio management, not specifically IT	MSP (Managing Successful Programmes)
		Prince2
		PMBoK (Project Management Body of Knowledge)
		IPMA Competence Baseline
Incident management	Frameworks that focuses on the management of events, incidents and vulnerabilities, specifically IT	ISO/IEC 27035
		ITIL
		COBIT
		ISO/IEC 27001

Adapted from Paradia, 2012.

to satisfy operational requirements. From a MSSP perspective, it might be a good idea to seek certification against a standard which will improve their marketability, and also provide assurance to clients – negating the need for clients to conduct their individual security tests on the SOC. Once the high-level requirements are determined, the cyber security centre requirements analysis is done which in turn leads to a functional specification with MoEs, and a technical specification for the SIEM tool with its MoPs coupled to it. The high-level process to build a cyber security centre is shown in Figure 5.

Some guidance in terms of the people, processes and tools framework:

- **People:**
  - The organisation has to ensure that it has the correct skills and certifications.
  - Training modules specific for the services and functions identified should be considered, e.g. TERENA TRANSITS I and II.
  - Continuous growth and training of people with a clear growth path should be in place, e.g. an analyst need to know what is required to advance to a Level 1 Engineer.
  - Analysts will have to be able to recognise attacks and anomalies from various different sources, e.g. VPN, active directory and database logs. For the building of an analyst capability, a framework such as the Skills Framework for the Information Age (SFIA) can be used.
  - Analysts must have a very clear understanding of threat detection and mitigation processes, but also of business processes affected by the threat detection and mitigation processes, e.g. threats identified against

Figure 5. Cyber security centre high-level development framework

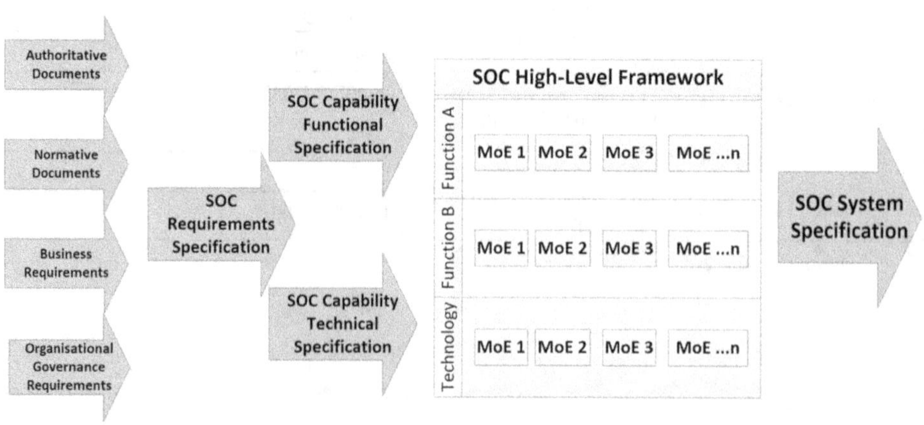

a financial system will have a direct business impact on the finance department. Enterprise architecture principles are helpful in this regard.
    - Some analysts require an understanding of big data concepts and data science in order to analyse a diverse set of internal and external data. This will provide analysis over larger data sets and time periods.
- **Processes:**
    - Services and functions need to be identified and defined using an internationally accepted best practice, such as ENISA. Processes need to be identified and assigned to fulfil the identified services.
    - Threat detection and mitigation processes must be clearly defined within the context of the existing business processes. Enterprise architecture can be helpful and a unified modelling language such as Business Process Modelling and Notation (BPMN) could be used.
    - Processes must be regularly tested and a maturity model should be developed or an existing maturity model should be used to measure the effectiveness of these processes.
    - Key performance indicators must be assigned to the processes to measure effectiveness.
- **Tools:**
    - Tools and technology should support the business requirements, functions and services of the cyber security centre.
    - Technology should be selected using a formal process, such as a systems engineering process, where user requirements, functional requirements and technical requirements are determined with the respective measures of effectiveness and measures of performance.
    - Assets which need to be monitored should be assigned a criticality rating.
    - The identified technology licensing model should be fit for the intended use, e.g. some vendors licence on events per second, whilst other vendors license on type of and number of events monitored. This will build on the technical requirements identified through the system engineering process.
    - Existing organisation legal and regulatory requirements must be considered carefully since these can have an impact on the centre, e.g. data retention requirements will have an impact on storage capability.

Each cyber security centre will have to make a decision in terms of the priority of the components of the people, processes and tools framework. Of the three components, one component needs to act as the constraint (i.e. this is fixed and cannot be changed), one component needs to be enhanced (i.e. this can be optimised in support of the constraint), and one component needs to be accepted as is (i.e. this can be traded off to support the constraint). For example, if the cyber security centre's technology component acts as the constraint (i.e. specific tools and technologies are already in place and thus needs to be incorporated into the centre as a means to save costs), the people component can be enhanced by training employees in the specific tools and technologies that are in place. The processes component will have to be accepted as is, since the tools, technologies and training dictate the processes that need to be implemented. Table 6 presents this example in the priority matrix.

Integrated Model

SOCs, CSIRTs and SICs can play a pivotal role in the monitoring of, and response to threats, attacks and vulnerabilities in organisations, including governments. While the focus of a SOC is on the monitoring of technical security controls and critical assets, and the response to attacks and threats, CSIRTs' main focus is on response and incident management. A SIC is an evolved SOC and as such incorporates big data into the existing SOC model. For the purposes of this model, SIC will be regarded as a type of SOC. One postulation is that a CSIRT is a highly specialised sub-capability of a SOC, whereas another postulation could be that a SOC serves as an input mechanism into CSIRTs and CERTs. This leads to an integrated model for the establishment of an initial cyber security centre.

When considering the process aspect of these cyber security centres, it is a good idea to leverage off the existing organisational processes, and map the cyber security centre processes to the existing organisational processes where possible. This avoids duplication and saves money and reduces effort in doing so. A good example is to map the cyber security centre incident management process to the organisational incident management process (if the organisation has one). The organisational incident management process would in all likelihood have a tool or technology supporting the process which the cyber security centre can leverage. In the integrated model an approach to service integration is proposed on the services applicable to SOCs, CSIRTs and SICs. The model is generic to enable the application to all service

Table 6. Example priority matrix

	People	Processes	Tools
Constrain (Fixed)			x
Enhance (Optimise)	x		
Accept (Trade-off)		x	

instances. The three criteria which must be met for an organisation to be classified as a cyber security centre are (Shirey, 2014; Zimmerman, 2014):

- A mechanism for clients or constituents to report cybersecurity incidents.
- A mechanism to provide incident handling assistance to constituents.
- A mechanism to disseminate incident related information to constituents and other third parties.

By integrating the respective services and functions from the SOC, CSIRT and SIC, all three criteria can be met in the instance of a cyber security centre. Linking to the people, processes and tools framework, the respective services can be extended to ensure that all service elements are catered for. The people, process and technology framework underpins the ITIL service delivery domain (Leopoldi, 2005).

As an operating model, the plan-build-run model is used. Using this next-generation operating model (KPMG, 2014) allows for the breaking down of traditional technology silos and rather focus on ICT functions, thus allowing for growth (Agarwal, Bommadevara & Weinberg, 2013). It is the authors' opinion that following such an approach will provide the best fit for the cyber security centre. The people, process and technology framework will ensure that all service requirements are met, while operation of the integrated model is governed by the plan-build-run model. The cyber security centre infrastructure needs to be secured, and all governance requirements of the organisation need to be met. These are normally in the form of legal requirements, regulations and others normative documents. The complete high-level model is shown in Figure 6.

CONCLUSION

The chapter provided an overview of three main cyber security centre types: SOCs, CSIRTs and SICs, and the key characteristics, benefits, and capabilities of each. The chapter outlined the characteristics, an approach to build, a proposed framework and possible challenges for each of the identified cyber security centre types.

The chapter further provides guidance in terms of cyber security threat detection and mitigation, and proposes a holistic cyber security centre framework in which the characteristics of a SOC, CSIRT and SIC can be combined. Despite the potential of such centres, there is sometimes resistance to establishing them due to a number of possible challenges. For example, strong governance and management is required to ensure the successful implementation and operation of cyber security centres. The chapter adopts the people, processes and tools framework to ensure all

Figure 6. Cyber security centre services integration high-level model
Source: Jacobs, Von Solms & Grobler, 2016.

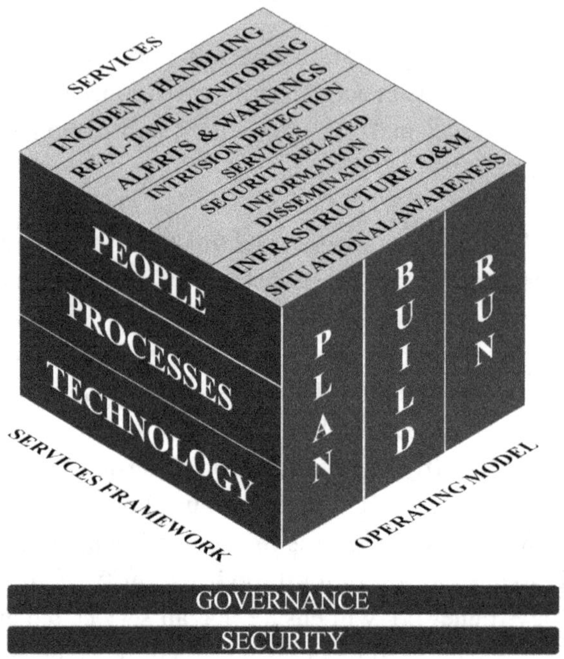

aspects of the cyber security centre requirements are covered within the proposed integrated model.

Cyber security centres provide services that promote the efficient detection of threats or incidents, and effective response to any confirmed incidents or breaches. In particular, the monitoring and real-time correlation of large data sets that mature cyber security centres conduct is instrumental in combatting the rapid evolution of cyber threats and the growing prevalence of advanced persistent threats.

REFERENCES

Agarwal, H., Bommadevara, N., & Weinberg, A. (2013). *Using a Plan-Build-Run Organisational Model to Drive IT Infrastructure Objectives.* Retrieved from http://www.mckinsey.com/~/media/mckinsey/dotcom/client_service/BTO/ PDF/ MOBT32_28-36_Plan-Build-Run_R5.ash

BAE Systems. (2015). *Intelligence-Led Threat Mitigation.* Retrieved from www.baesystems.com/ai

Ball, E. (2006). *Getting to Know ITIL*. Retrieved from http://www.theiia.org/intAuditor/itaudit/archives/ 2006/june/getting-to-know-itil/

Carnegie Mellon University SEI - CERT Coordination Centre. (2003). *Handbook for Computer Security Incident Response Teams (CSIRTs)*. Retrieved from http://www.sei.cmu.edu/library/abstracts/reports/ 03hb002.cfm

Chismon, D., & Ruks, M. (2015). *Threat Intelligence: Collecting, Analysing, Evaluating, MWR InfoSecurity, Centre for the Protection of National Infrastructure, and CERT-UK*. Retrieved from http://CPNI.gov.uk

CSA Big Data Working Group. (2013, September). *Big Data Analytics for Security Intelligence, Cloud Security Alliance*. Retrieved from http://www.cloudsecurityalliance.org/research/big-data

Dalziel, H. (2015). *How to Define and Build an Effective Cyber Threat Intelligence Capability*. Waltham, MA: Syngress.

Dickerson, C., & Mavris, D. N. (2010). *Architecture and Principles of Systems Engineering*. CRC Press.

ENISA. (2006). *A Step-By-Step Approach On How To Set Up a CSIRT*. Deliverable WP2006/5.4 (CERT-D1/D2).

ENISA. (2015). *CSIRT Structure*. Retrieved from https://www.enisa.europa.eu/activities/cert/support/ guide2/internal-management/structure

Gallego, R. (2015). *Security Intelligence Is Useful Only If Isolated From The Noise, Computer Weekly*. Retrieved from http://www.computerweekly.com/news/4500255079/Security-Think-Tank-Security-intelligence-useful-only-if-isolated-from-the-noise

Gandotra, V., Singhal, A., & Bedi, P. (2009). Threat Mitigation, Monitoring and Management Plan - A New Approach in Risk Management. *2009 International Conference on Advances in Recent Technologies in Communication and Computing*. Retrieved from http://ieeexplore.ieee.org/stamp/stamp.jsp?tp=&arnumber=5328897

Gartner. (2013). *Managed Security Service Provider (MSSP)*. Retrieved from http://www.gartner.com/it-glossary/mssp-managed-security-service-provider

Graves, T. (2012). *Service, Function and Capability*. Retrieved from http://weblog.tetradian.com/2012/09/ 22/service-function-capability-again/

Hewlett-Packard. (2013). *5G/SOC: SOC Generations*. Retrieved from http://www8.hp.com/us/en/software-solutions/software.html?compURI=1343719#!

IBM. (2013). *Protecting Your Critical Data With Integrated Security Intelligence, Software Thought Leadership White Paper*. Retrieved from http://www-01.ibm.com/common/ssi/cgi-bin/ssialias? htmlfid=WGW03023USEN

INSA Cyber Intelligence Task Force. (2014). *Operational Cyber Intelligence*. Arlington, VA: Intelligence and National Security Alliance. Retrieved from www.insaonline.org

INSA Cyber Intelligence Task Force. (2015). *Tactical Cyber Intelligence*. Arlington, VA: Intelligence and National Security Alliance. Retrieved from www.insaonline.org

ISACA. (2012). *COBIT 5 – A Business Framework for the Governance and Management of Enterprise IT*. Retrieved from http://www.isaca.org/cobit/pages/cobit-5-framework-product-page.aspx

ISO/IEC. (2009). *An Introduction to ISO 27001, ISO 27002.... ISO 27008*. Retrieved from http://www.27000.org/

ITWeb. (2015). *Vodacom talks cyber hunting*. Retrieved from http://www.itweb.co.za/index.php?option=com_content&view=article&id=143114:Vodacom-talks-cyberhunting&catid=234

Jacobs, J., & Rudis, B. (2014). *Data Driven Security*. Indianapolis, IN: Wiley.

Jacobs, P. (2015). *Towards a Framework For Building Security Operation Centres*. Rhodes University. Retrieved from http://contentpro.seals.ac.za/iii/cpro/DigitalItemViewPage.external?lang=eng&sp=1017932&sp=T&suite=def

Jacobs, P., Von Solms, S. H., & Grobler, M. M. (2016). E-CMIRC – Towards a Model For The Integration of Services Between SOCs and CSIRTs. *Proceedings of the 15th European Conference on Cyber Warfare and Security (ECCWS) 2016*.

Jiejin, H. (2009). *A Practical Approach to the Operation of Telecommunication Services Driven by the TMF eTOM Framework* (Master's Thesis). Universitat Poliècnica de Catalunya.

KPMG. (2012). *Security Operations Centre (SOC) Globalization*. Retrieved from http://www.kpmg.com/SG/en/IssuesAndInsights/ArticlesPublications/Documents/Advisory-CS-Security-Operations-Center-SOC-Globalization.pdf

KPMG. (2014). *Next Generation IT Operating Models*. Retrieved from http://www.kpmg.com/BE/ en/IssuesAndInsights/ArticlesPublications/Documents/Next-Generation-IT-Delivery-Models.pdf

Leopoldi, R. (2005). *The Role of ITIL in IT Governance - ITSM*. Retrieved from www.itsm.info/ITSM ITIL and IT Governance.pdf

Li, Q., & Clark, G. (2015). *Security Intelligence: A Practitioner's Guide to Solving Enterprise Security Challenges*. Indianapolis, IN: Wiley. doi:10.1002/9781119183617

MITRE. (2016). *Risk Mitigation Planning, Implementation, and Progress Monitoring, Systems Engineering Guide*. Retrieved from https://www.mitre.org/publications/systems-engineering-guide/ acquisition-systems-engineering/risk-management/risk-mitigation-planning-implementation-and-progress-monitoring

NIST. (2012). *NIST Publishes Draft Implementation Guidance for Continuously Monitoring an Organization's IT System Security*. Retrieved from http://www.nist.gov/itl/csd/monitoring-012412.cfm

Northcutt, S. (2009). *Security Controls*. Retrieved from http://www.sans.edu/research/security-laboratory/article/security-controls

Paradia, S. (2012). *Enterprise Service Operation Center*. Retrieved from http://www.slideshare.net/sameer paradia/it-enterprise-service-operation-center

Rothke, B. (2012). *Building a Security Operations Center (SOC)*. Retrieved from http://www.rsaconference.com/writable/presentations/file_upload/tech-203.pdf

RSA. (2013). *RSA Advanced Cyber Defense Program for Nextgen SOC, RSA Technical Brief, EMC²*. Retrieved from https://www.emc.com/collateral/data-sheet/rsa-advanced-cyber defense-program-nextgen-soc-ds.pdf

RSA. (2014, June). *Building an Intelligence-Driven Security Operations Centre, RSA Technical Brief, EMC²*. Retrieved from http://www.rsa.com

SearchSecurity. (2016). *Four Features to Look for in a Threat Intelligence Service, TechTarget*. Retrieved from http://searchsecurity.techtarget.com/ThreatIntelligence/Four-Features-to-Look-for-in-a-Threat-Intelligence-Service

Sensage. (2014). *A Practical Guide to Next-Generation SIEM*. Retrieved from http://www.sensage.com

Seward, M. (2016). *Addressing the New Threat Intelligence Problems of Scale and Relevance, Anomali and Information Week Dark Reading*. Retrieved from http://event.on24.com/r.htm?e=1137629&s=1&k=90E87A5A7DA758E4275E0D35A90B558C

Shirey, R. (2014). *Internet Security Glossary, Version 2*. IETF RFC 4949. Retrieved https://tools.ietf.org/html/rfc4949

Sqrrl. (2016). *A Framework for Cyber Threat Hunting*. Retrieved from http://www.sqrrl.com

Strategy&. (2012). *What is a Capability?* Retrieved from http://www.strategyand.pwc.com/global/home/ what-we-think/multimedia/video/mm-video_display/what-is-a-capability

Symantec. (2015). *Five Threat Intelligence Traps to Avoid, UBM Tech*. Retrieved from http://www.informationweek.com/whitepaper/risk-management-security/cybersecurity/five-threat-intelligence-traps-to-avoid/364973

Van Niekerk, B., & Jacobs, P. (2015). Towards a Secure Data Centre Model. *ISACA Journal, 3*. Retrieved from http://www.isaca.org/Journal/archives/2015/Volume-3/Pages/toward-a-secure-data-center-model.aspx

West-Brown, M., Stikvoort, D., Kossakowski, K. P., Killcrece, G., & Ruefle, R. M. (2003). Handbook for Computer Security Incident Response Teams (CSIRTs) (2nd ed.). CMU/SEI. Retrieved from http://resources.sei.cmu.edu/asset_files/Handbook/2003_002_001_14102.pdf

Zimmerman, C. (2014). *Ten Strategies of a World-Class Cybersecurity Operations Centre*. The Mitre Corporation. Retrieved from https://www.mitre.org/sites/default/files/publications/pr-13-1028-mitre-10-strategies-cyber ops-center.pdf

KEY TERMS AND DEFINITIONS

Computer Security Incident Response Team (CSIRT): Facility or structure that enables the facilitation of cyber security incident handling.

Cyber Intelligence: Actionable information regarding threats, the associated detection methods and countermeasures, and the relevance to the environment. This is generated from the analysis of a variety of internal and external sources.

Cyber Security: The protection of enterprise and/or national information infrastructure and services from malicious actors on the Internet.

Cyber Security Incident: Any event that is suspicious or malicious that affected, or has the potential to affect, the well-being of information, networks or systems.

Framework: Conceptual structure or skeleton used as a reference when planning or building a capability. It ensures that all aspects needed for the capability are considered.

Security Intelligence Centre (SIC): A next-generation SOC which has enhanced analysis capabilities to provide intelligence and effective response to threats and incidents.

Security Operations Centre (SOC): Facility or structure that enables the detection of attacks, threats and anomalies and respond to those. It most often uses a tool SIEM, and can be delivered in-house or as a service.

Threat: An actor or situation that has the potential or intent to cause harm.

Vulnerability: A weakness that a threat can exploit in order to cause harm.

Chapter 3

Methods and Tools of Big Data Analysis for Terroristic Behavior Study and Threat Identification:

Illegal Armed Groups during the Conflict in Donbas Region (East Ukraine) in Period 2014-2015

Yuriy V. Kostyuchenko
National Academy of Sciences of Ukraine, Ukraine

Maxim Yuschenko
National Academy of Sciences of Ukraine, Ukraine

ABSTRACT

Paper aimed to consider of approaches to big data (social network content) utilization for understanding of social behavior in the conflict zones, and analysis of dynamics of illegal armed groups. Analysis directed to identify of underage militants. The probabilistic and stochastic methods of analysis and classification of number, composition and dynamics of illegal armed groups in active conflict areas are proposed. Data of armed conflict – antiterrorist operation in Donbas (Eastern Ukraine in the period 2014-2015) is used for analysis. The numerical distribution of age, gender composition, origin, social status and nationality of child militants among illegal armed groups has been calculated. Conclusions on the applicability of described method in criminological practice, as well as about the possibilities of interpretation of obtaining results in the context of study of terrorism are proposed.

DOI: 10.4018/978-1-5225-1938-6.ch003

INTRODUCTION

In criminological practice and research of terrorism, there are many cases requiring application of sophisticated scientific instruments. Not only at the stage of analysis of evidences, but also at the preliminary stages, in particular, at the stage of crime identification.

Description of criminal activity and identification of a crime is a challenge in some cases, for example, in the areas of crisis, conflict, and fighting. This is due to significant limitations of existing information and data available.

In such cases it is necessary to use many different sources of information, including social networks, with adapted statistical approaches to assessment of this data. Correct statistical methods of data collection, analysis, filtering and regularization in such cases are critical.

Obtained with the mathematical methods, robust spatial-temporal distributions of data could be used to define the event and to identify a crime.

Social networks reflect the motivations of the behaviour of different groups of society and varied social environments (Wasserman, Galaskiewicz, 1994). So, because of the large scope, it is a good base for sociometrics and behavior analysis (Krause, Croft, James, 2007). However, formal numerical methods of social network data analysis are still not sufficiently developed for a wide range of important cases, in particular for crisis management (Lerbinger, 2012) and conflict analysis (Scott, 2012).

The task of extraction of structured distributions of data regularized by determining parameters from the large sets of non-structured data is solving. The algorithm based on specifics of dig data distribution, and on data source characteristics (e.g. group behavior). Realization of this approach aimed to detection of stable indicators of criminal and/or terroristic activity. At the same time the analyzed data and cyber activity, producing it might be legal in most cases (if propaganda is beyond our scope).

The proposed algorithm allows to analyze the content of social networks on the base of the set of selected indicators. These indicators allow to control the social dynamics of different social groups represented in social networks and analyze their behavior, including identification of evidences of terrorism.

The analysis of social dynamics and the behavior of social groups is becoming an important factor in the control of terrorist activities. Not only and not so much as for the identification of the criminals, but the identification of the crime, the identification of its social base, to determine the causes and driving forces of terrorism is a task to be solved by approaches proposed in this study.

In this paper, using the data on the conflict in some regions of Lugansk and Donetsk regions (Eastern Ukraine) shows the algorithm and method of data collection, filtering and regularization in tasks of child involvement in illegal military groups.

In other words, it shows how to use statistical algorithms in tasks of criminology and terrorism study for decision-making under deep uncertainty.

Antiterrorist operation - the armed conflict in the Donbas (Eastern Ukraine – see Figure 1) extends in spatially distributed urban agglomerations, e.g. with widely available and accessible information infrastructure and communications. The conflict zone is an area about 16,800 square kilometers with varying number of people: from 2.1 to 2.6 million up to 3.1 in maximum, depending on the intensity of fighting, economic situation and the season. In this area is available 990,620 telephone lines, 1,786,300 Internet lines (including optical fiber), there are 1,196,820 Internet users, 3,705,750 mobile phones serviced by 5 providers.

While traditional media functioning under external pressure and censorship, and often using as an "information weapon", social media, first of all, social networks have become an important source of information about the development of conflict. In this case, the question of the correct methods of collecting, processing and analyzing data from social networks is important in the context of analysis of the conflict.

Figure 1. Conflict localization map

METHODOLOGY

In the conflicts and crises the accurate, timely and adequate information acquiring is a daunting challenge. Usually, parties of the conflict use different tools of censorship, all public information is distorted, and access to the crisis area for observers is usually limited. In such situation, social networks became a good source of operational information.

Psychological features and characteristics of group behavior of social network users have determine the distribution of data and information flows (Lerman, 2013).

Usually, when we consider the conflicts of values, and also in many cases of racial, religious and ethnic conflicts, there are stable sets of motivations of group behavior, presented in the form of symbols. It serves as a justification for violence and are widely used in propaganda. These symbols are the value-markers of certain social groups. In social networks, these symbols are distributed in the form of a hashtag, and are convenient search criteria for gathering information (Sultana, Paul, & Gavrilova, 2015).

In each case, the information contained in social networks is incomplete, with high noise, distributed by interconnected clusters. So, we should apply robust statistical algorithms for collection, filtering and classification of information using a set of interrelated criteria. Furthermore, the classified information should be regularized to obtain spatial-temporal distributions with controlled reliability. As result we obtain a coherent description of a situation in which all data will be statistically related to each other and mutually verified. Only in this case our data will be meaningful and has a sense as evidence, rather than fragmentary information from unknown and unconfirmed sources (Caverlee, Cheng, Sui, et al, 2013).

The first stage is the search and collection of data by identifying hashtags on the area of conflict, location, group membership, age/date of birth.

Data Classification Approach

The accumulated massive of data should be classified. In other worlds, a decision making procedure should be applied to each entry of general community - the multidimensional dataset to extract a class of "infant participant of illegal armed groups". We propose to use for it a data classification approach based on the Bayes rule for minimum classification error in terms of maximum-a-posterior decision task in Markov random field model representation of multi-temporal, multi-source data (Duda, Hart, & Stork, 2012).

Let I_j $\{x_1, x_2,..., x_N\}$ be a given find record describing a community of illegal armed groups combatants, modeled as a set of N identically distributed n-variate random vectors of identifiers. We assume M classes $\omega_1, \omega_2,..., \omega_M$ to be present among a set of combatants and we denote the resulting set of classes by $\Omega=\{\omega_1, \omega_2,..., \omega_M\}$ and the class label of the k-th community member ($k=1, 2,..., N$) by $s_k \in \Omega$. By operating in the context of community classification, we assume that verified events, recognized as a training set to be available, and we denote the index set of the training community member by $T \subset \{1,2,...N\}$ and the corresponding true class label of the k-th training community member ($k \in T$) by s_k*.

When collecting all the feature vectors of the N member of community of combatants in a single ($N \cdot n$)-dimensional column vector $X= col[x_1, x2, ..., x_N]$ and all the community member labels in a discrete random vector $S = (s_1, s_2, ..., s_N) \in \Omega^N$, the Bayes rule approach (Duda, Hart, Stork, 2001) assigns to the community of combatants data X the label vector $\tilde{S}$, which maximizes the joint posterior probability $P(S \mid X)$:

$$\tilde{S} = \arg\max_{S\in\Omega^N} P(S \mid X) = \arg\max_{S\in\Omega^N}\left[p(X \mid S)P(S)\right] \tag{1}$$

where $p(X \mid S)$ and $P(S)$ are the joint probability density function (PDF) of the global feature vector X conditioned to the label vector S and the joint probability mass function (PMF) of the label vector itself, respectively. The Markov random field (MRF) approach offers a computationally tractable solution to this maximization problem by passing from a global model for the statistical dependence of the class labels to a model of the local general community properties, defined according to a given neighborhood system (Geman, S., Geman, D., 1993). Specifically, for each k-th member of general community, a neighborhood $N_k \subset \{1,2,...N\}$ is assumed to be defined, such that, for instance, N_k includes the four (first-order neighborhood) or the eight (second-order neighborhood) member surrounding the k-th member ($k=1, 2,..., N$). More formally, a neighborhood system is a collection $\{N_k\}_{k=1}^{N}$ of subsets of member such that each member is outside its neighborhood (i.e., $k \notin N_k \forall k = 1,2,...N$) and neighboring member are always mutually neighbors (i.e., $k \in N_h$ if and only if $h \in N_k \forall k$, $h = 1,2,...N, k \neq h$). This simple discrete topological structure attached to the general community data is exploited in the Markov random field framework to model the statistical relationships between the class labels of spatially distinct member and to provide a computationally affordable solution to the classification problem of (1). Specifically, we assume the feature vectors $x_1, x_2,..., x_N$ to be conditionally independent and identically distributed with probability density function $p(x \mid s)$ ($x \in \Re^n, s \in \Omega$), that is (Geman, S., Geman, D., 1993):

$$p(X \mid S) = \prod_{k=1}^{N} p(x_k \mid s_k) \quad (2)$$

and the joint prior probability mass function $P(S)$ to be a Markov random field with respect to the abovementioned neighborhood system. The probability distribution of each k-th general community label, conditioned to all the other general community labels, is equivalent to the distribution of the k-th label conditioned only to the labels of the neighboring members (*k=1, 2,..., N*):

$$P\{s_k = \omega_i \mid s_h : h \neq k\} = P\{s_k = \omega_i \mid s_k : h \in N_k\}, i = 1, 2, \ldots N \quad (3)$$

The probability mass function of *S* is a strictly positive function on Ω^N, so, $P(S) > 0 \forall S \in \Omega^N$.

The Markov assumption expressed by (3) allows restricting the statistical relationships among the general community labels to the local relationships inside the predefined neighborhood, thus greatly simplifying the contextual model for the label distribution as compared to a generic global model for the joint probability mass function of all the general community labels.

Basing on described approach, the Ho–Kashyap method (Ho, Kashyap, 1965) has been applied to classification of the general community.

As the result of classification we obtain a dataset with all records that meet the specified condition. For example, distribution of members of community "infant participant of illegal armed groups" with age, sex, social status, accessory, spatial and temporal marks inside the general community of members of illegal armed groups' combatants.

Data Regularization Algorithm

Further, we need regularized spatial-temporal distribution of classified records. It is necessary to avoid duplication and fix falsification, inter-verify data, trace robust trends in the distributions, so get a basis for interpretation of the data. We propose to use a two-stage procedure of data regularization.

The method proposed is based on non-linear kernel-based principal component algorithm (KPCA) modified according to specific of data. Using this method the set of selected records has been analyzed. The robust technique of data regularization for normalization of data reliability is proposed. The technique utilizes data from different sources, different nature, and with different metrics. This approach allows to calculate regularized distributions in units invariant toward data properties and quality. We can analyze simultaneously different types of data, regardless spatial and temporal scales and heterogeneities using this approach.

Correct statistical analysis requires the set of data x_i with controlled reliability (above mentioned "training set"), which reflects distribution of investigated parameters during whole observation period (taking into account variances of reliability of data x_t). Set of data x_t ($x_t \in R^m$) consists of multi-source data, including data with sufficient reliability x_j ($x_j \in R^m$), where $j = 1, \ldots N$. Problem of determination of controlled quality and reliability spatial-temporal distribution of investigated parameters x_i might be solved in framework of tasks of multivariate random processes analysis and multidimensional processes regularization (Kostyuchenko, Movchan, Kopachevsky, et al, 2015).

Required regularization may be provided by 2-stage procedure. If we able to formulate stable hypothesis on distribution of reliability of data in the framework of defined problem we may to propose relatively simple way to determine investigated parameters distributions $x_t^{(x,y)}$ towards distributions on measured sites x^m_t basing on (Kostyuchenko & Movchan, 2015):

$$x_t^{(x,y)} = \sum_{m=1}^{n} w_{x,y}(\tilde{x}_t^m) x_t^m \tag{4}$$

where weighting coefficients $w_{x,y}(\tilde{x}_t^m)$ determined as:

$$\min \left\{ \sum_{m=1}^{n} \sum_{x_t^m \in R^m} w_{x,y}\left(\tilde{x}_t^m\right) \left(1 - \frac{x_t^m}{\tilde{x}_t^m}\right)^2 \right\} \tag{5}$$

according to (Kostyuchenko & Movchan, 2015). Here m – number of records; n – number of sources/series; x^m_t – distribution of data; R^m – set (aggregate collection) of data; $\tilde{x}_t^m$ - mean distribution of searching parameters.

This is the simple way to obtain a regular spatial distribution of analyzed parameters, on which we can apply further analysis, in particular temporal regularization. At the same time, this is the first stage of regularization. This algorithm may be interpreted as the general form of Kolmogorov regularization procedure (Ermoliev, Makowski, & Marti, 2012).

Further, second stage of regularization should take into account both data distribution temporal non-linearity (caused by imperfection of available multi-source statistics) and features of temporal-spatial heterogeneity of data distribution caused by systemic complexity of studied community. According to (Kostyuchenko, &

Movchan, 2015) the kernel based non-linear approaches are quite effective for analysis of such types of distributions.

Proposed method is based on modified kernel principal component analysis (KPCA). In the framework of this approach the algorithm of non-linear regularization might be described as following rule (Kostyuchenko, Movchan, Kopachevsky, et al, 2015):

$$x_i = \sum_{i=1}^{N} \alpha_i^k \tilde{k}_t(x_i, x_t) \tag{6}$$

In equation (6) the coefficients α selected according to optimal balance of relative validation function and covariance matrix, for example as (Kostyuchenko, 2015):

$$C^F v = \frac{1}{N}\sum_{j=1}^{N} \Phi\left(x_j\right)\Phi\left(x_j\right)^T \cdot \sum_{i=1}^{N} \alpha_i \Phi\left(x_i\right) \tag{7}$$

where non-linear mapping function of input data distribution Φ determined as:

$$x_i = \sum_{i=1}^{N} \alpha_i^k \tilde{k}_t(x_i, x_t) \tag{8}$$

$$\sum_{k=1}^{N} \Phi(x_k) = 0 \tag{9}$$

and $\tilde{k}_t$ - is mean values of kernel-matrix

$$\boldsymbol{K} \in R^N \left(\left[\boldsymbol{K}\right]_{ij} = \left[k(\boldsymbol{x}_i, x_j)\right]\right).$$

Vector components of matrix determined as:

$$\boldsymbol{k}_t \in R^N ; [\boldsymbol{k}_i]_j = \left[k_t(\boldsymbol{x}_t, \boldsymbol{x}_j)\right].$$

Matrix calculated according to modified rule:

$$k_t(x_i, x_t) = \left\langle \rho_{j,t}^{x_j} \left(1 - \rho_{j,i}\right)^{x_j} \right\rangle,$$

where ρ – empirical parameters, selected according to the classification model of study community.

Using described algorithm, it is possible to obtain regularized spatial-temporal distribution of investigating parameters over the whole observation period with rectified reliability and controlled uncertainty (Kostyuchenko, 2015, 2016). As the result we obtain a regular spatial-temporal distribution of child soldiers registered, prepared for the interpretation.

DATA AND RESULTS

By the requested hashtag over 21,500 profiles in social networks (VK, Facebook, Instagram, Twitter) in five languages (Russian, Ukrainian, English, Serbian and German) has been analyzed. Following hashtags were used: #донбасс(Donbas), #новороссия(NovoRossia), #юныегероиновороссии(NovoRossiaYoungHeroes) #героиновороссии(NovorossiaHeroes) #память(Memory). More than 480 000 entries had satisfied to determined criteria. After filtering by age (12-18 years) / date of birth (after 1998) and of belonging to an illegal military group, has been collected about 5400 entries.

A classification procedure (1)-(3) has been applied to these entries. After it data has been regularized by a two-stage procedure (4)-(9).

The used method allowed to obtain an overview of distribution of key characteristics of illegal armed groups in the conflict zone. There were identified from 40 to 60 separated armed groups during the observation period. All of them are significantly varied by weapons, equipment, ideology (Soviet patriots, Russian imperialists, neo-Stalinists, anarchists, neo-Nazi, Christian orthodox extremists), and united only anti-Western ideas and the pro-Russian rhetoric.

They acted quite independently. The opinion of the nominal leaders of the region determined their actions no more that they are dependent on the supply of arms, ammunition, fuel and food from Russia via the central leadership. However, to the end of the observation period, the number of autonomous groups has decreased, and their centralization is increased.

At the beginning of 2016 a number of armed groups was about 30. During the active phase of conflict in April 2014 – December 2015 the total number of illegal armed group members ranged from 19.5 to 52 thousand, and to the end of 2015 it amounted to 35.7 thousand. Noticeable is heterogeneity and uncertainty of this population: more than 60% of the militants are citizens of Russia and other countries, about 15% are Russian military personnel.

For a detailed study of the general population, it is necessary to use additional tools based on stochastic methods of assessment (Ermoliev, Makowski, & Marti, 2012). This makes possible to analyze distributions of socioeconomic parameters of members of illegal armed groups.

In this case, we limited ourselves to a narrow concrete case of child soldiers. Calculated distributions of the study parameters are demonstrating following results.

Data were obtained based on the information resources of local communities, their supporters groups, homepages of militants and their families in social networks, and media publications. This allows us to estimate the number of children, their distribution and dynamics, age and gender characteristics, and in most cases, to identify their origin and identity, including determination of residency and schools, to restore the history of their participation in armed conflict.

First case of involvement of children into illegal armed group was detected in town Slovyansk in end of April 2014. Underage personnel were recruited into pro-Russian troop headed by Igor Stryelkov (a.k.a. Girkin), who captured the town. Then, during the fighting in Slovyansk in June 10-30, 2014, the first 2 child soldiers were killed and a few kids were injured. The estimated (by calculation using procedures (1)-(3), (6)) total number of people under the age of 18, who was involved in the conflict as military personnel on the side of illegal groups, during the events in Slovyansk is about 35 - 40.

Later, during 2014 the number of children in the illegal military groups gradually increased. Almost every group had underage participants, which in many cases were involved into the fighting.

The total number of underage soldiers in terrorist groups during the conflict in Donbas in the period April 2014 - January 2015 could be estimated at 210-250 persons. Their age ranges from 13-18, mostly 16-17 years, but registered even cases of 12-year-old kids (Figure 2).

Most (90%) underage members of illegal armed groups in 2014 and in 2015 are boys with an insignificant tendency to increase their part (Figure 3).

In 2015 the number of child soldiers began to decline. The reason for this is likely to organizing the chaotic terrorist groups and reducing its support from Russian Federation. However, local children continue to cooperate with militants using them as a source of income during the economic degradation of the territory. During 2014-2015 the direct subsidy from Russia of conflict region was about 88% of their total budget, while local economic activity (including illegal) accounted for only 12%. At the same time in 2015 losses of underage soldiers are increasing.

Question of involvement of children in the illegal armed groups is also important. In 2014, especially in the period April-October, the majority (80%) of underage

Figure 2. The age distribution of underage members of illegal armed groups in 2014 and 2015

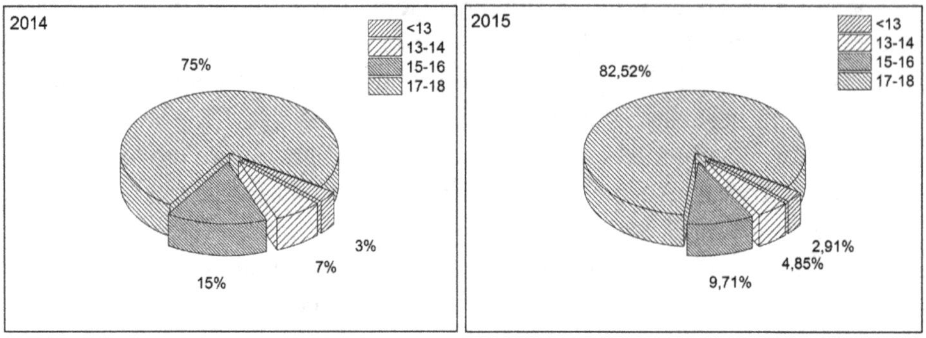

Figure 3. The gender distribution of underage members of illegal armed groups in 2014 and 2015

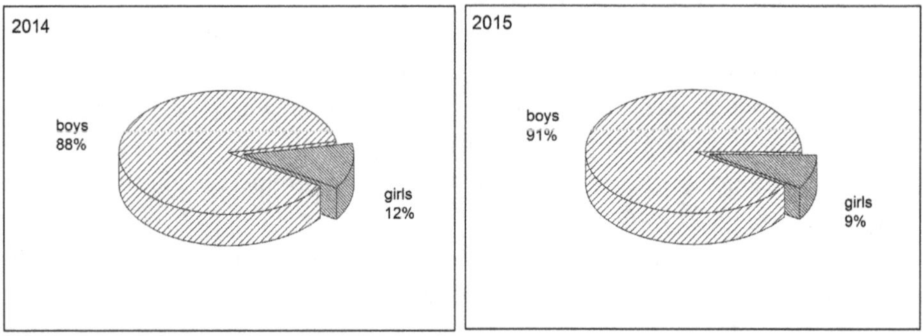

soldiers were kids of members of these groups (Figure 4). Their parents (mainly fathers) were also detected among the members of armed groups. About 10% were children who have no parental care (registered by the regional social service or with permanent residency in other localities), and 10% - children from local families who accidentally joined armed groups (kids with local addresses and registered in local schools). Ways of recruitment of children are usually the military-patriotic and sport clubs where children were indoctrinated by specific pro-Russian and pro-Soviet ideas. Soviet symbols of WWII were widely used in propaganda among child.

In 2015 the situation is changing: the proportion of kids of members of armed groups is reduced to 25%, and the proportion of children without parental care increased to 50%. The number of children from ordinary families also increased to 25%. This is due to changes both in the structure of local communities and in the economy of the territory.

Figure 4. Distribution of social status and origin of underage members of illegal armed groups in 2014 and 2015

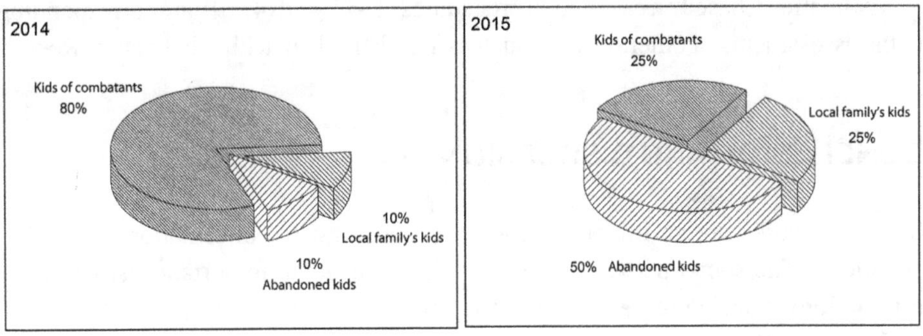

In 2015 at the background of the observed influx of mercenaries from Russia, the number of local residents among illegal armed groups as well as local child soldiers is decreased (Figure 5).

With a significant reduction in intensity of fighting, over 2015 children are mainly used by militants for intelligence activity, housekeeping and security functions.

The total number of underage soldiers at the end of 2015 in the Russian-supported illegal armed groups could be estimated at 70-90 persons. This number is not constant, children who are local residents, are not a part of the group permanently. The most stable groups of children - up to 50% of them - are Russian citizens.

Thus, as the data demonstrate, for the period 2014-2015 during the armed conflict in the Donbas (Eastern Ukraine) on the side of the Russian-supported illegal armed groups about 300 underage soldiers (under 18 years) has been involved.

Figure 5. Distribution of citizenship of underage members of illegal armed groups in 2014 and 2015

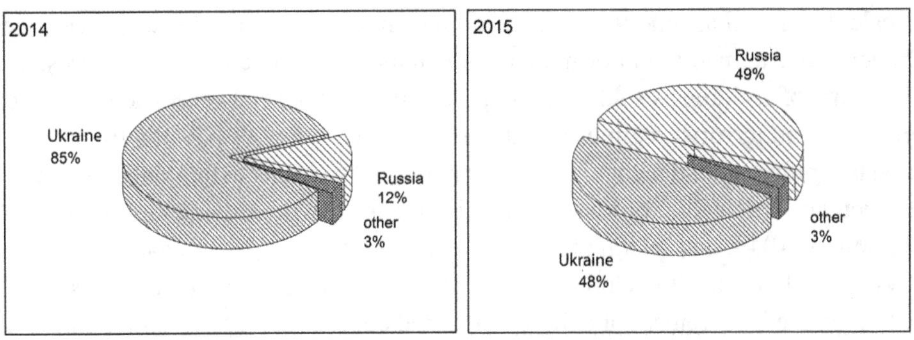

About 150-180 child soldiers directly participated in the combats. Up to 50 minors were killed during the fighting.

Estimations, based on calculation procedures (1)-(3) and (4)-(9), are corresponding to the assessments of other sources, such as The 2016 Trafficking in Persons Report.

CONCLUSION AND DISCUSSION

Social networks are an important and valuable source of information in conflict situations. The social network data analysis could be an important instrument in criminology, conflict research and terrorism study.

However, the specific of big data requires application of correct techniques of data collection processing and analysis. In this context, it is important not only to correctly collect and filter the data, but also to apply the regularization procedure, which will provide regular spatially-temporally distributed data, mutually verified, with controlled reliability.

For data filtering and classification has been proposed an approach based on the Bayes rule for minimum classification error in terms of maximum-a-posterior decision task in Markov random field model representation of multi-temporal, multi-source data. As the result of classification we obtain a dataset with all records that meet the specified condition. For collected, filtered and classified data regularization it has been proposed non-linear kernel-based principal component algorithm (KPCA) based method, modified according to specific of data. Using this algorithm was obtained regularized spatial-temporal distributions of investigating parameters over the whole observation period with rectified reliability and controlled uncertainty, prepared for the interpretation.

Proposed in this paper methodological framework requires some a priori information about the studied distributions. Consequently, methods based on this methodology require regional adaptations.

Thus, application of the proposed method resulting to distribution, which cannot be obtained by another way, at least during the active conflict. We can recognize a calculated data as unique, reliable, and meaningful and has a sense as evidence, rather than fragmentary information from unknown and unconfirmed sources.

Using of children as soldiers and military personnel is a widespread practice of criminal and terrorist regimes worldwide. It is clearly condemned by the international community and recognized as a war crime. According to international law (in particular, Article 77.2 of the Additional Protocol I to the Geneva Conventions of 12 August 1949, and relating to the Protection of Victims of International Armed Conflicts, 1977, and the Optional Protocol to the Convention on the Rights of the Child on the involvement of children in armed conflict, 2002), the participation of children under 15 years in army is prohibited, as well as the recruitment of persons under 18 years to paramilitary and guerrilla nongovernmental groups with any goals. Actually, child soldiers are separate, independent indicator of terrorism. According to international statistics, more than 300,000 children and teenagers were used in the last 10 years as a soldier, a partisan military personnel to support the armed forces in the world. The most famous historical cases of mass use of minors in armed conflict are a children's group "The Liberation Tigers of Tamil Eelam" in Sri Lanka, the young militants of Hamas and "Islamic Jihad" in the Palestinian Authority, "Lord's Resistance Army" in Uganda, left-wing guerrilla groups in Colombia and other.

Unfortunately, in the framework of the conflict in Donbas, the practice of using children as soldiers and military personnel also widely used – both by Russian military troops and Russian supported illegal armed groups mixed from Russian mercenaries and local collaborators. Mass systematic use of persons under 18 as soldiers and military personnel during the conflict proves that we are dealing with a classic terrorist organization, controlled, managed and supported by Russia.

ACKNOWLEDGMENT

The authors are deeply grateful to the anonymous referees for constructive suggestions that resulted in important improvements to the paper, to colleagues from the American Statistical Association (ASA) for their critical and constructive comments and suggestions.

REFERENCES

Caverlee, J., Cheng, Z., Sui, D. Z., & Kamath, K. Y. (2013). Towards Geo-Social Intelligence: Mining, Analyzing, and Leveraging Geospatial Footprints in Social Media. *IEEE Data Eng. Bull.*, *36*(3), 33–41.

Duda, R. O., Hart, P. E., & Stork, D. G. (2012). *Pattern classification*. John Wiley & Sons.

Ermoliev, Y., Makowski, M., & Marti, K. (2012). *Managing Safety of Heterogeneous Systems*. Heidelberg, Germany: Springer. doi:10.1007/978-3-642-22884-1

Geman, S., & Geman, D. (1993). Stochastic relaxation, Gibbs distributions and the Bayesian restoration of images. *Journal of Applied Statistics*, *20*(5-6), 25–62. doi:10.1080/02664769300000058

Ho, Y. C., & Kashyap, R. L. (1965). An algorithm for linear inequalities and its applications. *IEEE Transactions on Electronic Computers*, *5*(5), 683–688. doi:10.1109/PGEC.1965.264207

Kostyuchenko, Yu. V. (2015). Geostatistics and remote sensing for extremes forecasting and disaster risk multiscale analysis. In S. Kadry & A. El Hami (Eds.), Numerical Methods for Reliability and Safety Assessment: Multiscale and Multiphysics Systems. Springer International Publishing. doi:10.1007/978-3-319-07167-1_16

Kostyuchenko, Yu. V. (2016). Risk Perception Based Approach to Analysis of Social Vulnerability. In Risk Perception: Theories and Approaches. Nova Publ.

Kostyuchenko, Yu. V., & Movchan, D. (2015). Quantitative parameter of risk perception: can we measure a geoethic and socio-economic component in disaster vulnerability? In Geoethics: the Role and Responsibility of Geoscientists. Geological Society. doi:10.1144/SP419.10

Kostyuchenko, Yu. V., Movchan, D., Kopachevsky, I., & Bilous, Y. (2015). Robust Algorithm of Multi-Source Data Analysis for Evaluation of Social Vulnerability in Risk Assessment Tasks. In *Proc. of SAI IntelSys*. doi:10.1109/IntelliSys.2015.7361256

Krause, J., Croft, D. P., & James, R. (2007). Social network theory in the behavioural sciences: Potential applications. *Behavioral Ecology and Sociobiology*, *62*(1), 15–27. doi:10.1007/s00265-007-0445-8

Lerbinger, O. (2012). *The crisis manager*. Routledge.

Lerman, K. (2013). Social Informatics: Using Big Data to Understand Social Behavior. In Handbook of Human Computation (pp. 751-759). Springer New York.

Scott, J. (2012). Social network analysis. *Sage (Atlanta, Ga.)*.

Sultana, M., Paul, P. P., & Gavrilova, M. (2015). Social Behavioral Biometrics: An Emerging Trend. *International Journal of Pattern Recognition and Artificial Intelligence*, *29*(08), 1556013. doi:10.1142/S0218001415560133

The 2016 Trafficking in Persons Report. (2016). U.S. Department of State Publications, Office of the Under Secretary of Civilian Security, Democracy and Human Rights.

Wasserman, S., & Galaskiewicz, J. (Eds.). (1994). *Advances in social network analysis: Research in the social and behavioral sciences* (Vol. 171). Sage Publications.

Section 2

The Meaning of Security and Cyber Risk

Chapter 4
Analysis of Cyber-Attacks against the Transportation Sector

Brett van Niekerk
University of KwaZulu-Natal, South Africa

ABSTRACT

For many countries the physical transport infrastructure is critical to the economy, with ports forming a gateway for the majority of trade, and rail and road used to distribute goods. Airlines are crucial to the tourism industry. Whilst the focus of cyber-defense is on financial networks and the power grid, recent incidents illustrate that the transport infrastructure is also susceptible to cyber-attacks. The chapter provides an overview of cyber-security incidents related to the transportation sector, and analyses the reports of the incidents to illustrate the prevalence of threat types and impact. The chapter then discusses some efforts to mitigate the threats in terms of regulations, threat intelligence and information sharing, and awareness training.

INTRODUCTION

Cyber-defense has focused on the financial and electric power grids due to the potentially catastrophic outcomes should a major cyber-attack on those infrastructures cause widespread damage. More recently, more focus has been applied to the cyber-security of the physical transport sector. Whilst this infrastructure may not be considered as critical as the power or financial systems, international trade and

DOI: 10.4018/978-1-5225-1938-6.ch004

employees travelling to work depend on the transportation. Military deployments are also dependent on the infrastructure, although it may not be interconnected with the civilian versions. Cyber-attacks on this sector therefore could have severe implications for national economies and international trade.

Hughes, quoted in Stelter (2015), indicates that the threats faced by a nation were different when transport infrastructure was built. Similar to other critical infrastructure, many of the IT systems for the transport sector were designed assuming that the Internet and those online could be trusted, therefore there is a lack of inherent cyber-security. Combining this vulnerability with the fact that the World Economic Forum (2015) listed cyber-attacks and critical infrastructure failure in the top risks, it is evident that there is a need to investigate cyber-attacks targeting the transportation sector.

This study will follow a similar methodology to that of Miller and Rowe (2012). In their study, they classified attacks against industrial control systems (ICS) according to impact and attack vector. For this study, the incidents related to the transportation sector will be assessed according to the threat type (attacker) and the impact. The source of the incident descriptions are from primarily from online news reports and from existing analysis where it is available. A thematic (qualitative) analysis of the reports is used to identify the threat types and impact for each incident. For the purposes of this chapter, the threats types include:

- Individual hackers,
- Disgruntled insiders (and a category for a combination of hackers and insiders),
- State-sponsored attacks, cyber-criminals, and
- Malware.

Researchers are also considered even though they are not necessarily a threat, however they do expose system vulnerabilities. The impact categorizations include:

- Disruption of operations,
- Data loss,
- Financial theft,
- Illegitimate control of networks or systems,
- Unauthorized access to information or systems, and
- Proof of concept of vulnerabilities and/or attack methods that researchers develop.

The incidents under consideration are described in the next section. Thereafter the analysis provides the tabulated number of occurrences for each threat type and impact per sector, as well as the impacts for each threat type. A timeline of the number of incidents is also provided. The chapter will then go on to describe attempts being made to mitigate cyber-threats against the sector.

CYBER-ATTACKS ON THE TRANSPORTATION SECTOR

There are multiple reasons for conducting cyber-attacks against the transportation sector. Due to the reliance of trade on the sector, an attack could be used to affect trade in general, or even target a specific commodity. Due to the interdependence of the various transport infrastructures, there are a variety of targets to impact on the trade: railways or roads could be targeted to prevent goods reaching the ports, and disrupting the ports themselves would hinder any import or export. Airports can be targeted to affect tourism. Similarly, disrupting operations could delay military deployments. Cyber-attacks could potentially be seen as the modern version of a naval blockade, however possible at a fraction of the cost. Cyber-criminals would gain from fraud due to the large financial transfers involved.

The Symantec Internet Security Threat Reports provide an overview of the transportation sector (combined with other sectors), as shown in Table 1. It is further reported that just under 52% of email to the sector is spam, and there were 2.7 spear phishing attacks per organization with nearly 11% of their customers being targeted (Symantec, 2016a). This is the highest rate of targeted customers across all sectors. As the transportation sector is combined with other sectors in the reports, these will not be considered in the analysis.

The *Verizon Data Breach Investigations Report* (Verizon, 2016b) indicated 31 incidents related to the transportation sector: 35% were due to web applications, 26% denial-of-service (DoS), 16% cyber-espionage, and 10% crime-related malware. The Industrial Control System Computer Emergency Response Team (ICS-CERT)

Table 1. Overview of incidents and threats recorded by Symantec

Year	2013	2014	2015
No. of Incidents	6	9	6
No. of Identities Exposed	1,460,340	~825,039*	156,959
Avg. Identities per Incident	243,390	91,671	~26,160*
Rate of Malicious Code in Emails	1 in 480.5	1 in 289.0	1 in 338
Rate of Phishing Emails	1 in 2,226.1	1 in 2,172.9	1 in 2,948

Sources: Symantec (2014; 2015; 2016a; 2016b).

* These figures were not provided in the reports and were calculated.

in the US recorded 23 incidents in the transportation sector for the 2015 fiscal year, and conducted nine onsite assessments. The number of assessment was lower, as ten were conducted per year for the 2012 to 2014 fiscal years (ICS-CERT, 2016).

The remainder of this section will describe various incidents and research activities related to the transportation sector. The section is divided into five main aspects of transportation: maritime, rail, road, air and pipelines. These descriptions will then be used to assess the prevalence of impacts and threat types.

Maritime Sector

There is limited information regarding cyber-attacks in the maritime and transport industry, possibly due to companies not reporting incidents to avoid reputation damage (Wagstaff, 2014). This section describes the known cyber security incidents that have affected the maritime sector, the majority of which having occurred since 2013. The earliest incident was in 2001, when operations at the Port of Houston were severely disrupted due to the computer systems crashing as a result of a DoS attack. The port itself was not a target, however the servers were vulnerable and had been compromised by an individual in order to conduct the attack (McCue, 2003). This incident illustrates the impact of DoS on port operations.

Criminal syndicates managed to gain illegitimate access to Australian Customs cargo systems in 2012, with the purpose of checking if any of their containers were considered as suspicious, indicating if it was safe to collect them or not (CyberKeel, 2014). In 2013 it was discovered that the IT systems in port of Antwerp in Belgium had been compromised for approximately two years by an organized crime gang who were smuggling drugs through the use of remote access devices (Dunn, 2013). The access they had to the systems allowed them to easily identify the location of containers in which the drugs were hidden, and sometimes even control the location of the containers (King, 2014; Ship Technology, 2013; Wagstaff, 2014). Such access could be used to disrupt operations rather than aid smuggling, as reportedly occurred in 2011 when a major Iranian shipping line suffered a cyber-attack that corrupted data, hindered internal communications and disrupted operations (CyberKeel, 2014). In 2015 it was reported that a disgruntled employee at a European port jammed the GPS signals that it relied it, disrupting operations for 12 hours (Knox, 2015).

A cyber espionage campaign using the "Icefog" APT targeted the shipbuilding and maritime sectors in Japan and South Korea amongst other target classes. The APT was discovered by Kaspersky Labs' Global Research & Analysis Team in 2013, and indications were that it had been operating since 2011. Earlier versions copied data to the command and control servers automatically, however later versions allowed the attackers to interact with the compromised systems (Kaspersky Lab, 2013).

In 2013 fake emails with PDF attachments were used by Nigerian scammers to deploy tools on target systems, which hackers could then use to steal credentials which may be used in future scams. The oil and gas maritime transport sector was the prime target of the fake emails (Kovacs, 2015). CyberKeel and partners investigated cyber-attacks aimed at deceiving targets into the shipping industry into transferring large amounts of money into fake bank accounts (CyberKeel, 2014). In December 2015 it was reported that similar fraudulent emails had been used to target US shipping companies (US Coast Guard, 2015). Attempts to compromise the webpages of maritime organizations were also reported (US Coast Guard, 2016a). This is concerning as Cyberkeel (2014) conducted research which uncovered vulnerabilities leaving container carriers exposed.

Sea-going vessels have also been targeted: it was reported in 2009 that a disgruntled employee had disabled the leak detection systems of three oil rigs (Kravets, 2009). In 2014 it was reported that two oil rigs had been affected by cyber-incidents. A floating oil rig off Africa was rendered inoperable by hackers who tilted it, which took a week to identify the cause and resolve the issue. An oil rig was considered unseaworthy for 19 days due to a severe malware infection (CyberKeel, 2014; Wagstaff, 2014). During 2013 an oil rig off Mexico was infected with malware, which disrupted its navigation systems and resulted it in drifting off the drill site (Knox, 2015; Swanbeck, 2015). This is not the first time a sea-going vessel has been infected; in 2009 it was reported that three-quarters of the Royal Navy's warships had their NavyStar/N* systems infected by a version of Conficker (Kirk, 2009; Page, 2009).

Vulnerabilities in navigation and related systems have been identified through a series of tests. Global positioning systems (GPS) have been shown to be vulnerable to attack. Whilst it is simpler to interfere and jam GPS signals, thereby denying the service, it may also be possible to create false signals that results in an aircraft or ship to divert its course do to false readings (Emspak, 2012). This was demonstrated in when a 2013 a luxury yacht was successfully diverted off course using fake GPS signals (Israeli Homeland Security.com, 2013). Researchers have shown it is possible to hack and interfere with the automatic identification system (AIS) that is installed on cargo vessels. Examples they provided include spoofing a man in the water distress signal, disabling AIS transponders, triggering a collision warning alert, and interfere with weather forecasts (Balduzzi, Wihoit & Pasta, 2013). The Electronic Chart Display and Information System (ECDIS) display information from a wide range of navigational systems and sensors, and vulnerabilities have been discovered in the system providing unauthorised access and subsequently allowing attackers to affect the ship's navigational network (Dyryavyy, 2014). In 2014 a researcher demonstrated vulnerabilities with satellite communications equipment used in the maritime, aerospace and military sectors (Santamarta, 2014). Vulnerabilities in the Voyage Data Recorders have also been reported (US Coast Guard, 2016b). In 2013 researchers illustrated the impact of vulnerabilities in control systems of oil rigs, where they could activate or deactivate pumps (Swanbeck, 2015).

Another incident in 2014 saw the US weather service and associated satellites being affected for two days, after the National Oceanic and Atmospheric Administration (NOAA) realized it was targeted by a possible state-sponsored cyber-attack and had to institute incident response procedures. The response included shutting off access to data important to shipping (Flaherty, Samenow, & Rein, 2014). Reports surfaced that systems of the Indian Navy were penetrated in an attempt to track the two aircraft carriers. The attempt, linked to a Chinese APT group, was thought to be successful for a period (Maritime Executive, 2016).

In 2009 it was reported that Somali pirates were receiving information regarding shipping from the UK (Webb, 2009). Openly available information, such as the information broadcast by AIS, can be viewed online and could aid pirates in identifying ships and their location (Wagstaff, 2014). It was discovered that pirate groups had employed hackers to gather intelligence on vessel cargo, allowing them to target specific vessels and quickly retrieve the desired cargo. The targeted company used an in-house developed content management system for managing shipping inventories, which was compromised by the hackers (Prabhu, 2016; Verizon, 2016a).

Rail

Similar to the maritime sector, malware and hackers can cause disruption to rail networks. In 2003 the Sobig virus caused system shutdowns at CSX Corporation, ultimately resulting in train delays (Miller & Rowe, 2012). A Polish teen gained constructed a basic electronic interface which allowed him to trigger switching points of tram lines using a basic remote control. After gaining access to the tram lines he triggered the device, resulting in the derailment of some carriages in 2008 (Ismail, Sitnikova & Slay, 2015; Leyden, 2008). During December 2011 an unnamed railway in the US experienced delays after a network intrusion affected signals over two days. On the first day this resulted in a fifteen minute delay in the schedules. At first it was thought that this was a targeted attack, however there are claims that this was a random incident (Ragan, 2012; Sternstein, 2012). In this case information was shared with the relevant organizations to handle the cyber-attack, and it was indicated that the government processes in place were sufficient (Ragan, 2012; Sternstein, 2012).

Researchers have indicated that railways often have outdated systems which are now networked and vulnerable to cyber-attack. These include communication and collision avoidance systems, and possible linkages between the passenger entertainment systems and the operational systems of the trains (Pauli, 2016).

Road

There are limited examples of cyber-attacks against road infrastructure, however an example of road traffic being disrupted is a malware infection of a major Israeli toll road. The camera system was affected, with a 20 minute shut down occurring on the 8 September 2013 and another eight hour shutdown on the following day (Ashford, 2013).

Researchers demonstrated that vehicles could potentially be hacked and remotely interfered with. In 2015 this was tested with a moving vehicle (Greenberg, 2015). As some point out, it is highly unlikely that a random car could be hacked, however the potential is there given that numerous experiments have shown different models are potentially vulnerable (Pogue, 2016). Remotely controlled cars could be used to cause accidents and block major roadways, therefore there is a possibility of using them to target the transportation infrastructure.

Traffic lights and signs are also vulnerable. In June 2014 an alert was made regarding the traffic signs (Kovacs, 2014). Subsequently there have been numerous reports of electronic traffic signs being hacked, and online humor websites provide over 30 images of hacked signs (stuffyoushouldknow, 2015; worldwideinterweb, 2015). Researchers have also shown that traffic lights are vulnerable (Ghena *et al*, 2014; Prigg, 2014). Hacking and controlling traffic lights could result in accidents and sever blockages, particularly in city centers.

Aviation Sector

There are a number of cyber-related incidents which have resulted in operational disruptions, some preventing flights being delayed or cancelled. A juvenile hacker targeting a telecommunications provider ultimately affected the airport in Worcester, MA in 1997. This resulted in the airport losing communications with a number of safety functions and the control tower for six hours (Ismail, Sitnikova & Slay, 2015; Miller & Rowe, 2012). During the 2009 Conficker worm outbreak, a French military airfield was affected, preventing aircraft from taking off (Willsher, 2009). In June 2015 ten flights from Polish airline LOT were cancelled and twelve were delayed due to a DDoS attack that targeted the airline's systems that issued flight plans (BBC, 2015; Brook, 2015). Officials were quaoted as stating that the attacks were a concern to other sectors (BBC, 2015; Brook, 2015). Sweden air traffic control at three airports experienced outages in November 2015, which resulted in flights being cancelled. This was initially claimed to be a result of a solar-storm, however subsequent reports claim that the outages may have been due to a state-sponsored cyber-attack, with warnings being provided to NATO and the Nordic nations (Leyden, 2016; Stormark, 2016). Other reported incidents include hackers penetrating air traffic control in Alaska and the US Air Force, stolen personal information on Federal Aviation Administration employees and gaining control of their networks (Claburn, 2009). Malware has also affected control systems. In October 2011 reports surfaced that systems used to pilot and control the US military drones were infected with malware (Roberts, 2011; Schachtman, 2011).

A Chinese man was charged in 2014 with hacking Boeing and Lockheed Martin, and stealing information regarding the designs for military aircraft. The initial hacks and transmission of data occurred between 2009 and 2010. The arrested perpetrator owned his own aviation company based in Canada (Chabrow, 2014; Walker, 2014)

As with other sectors, the airline industry is affected by online crime. South African Airways was amongst the companies affected by a wave of phishing attacks in 2013 (Fraud Alert, 2013). In 2015 RyanAir lost approximately $5 million which was fraudulently transferred to a Bank in China (Treanor, 2015). A similar attack in January 2016 saw an aircraft parts manufacturer, FACC, supplying Boeing lose $55 million as a result of being targeted by online fraudsters. After making the announcement of the hack, shares dropped by 17% (Cluley, 2016). The CEO and CFO of an aircraft parts manufacturing company in Europe were dismissed after falling victim to a whaling scam, costing the company an estimated €40 million (Reeve, 2016).

As mentioned in the maritime section, Santamarta (2014) demonstrated vulnerabilities with satellite communications equipment. This illustrates that vulnerabilities may be common across the transportation sub-sectors, and that information sharing will be beneficial. Another researcher who raised concerns regarding vulnerabilities in aircraft was later banned from flying in 2015 after tweeting as a joke about hacking satellite communication systems. In the resultant investigations, he claims to have hacked an aircraft through the in-flight entertainment system and affected an engine, causing the aircraft to veer sideways (Zetter, 2015).

Pipelines

In 1982 a Trans-Siberian pipeline exploded; reports later surface alleging that this was due to a logic bomb inserted into the control system design by the CIA. The explosion was reported to be 3kT TNT equivalent (Miller & Rowe, 2012; Weiss, 2008; Andress & Winterfield, 2011). This is not necessarily what one would class as a cyber-attack, however it indicates the importance of validating designs and supply chain security.

In 1999 attackers, with aid of a disgruntled insider, used backdoors to access and take control of systems managing the flow in the pipelines at Gazprom (Miller & Rowe, 2012). Twenty-three US pipeline operators were targeted from December 2011 to June 2012 in what appeared to be state-sponsored attacks, and sensitive operational documentation was exfiltrated (Clayton, 2013).

In August 2008 the Baku-Tbilisi-Ceyhan pipeline experienced an explosion, however none of the sensors or security cameras detected the flames. In 2014 it was reported that an infra-red camera not on the network detected two men with laptops, claimed to be Special Forces troops. The timing coincided with what appeared to be a computer intrusion. The reported vulnerability was the security cameras, which allowed the attackers to introduce malware onto the alarm system, and intentionally over-pressurize the pipeline, which was out of operation for three weeks (Robertson & Riley, 2014). Lee, Assante and Conway (2014) analyze the reports and comment that due to lack of details and corroboration, the credibility of the reports of the system penetration is reduced, despite confirmation of the explosion occurring. However, it will be considered as it is a suspected state-sponsored attack.

ANALYSIS OF INCIDENTS

This section will provide an analysis of the descriptions from the previous sections, to determine the prevalence of the threat and impact types. Forty-nine incidents were described above, spanning over 34 years. To illustrate the increasing frequency of cyber-incidents related to the transportation infrastructure, Figure 1 provides a timeline. From this it can be seen that well over a half of the incidents occurred since 2013, and just over a quarter from 2008 to 2012. The overall numbers are not that large, however these are the incidents that are considered newsworthy, and appear in the press and case studies. It is likely that numerous other incidents have occurred which are not commonly known. As laws are passed that require organizations to notify government or stakeholders of cyber-incidents, it is likely that more incidents will come to light. For the purposes of this analysis, the traffic sign hacking was considered as a single incident given that there is no clear indication of the number of perpetrators involved, or the exact timing of the signs being tampered with. Likewise, the research surrounding car hacking is only considered as one incident.

Table 2 indicates the threat type per sector. Overall, individual hackers and cyber-criminals are prevalent, with both the maritime and aviation sectors being the worst affected. The researchers are also noticeable; again the maritime sector is most affected. This could be as a result of the numerous navigation and communication aids that sea-going vessels incorporate into their daily operations, providing researchers with a number of different systems to experiment with. Of the state-sponsored attacks, the maritime and pipeline sectors appear to be the worst affected. This could be a result of the national importance of oil and trade, upon which most transportation and power generation is reliant. Consequently, this is also of strategic importance in a military context, as removing oil sources can hinder military deployments and operations. Malware is also noticeable, however these are older worm variants. As

Figure 1. Timeline of incidents related to the transportation infrastructure

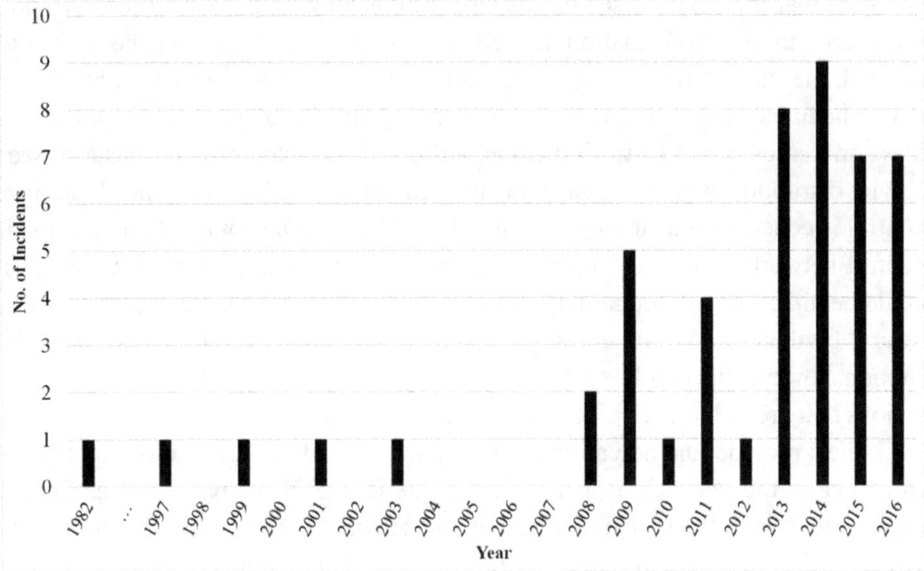

Table 2. Threat types per transport sector

	Maritime	Air	Road	Rail	Pipeline	Total
Individual hacker	5	5	1	2		13
Disgruntled insider	1					1
Hackers with aid of insider					1	1
Possible state-sponsored Cyber-espionage/warfare	3	1			3	7
Cyber-crime	7	4				11
Malware	3	2	1	1		7
Research	6	1	2	1		10
Total	25	13	4	4	4	

of yet, there have been no major reports of the more recent ransomware such as crypto-locker. Ransomware can pose a serious threat to operations should it manage to encrypt major systems for air traffic control or a port terminal operating system. The incidents of malware affecting military systems (airfield, warships, and drone control systems) illustrates the potential for targeted malware to be used in cyber-warfare at a tactical level. The state-sponsored cyber-espionage/warfare ranked equivalent to the malware. The insider threat did feature, but was not as prevalent as would usually be expected.

Table 3 illustrates the primary impact of the incidents for each sector. It should be noted that the aviation sector has an additional count here as an incident provided both theft of information as well as illegitimate control over the networks. Overall, the most common impact is operational disruptions, followed by loss of data. The impact to the maritime sector is mostly split amongst the disruption, data loss and the research identifying vulnerabilities. The aviation sector has also seen service disruption, which is equal with the financial impact. However, the illegitimate control is concerning as it includes aircraft, military drone controls, and air traffic control networks. This could potentially be used to aid terrorism in that aircraft could be hijacked without gaining access to the cockpit, or misdirect drone attacks against civilian or government targets. Railways and pipelines have also seen disruption, however the pipeline reports indicate possible physical destruction through actions that are at least supported by a cyber component.

Table 4 provides an indication of the impacts caused by the various threat types. In this case, the research aspect indicates one less as Santamarta's research was applicable to both the maritime and aviation sectors. The top combination is individual hackers causing disruption, followed by cyber-criminals targeted finances and malware causing disruptions. The suspected state-sponsored attacks tend slightly towards data theft over service disruption, indicating cyber-espionage. This is not surprising as both may provide an economic advantage to the attacker. Data theft enables decision makers to counter a competing country's planned actions, and disrupting supply reduces their effectiveness as an economic competitor. The results in terms of service disruptions, loss of confidentiality (with possible cyber-espionage), and cyber-crime is consistent with the results of the The *Verizon Data Breach Investigations Report* (Verizon, 2016b).

Table 3. Primary impact per transport sector

	Maritime	Air	Road	Rail	Pipeline	Total
Denial of service / operation disruption	8	4	1	3	2	18
Loss of data confidentiality	7	2			1	10
Financial	3	4				7
Potential Illegitimate control		3	1		1	5
Unauthorized access	1					1
Proof of concept / research	6	1	2	1		10
Total	25	14	4	4	4	

Table 4. Impacts per threat type

	Individual Hacker	Disgruntled Insider	Hackers with Aid of Insider	Cyber-Espionage/ Warfare	Cyber-Crime	Malware	Research
Denial of service / operation disruption	8	1		3		6	
Loss of data confidentiality	2			4	4		
Financial					7		
Potential Illegitimate control	2		1			1	
Unauthorized access	2						
Proof of concept / research							9

With both individual hackers and malware able to cause disruptions the nature of the vulnerabilities in critical systems are evident. Given the prevalence of the service disruption and loss of confidentiality, it can be concluded that the transportation sector is vulnerable to cyber-warfare and cyber-terrorism, where large disruptions can occur. Varying (or a complete lack of) regulations and standards for cyber-security in the transport sector at a national and inter-national level can be considered as partly to blame. Whilst some nations are attempting to drive the improvement of cyber-security in these sectors, reports indicate that there are still gaps that need to be closed. In addition, the unwillingness of organisations to share information with government, and sometimes unwillingness of nations to share intelligence with each other can also be a contributor. The following section discusses the efforts to mitigate the cyber-threat in the transportation sector.

EFFORTS FOR THREAT MITIGATION

Three main areas for aiding in threat mitigation will be considered: the regulatory environment to provide a minimum level of cyber-security, threat intelligence and information sharing to aid in detecting attacks and minimizing impact, and awareness which is required to support the other efforts.

Regulations

Regulations, legislation, and policies which are enforceable will provide the sector with a set of controls and requirements to ensure a minimum level of cyber-security across the sector. This will aid in minimizing the impacts of the cyber-threat. However, sector-specific regulations are scarce, as is discussed in this section.

Reports by the European Network Information Security Agency (ENISA), the US Government Accountability Office (GAO) and Kramek indicate a number of vulnerabilities in the maritime sector. The ENISA (2011) report indicated that there was a lack of cyber-security awareness in the maritime sector, no holistic approach to cyber-security, cyber-security was not considered in port security regulations, and that there was little or no incentive to effectively address cyber-security in the sector. Kramek (2013) also highlighted a lack of awareness in the US, and indicated that while a few ports had made notable progress, that this was inconsistent across the country. The GAO (2014) report supported these findings, indicating that the relevant risk assessments and security regulations did not consider cyber-security and that there was inconsistent information sharing regarding cyber-security. Since those reports, efforts have been made to provide guidance on cyber-security in the shipping industry. One set of efforts to produce cyber-security policies in the maritime sector is the Guidelines on Cyber Security Onboard Ships (BIMCO, CLIA, IS, INTERCARGO & INTERTANKO, 2016). The guidelines provide an overview of threats, security controls, and systems that could potentially be targeted, and make specific reference to the National Institute of Standards and Technology's Cyber Security Framework (NIST, 2014). According to Miller, quoted in Tomlinson (2016), the biggest limiting factor of the effort is that it is only guidelines, and there is no mandatory minimum controls that are enforceable. A whitepaper on ICS security specifically in the maritime sector has also been released (Wallischeck, 2013).

Similarly to the maritime sector, the GAO released two reports on cyber-security for the aviation sector. The first was released in January 2015, and focused on the Federal Aviation Administration (FAA) needing to mitigate vulnerabilities in air traffic control systems (GAO, 2015a). The second report, released in April 2015, focused more on the risks posed by the modernization of the FAA and the aviation sector, particularly where networking technologies were being implemented both onboard aircraft as well as ground facilities (GAO, 2015b). In 2016 US Senator Markey provided a Bill known as the "Cybersecurity Standards for Aircraft to Improve Resilience Act of 2016'' or the ''Cyber AIR Act". The bill proposed that organizations in the aviation sector be mandated to disclose cyber-attacks targeting them, and to identify and mitigate vulnerabilities in the US aviation sector (Markey, 2016). In June 2016 a "preliminary agreement" was reached regarding cyber-security for airlines (Dark Reading, 2016).

As with the Guidelines on Cyber Security Onboard Ships, Carr (2015) also cites the NIST Cybersecurity Framework as a guideline for pipeline cyber-security. A number of regulations for the oil and gas sector can be applied to pipelines due to the overlap. An example of this is the Control Systems Cyber Security Guidelines for the Natural Gas Pipeline Industry (INGAA, 2011) and the Roadmap to Secure Control System in the Energy Sector (Eishenhauer, Donnelly, Ellis & O'Brien, 2006).

Other broader guidance and regulation for the transport sector include the Department of Homeland Security's (2010) national infrastructure protection plan for the transport sector, which includes cyber-security aspects, and a report by the Roadmap to Secure Control System in the Transportation Sector Working Group (2012). The fact that this report was released six years after its equivalent in the energy sector illustrates the challenges for guidance specific to the transportation sector.

To a large degree, operational systems in the sector can be considered similar to ICS or supervisory, control and data acquisition (SCADA) systems. The guidance provided for ICS/SCADA may be suitable here. These include NIST's Special Publication 800-82: Guide to Industrial Control Systems (ICS) Security (Stouffer, Falco & Scarfone, 2013), and the Centre for the Protection of National Infrastructure's Good practice guide Process Control and SCADA Security (CPNI, 2008).

There should not only be a focus on protection measures, but also recovery and incident response measures. Despite implementing cyber-security, it is possible that an attack is successful. It is therefore important to ensure business continuity and recovery regulations are followed in order to minimize the organizational and economic damage resulting from the attack.

As is evident, there is a lack of sector-specific regulations, particularly those that can be enforced internationally. At national and international levels more incentives should be provided to align to cyber-security guidance relevant to the transport sector. Awareness of the issues needs to be raised to aid buy-in of the available guidance, while enforceable regulations are continued to be developed.

Threat Intelligence

As indicated in the 2011 US railway incident, information sharing and communications aided in the effective collaboration to investigate and resolve the incident (Ragan, 2012; Sternstein, 2012). For the case of the Swedish air traffic control outages, the warnings provided to NATO and Nordic nations was the cause of speculation that the outages were in fact caused by cyber-attacks and not the official report of solar storms (Leyden, 2016; Stormark, 2016). However, in such a case this would allow the recipients of the alerts to become more vigilant, and if specific signatures of the attack were communicated, they may be able to detect and minimize the impact of similar attacks. The Polish airline attack, in particular the comment that it was of concern to other sectors (BBC, 2015; Brook, 2015), provides further motivation for the sharing of threat intelligence.

In the US, a number of Information Sharing and Analysis Centers (ISAC) have been implemented to facilitate the information sharing within and across the various sectors. The National Council of ISACs have a number of members which are related to the transport sector, including:

- Aviation ISAC (www.a-isac.com).
- Maritime ISAC (www.maritimesecurity.org).
- Oil & Natural Gas ISAC (www.ongisac.org).
- Surface Transportation, Public Transportation and Over-The-Road Bus ISACS (www.surfacetransportationisac.org).

In South Africa, the National Cybersecurity Policy Framework (State Security Agency, 2015) and the draft Cybercrimes and Cybersecurity Bill (Department of Justice and Correctional Services, 2015) mandate sector based response teams coordinated by a central cyber-security facility. In the US the ICS-CERT provides reports and alerts, many of which will be relevant to the transportation sector.

Awareness Training

As indicated in the regulation section, a number of reports have indicated that awareness of cyber-security issues in the sector is lacking. It is therefore important that when developing initiatives, that the relevant stakeholders at national and executive levels are engaged to gain support. At operational and management levels, buy-in is also required to ensure that the regulations and standards are implemented and the controls are maintained effectively.

National and regional exercises can be conducted to provide training for business and local government leaders and assesses the readiness to respond to significant cyber-attacks. At managerial and operational levels, similar exercises can be conducted with a more technical focus on how to recover actual systems, or to brainstorm possible vulnerabilities and attack techniques.

Such exercises are conducted in the US and Europe, such as Cyber Storm II. These should be conducted in all major transport points, to ensure there is a minimum level of response readiness to minimize any impact on global trade routes.

CONCLUSION

There are an increasing number of cyber-related incidents targeting the transport sector. From the incidents considered, over a half have occurred since 2013. The most common perpetrator was individual hackers, and the most common impact was a disruption of services. Sector-specific regulations and legislation will aid, however there is a lack of this. Whilst efforts are being made, they are inconsistent internationally, and are often not enforceable. The development of international regulations that are enforceable is important to aid in encouraging cyber-security be considered in the sector. Information sharing and threat intelligence will provide warnings to the transportation sector, and allow for attacks to be detected and identified earlier, minimizing the impact. The development of sector-based response teams or cyber-security centers should be a priority.

REFERENCES

Andress, J., & Winterfield, S. (2011). *Cyber Warfare: Techniques, Tactics and Tools for Security Practitioners*. Waltham: Elsevier. doi:10.1016/B978-1-59749-637-7.00001-0

Ashford, W. (2013, October 28). Cyber attack shuts down Israeli toll road tunnel. *Computer Weekly*. Retrieved from http://www.computerweekly.com/news/2240207924/Cyber-attack-shuts-down-Israeli-toll-road-tunnel

Balduzzi, M., Wihoit, K., & Pasta, A. (2013). Hey Captain, where's your ship? Attacking vessel tracking systems for fun and profit. *11th Annual Hack in the Box (HITB) Security Conference in Asia*. Retrieved from https://conference.hitb.org/hitbsecconf2013kul/materials/D1T1%20-%20Marco%20Balduzzi,%20Kyle%20Wilhoit%20Alessandro%20Pasta%20-%20Attacking%20Vessel%20Tracking%20Systems%20for%20Fun%20and%20Profit.pdf

BIMCO. CLIA, IS, INTERCARGO & INTERTANKO. (2016, January). *The Guidelines on Cyber Security Onboard Ships*. Retrieved from www.bimco.org

British Broadcasting Corporation (BBC). (2015, June 21). *Polish LOT aeroplanes grounded by computer hack*. Retrieved from http://www.bbc.com/news/world-europe-33219276

Brook, C. (2015, June 22). Polish planes grounded after airline hit with DDoS attack. *Threatpost*. Retrieved from https://threatpost.com/polish-planes-grounded-after-airline-hit-with-ddos-attack/113412

Carr, R. (2015, November 2). Empowering pipeline SCADA cybersecurity. *Tripwire*. Retrieved from http://www.tripwire.com/state-of-security/featured/empowering-pipeline-scada-cybersecurity/

Center for the Protection of National Infrastructure (CPNI). (2008). *Process control and SCADA security, Good practice guide 1: Understand the business risk*. Retrieved from http://www.cpni.gov.uk/Documents/Publications/2008/2008024-GPG_SCADA_Business_Risk.pdf

Chabrow, E. (2014, July 13). Details emerge of Boeing hack. *GovInfoSecurity*. Retrieved from http://www.bankinfosecurity.com/details-emerge-boeing-hack-a-7053/op-1

Claburn, T. (2009, May 7). Air traffic control system repeatedly hacked. *Dark Reading*. Retrieved from http://www.darkreading.com/risk-management/air-traffic-control-system-repeatedly-hacked/d/d-id/1079364

Clayton, M. (2013, April 19). Exclusive: Cyberattack leaves natural gas pipelines vulnerable to sabotage. *Christian Science Monitor*. Retrieved from http://www.csmonitor.com/Environment/2013/0227/Exclusive-Cyberattack-leaves-natural-gas-pipelines-vulnerable-to-sabotage

Cluley, G. (2016, January 21). Hackers steal $55 million from Boeing supplier. *Tripwire*. Retrieved from http://www.tripwire.com/state-of-security/security-data-protection/boeing-supplier-hacked-claims-55-million-worth-of-damage-as-stock-price-falls/

CyberKeel. (2014, October 15). *Maritime cyber-risks*. Retrieved from www.cyber-keel.com

Dark Reading. (2016, June 14). *Preliminary Agreement Reached On Airline Cyber-security*. Retrieved from http://www.darkreading.com/risk/preliminary-agreement-reached-on-airline-cybersecurity/d/d-id/1325904

Department of Justice and Correctional Services. (2015). *Cybercrimes and Cyber-security Bill*. Pretoria: Government of South Africa.

Dunn, J. E. (2013, October 16). Hackers planted remote devices to smuggle drugs through Antwerp port, Europol reveals. *Techworld*. Retrieved from http://news.techworld.com/security/3474018/hackers-planted-remote-devices-to-smuggle-drugs-through-antwerp-port-europol-reveals/

Dyryavyy, Y. (2014). *Preparing for cyber Battleships – electronic chart display and information system security*. NCC Group. Retrieved from https://www.nccgroup.com/media/481230/2014-03-03_-_ncc_group_-_whitepaper_-_cyber_battle_ship_v1-0.pdf

Eishenhauer, J., Donnelly, P., Ellis, M. & O'Brien, M. (2006, January). *Roadmap to Secure Control System in the Energy Sector*. Columbia, MD: Energetics Incorporated.

Emspak, J. (2012, May 7). GPS vulnerable to dangerous hacks and spamming. *Discovery Channel*. Retrieved from http://www.msnbc.msn.com/id/47324401/ns/technology_and_science-security/#.T7BXxoGvPLU

European Network and Information Security Agency (ENISA). (2011, November). *Analysis of cyber security aspects in the maritime sector*. Retrieved from http://www.enisa.europa.eu/activities/Resilience-and-CIIP/critical-infrastructure-and-services/dependencies-of-maritime-transport-to-icts/cyber-security-aspects-in-the-maritime-sector-1

Flaherty, M. P., Samenow, J., & Rein, L. (2014, November 12). Chinese hack U.S. weather systems, satellite network. *The Washington Post*. Retrieved from http://www.washingtonpost.com/local/chinese-hack-us-weather-systems-satellite-network/2014/11/12/bef1206a-68e9-11e4-b053-65cea7903f2e_story.html

Fraud Alert. (2013, December 18). *SAA warns against email phishing scam*. Retrieved from http://fraudalert.co.za/saa-warns-against-email-phishing-scam/

Ghena, B., Beyer, W., Hillaker, A., Pevarnek, J., & Halderman, J. A. (2014, August 19). Green lights forever: analyzing the security of traffic infrastructure. *Proceedings of the 8th USENIX Workshop on Offensive Technologies (WOOT '14)*. Retrieved from https://www.usenix.org/conference/woot14/workshop-program/presentation/ghena

Government Accountability Office (GAO). (2014, June). *Maritime critical infrastructure protection: DHS needs to better address port cybersecurity*. Retrieved from www.gao.gov/assets/670/663828.pdf

Government Accountability Office (GAO). (2015a, January). *FAA Needs to Address Weaknesses in Air Traffic Control Systems*. Retrieved from http://www.gao.gov/assets/670/668169.pdf

Government Accountability Office (GAO). (2015b, April). *FAA Needs a More Comprehensive Approach to Address Cybersecurity As Agency Transitions to NextGen*. Retrieved from http://www.gao.gov/assets/670/669627.pdf

Greenberg, A. (2015, July 21). Hacker remotely kill a jeep on the highway. *Wired.* Retrieved from http://www.wired.com/2015/07/hackers-remotely-kill-jeep-highway/

ICS-CERT. (2016). *NCCIC/ICS-CERT Year in Review: FY2015.* Washington, DC: Department of Homeland Security.

INGAA. (2011, January 31). *Control Systems Cyber Security Guidelines for the Natural Gas Pipeline Industry.* Washington, DC: INGAA.

Ismail, S., Sitnikova, E., & Slay, J. (2015). SCADA systems cyber security for critical infrastructures: case Studies in the transport sector. In *Proceedings of the 10th International Conference on Cyber Warfare and Security (ICCWS 2015),* (pp. 425-433).

Israeli Homeland Security.com. (2013, August 7). *$80 million yacht hijacked by students spoofing GPS signals.* Retrieved from http://i-hls.com/2013/08/80-million-yacht-hijacked-by-students-spoofing-gps-signals/

Kaspersky Lab. (2013). *The 'Icefog' APT: a tale of cloak and three daggers.* Retrieved from http://kasperskycontenthub.com/wp-content/uploads/sites/43/vlpdfs/icefog.pdf

King, R. (2014, July 9). Hacking groups target shipping ports. *Wall Street Journal.* Retrieved from http://mobile.blogs.wsj.com/cio/2014/07/09/hacking-groups-target-shipping-ports/?mg=blogs-wsj

Kirk, J. (2009, January 19). *Virus attacks Ministry of Defence.* Retrieved from http://www.cio.co.uk/news/3460/virus-attacks-ministry-of-defence/

Knox, J. (2015, June 15). *Coast Guard Commandant on Cyber in the maritime domain, US Coast Guard.* Retrieved from http://mariners.coastguard.dodlive.mil/2015/06/15/6152015-coast-guard-commandant-on-cyber-in-the-maritime-domain/

Kovacs, E. (2014, June 6). Default password exposes digital highway signs to hacker attacks. *Security Week.* Retrieved from http://www.securityweek.com/default-password-exposes-digital-highway-signs-hacker-attacks

Kovacs, E. (2015, May 18). Nigerian fraudsters attack oil logistics firms without using malware: report. *Security Week.* Retrieved from http://www.securityweek.com/nigerian-fraudsters-attack-oil-logistics-firms-without-using-malware-report

Kramek, J. (2013, July 3). *The critical infrastructure gap: U.S. port facilities and cyber vulnerabilities, Center for 21st Century Security and Intelligence.* Brookings Institute. Retrieved from http://www.brookings.edu/~/media/research/files/papers/2013/07/02-cyber-port-security-kramek/03-cyber-port-security-kramek.pdf

Kravets, D. (2009, March 18). Feds: hacker disabled offshore oil platforms' leak-detection system. *Wired*. Retrieved from http://www.wired.com/2009/03/feds-hacker-dis/

Lee, R. M., Assante, M. J., & Conway, T. (2014, December 20). *ICS CP/PE (Cyber-to-physical or process effects) case study paper – media report of the Baku-Tbilisi-Ceyhan (BTC) pipeline cyber attack*. SANS Institute. Retrieved from https://ics.sans.org/media/Media-report-of-the-BTC-pipeline-Cyber-Attack.pdf

Leyden, J. (2008, January 11). Polish teen derails tram after hacking train network. *The Register*. Retrieved from http://www.theregister.co.uk/2008/01/11/tram_hack/

Leyden, J. (2016, April 12). Sweden 'secretly blames' hackers – not solar flares – for taking out air traffic control. *The Register*. Retrieved from http://www.theregister.co.uk/2016/04/12/sweden_suspects_russian_hackers_hit_air_traffic_control/

Maritime Executive. (2016, June 9). *Hackers attempt to track Indian aircraft carriers*. Retrieved from http://maritime-executive.com/article/hackers-attempt-to-track-indian-aircraft-carriers

Markey, E. (2016). *Cybersecurity Standards for Aircraft to Improve Resilience Act of 2016*. Retrieved from http://www.markey.senate.gov/imo/media/doc/Cyber%20AIR%20Act%204-7-16.pdf

McCue, A. (2003, October 7). 'Revenge' hack downed US port systems. *ZDNet*. Retrieved from http://www.zdnet.com/article/revenge-hack-downed-us-port-systems/

Miller, B., & Rowe, D. C. (2012, October 11-13). A survey of SCADA and critical infrastructure incidents. *ACM Special Interest Group on Information Technology Education (SIGITE) Research in IT Conference*.

National Institute of Standards and Technology (NIST). (2014, February 12). *Framework for Improving Critical Infrastructure Cybersecurity*. Retrieved from http://www.nist.gov/cyberframework/

Pauli, D. (2016, January 4). Irked train hackers talk derailment flaws, drop SCADA password list. *The Register*. Retrieved from http://www.theregister.co.uk/2016/01/04/irked_train_hackers_talk_derailment_flaws_drop_scada_password_list/

Pogue, D. (2016, February 22) Why car hacking is nearly impossible. *Scientific American*. Retrieved from http://www.scientificamerican.com/article/why-car-hacking-is-nearly-impossible/

Prabhu, V. (2016, March 2). Pirates hack shipping company to find their target ships for raids. *Techworm*. Retrieved from http://www.techworm.net/2016/03/pirates-hacked-shipping-company-gather-intel-target-ships.html

Prigg, M. (2014, August 20). How to get green lights all the way to work: Hackers reveal how simple it is to control traffic lights in major cities using just a laptop. *Daily Mail*. Retrieved from http://www.dailymail.co.uk/sciencetech/article-2730096/How-green-lights-way-work-Hackers-reveal-simple-control-traffic-lights-major-cities-using-just-laptop.html

Ragan, S. (2012, January 25). Railway network disrupted after cyber attack. *Security Week*. Retrieved from http://www.securityweek.com/railway-network-disrupted-after-cyber-attack-report-says

Reeve, T. (2016, May 16). CEO sacked after aircraft company grounded by whaling attack. *SC Magazine*. Retrieved from http://www.scmagazine.com/ceo-sacked-after-aircraft-company-grounded-by-whaling-attack/article/499258/

Roadmap to Secure Control System in the Transportation Sector Working Group. (2012). *Roadmap to Secure Control System in the Transportation Sector*. Washington, DC: Department of Homeland Security.

Roberts, P. (2011, October 12). Air Force struggled for weeks with malware in drone fighter systems. *ThreatPost*. Retrieved from http://threatpost.com/en_us/blogs/report-air-force-struggled-weeks-malware-drone-fighter-systems-101211

Robertson, J., & Riley, M. (2014, December 10). Mysterious '08 Turkey pipeline blast opened new cyberwar. *Bloomberg*. Retrieved from http://www.bloomberg.com/news/articles/2014-12-10/mysterious-08-turkey-pipeline-blast-opened-new-cyberwar

Santamarta, R. (2014). SATCOM terminals hacking by air, sea, and land. *BlackHat USA 2014*. Retrieved from https://www.blackhat.com/docs/us-14/materials/us-14-Santamarta-SATCOM-Terminals-Hacking-By-Air-Sea-And-Land-WP.pdf

Shachtman, N. (2011, October 8). Computer virus hits US Predator and Reaper drone fleet. *ARS Technica*. Retrieved from http://arstechnica.com/business/news/2011/10/exclusive-computer-virus-hits-drone-fleet

Ship Technology. (2013, November 28). *Web of intrigue: protecting ports against cyber-terrorism*. Retrieved from http://www.ship-technology.com/features/feature-cybersecurity-port-computer-hackers-us-belgium/

State Security Agency. (2015). *The National Cybersecurity Policy Framework*. Pretoria: Government of South Africa.

Stelter, L. (2015, February 2). Securing the Nation's ports against cyberterrorism. *InHomelandSecurity.com*. Retrieved from http://inhomelandsecurity.com/securing-the-nations-ports-against-cyberterrorism/

Sternstein, A. (2012, January 23). Hackers manipulated railway computers, TSA memo says. *NextGov*. Retrieved from http://www.nextgov.com/cybersecurity/2012/01/hackers-manipulated-railway-computers-tsa-memo-says/50498/

Stormark, K. (2016, April 12). *Sweden issued cyber attack alert as its air traffic reeled*. Retrieved from https://www.aldrimer.no/sweden-issued-cyber-attack-alert-as-its-air-traffic-reeled/

Stouffer, K., Falco, J., & Scarfone, K. (2013). *Guide to Industrial Control Systems (ICS) Security. NIST Special Publication 800-82 Revision 1*. Gaithersburg, MD: NIST.

Swanbeck, S. (2015, June 22). *Coast Guard Commandant Addresses Cybersecurity Vulnerabilities on Offshore Oil Rigs, CSIS-Tech.org*. Retrieved from http://www.csis-tech.org/blog/2015/6/22/coastguard-commandant-addresses-cybersecurity-vulnerabilities-in-offshore-oil-rigs

Symantec. (2014, April). *2014Internet Security Threat Report, Volume 19, Supplemental Data*. Retrieved from http://www.symantec.com/content/en/us/enterprise/other_resources/b-istr_appendices_v19_221284438.en-us.pdf

Symantec. (2015, April). *2015Internet Security Threat Report, Volume 20, Supplemental Data*. Retrieved from https://www4.symantec.com/mktginfo/whitepaper/ISTR/21347931_GA-internet-security-threat-report-volume-20-2015-appendices.pdf

Symantec. (2016a, April). *2016Internet Security Threat Report, Volume 21*. Retrieved from https://www.symantec.com/security-center/threat-report

Symantec. (2016b, April). *2016Internet Security Threat Report, Volume 21, Supplemental Data*. Retrieved from https://www.symantec.com/security-center/threat-report

Tomlinson, K. (2016, January 6). *Even cruise ships can be hacked—and what is being done about it*. Archer Energy Solutions. Retrieved from http://www.archerenergysolutions.com/even-cruise-ships-can-be-hacked-and-what-is-being-done-about-it/

Treanor, J. (2015, April 29). Ryanair confirms hackers stole almost $5m via Chinese bank. *The Guardian*. Retrieved from http://www.theguardian.com/business/2015/apr/29/ryanair-confirms-hackers-stole-almost-5m-via-chinese-bank-electronic-transfer

US Coast Guard. (2015, December 28). *Spoofed business e-mail used to try and defraud maritime organization.* Maritime Cyber Bulletin 002-16.

US Coast Guard. (2016a, January 4). *Maritime organization's websites targeted by actors attempting to exploit Joomla vulnerabilities.* Maritime Cyber Bulletin 003-16.

US Coast Guard. (2016b, January 5). *Shipboard voyage data recorder vulnerabilities.* Maritime Cyber Bulletin 004-16.

Verizon. (2016a). *Data breach digest. Scenarios from the field.* Retrieved from http://www.verizonenterprise.com/resources/reports/rp_data-breach-digest_xg_en.pdf

Verizon. (2016b). *Data breach investigations report.* Retrieved from http://www.verizonenterprise.com/resources/reports/rp_data-breach-digest_xg_en.pdf

Wagstaff, J. (2014, April 23). All at sea: global shipping fleet exposed to hacking threat. *Reuters.* Retrieved from http://www.reuters.com/article/2014/04/24/tech-cybersecurity-shipping-graphic-pix-idUSL3N0NG0GP20140424

Walker, D. (2014, July 14). Chinese man charged with hack of Boeing, Lockheed Martin aircraft data. *SC Magazine.* Retrieved from http://www.scmagazine.com/chinese-man-charged-with-hack-of-boeing-lockheed-martin-aircraft-data/article/360786/

Wallischeck, E. Y. (2013, July). *ICS Security in Maritime Transportation.* Cambridge, MA: John A Volpe National Transportation Systems Center.

Webb, J. (2009, May 11). Somali pirates using London contacts - Spain radio. *Reuters.* Retrieved from http://www.reuters.com/articlePrint?articleId=USLB570114

Weiss, G. (2008, April 14). *The Farewell dossier: Duping the Soviets, Central Intelligence Agency.* Retrieved from https://www.cia.gov/library/center-for-the-study-of-intelligence/csi-publications/csi-studies/studies/96unclass/farewell.htm

Willsher, K. (2009, February 7). French fighter planes grounded by computer virus. *The Telegraph.* Retrieved from http://www.telegraph.co.uk/news/worldnews/europe/france/4547649/French-fighter-planes-grounded-by-computer-virus.html

World Economic Forum. (2015). *Global Risks 2015* (10th ed.). Geneva: World Economic Forum.

Zetter, K. (2015, May 15). Feds say that banned researcher commandeered plane. *Wired.* Retrieved from http://www.wired.com/2015/05/feds-say-banned-researcher-commandeered-plane/

KEY TERMS AND DEFINITIONS

Cyber-Attack: Attacks conducted in cyber-space primarily aimed at disrupting information services or operations, steal information, or making political statements.

Cyber-Crime: Actions by online criminal groups targeting organizations and individuals for financial gain.

Cyber-Espionage: Nation-state sponsored or backed efforts to penetrate networks to steal sensitive national or industrial information.

Cyber-Security: Protecting information, systems and networks from cyber-threats.

Cyber-Threat: Actors or malicious code that seek to do harm or steal information through the use of cyber-space.

Cyber-Warfare: Nation-state sponsored or backed attacks to disrupt military, economic, or other national services.

ICS: Industrial control systems – systems used to control machinery, industrial processes and equipment.

SCADA: Supervisory control and data acquisition – a subset of ICS.

Threat Intelligence: Actionable intelligence that provides signatures or indicators of cyber-threats or compromised systems.

Chapter 5
English Speaking Countries and the Culture of Fear:
Understanding Technology and Terrorism

Maximiliano Emanuel Korstanje
University of Palermo, Argentina

ABSTRACT

The direct intervention or full-scare led wars are ideologically legitimized by the needs of bringing the ideals of American democracy, liberty, freedom and mobility. However, at the bottom, this globalized culture of fear hidden dark interests associated to exploitation. Paradoxically, these types of interventions suggest that terrorism needs the use of force, but in so doing, impotence and deprivation surface. Undoubtedly, Anglo and Latin worlds have created, according to their cultural matrices, diverse tactics to adapt to environment, as the form of understanding the future. While Anglo-countries developed a fascinating attraction to risk and future, the sense of predestination alludes to what today has not occurred yet. Technology only helps to mitigate the temporal effects of uncertainty triggered by the orientation to future.

DOI: 10.4018/978-1-5225-1938-6.ch005

INTRODUCTION

Undoubtedly, English speaking countries have developed high technology, digital information to pose as the main powers of contemporary world. Dotted with a machinery of war, adjoined to consolidated economies and democracies, England seems to be situated as a global empire, even collapsed its hegemony after 19s century. In fact, this chapter explores the connection of English speaking cultures with risk perception, a deep-seated matter which merits to be studied. Over years, England has successfully imposed not only to its colonies the right of trade and the language, but forged a culture of fear, which persists to date. To set an example, at time of keying the word disaster in google reaching almost 151.00.000 records. If we go to Spanish "desastre" this cipher is limited to 60.000 records. This evident contrast between Protestant and Catholics seems to be one of the aspects that captivated our attention in this essay review. At some extent, the following tables summarizes how prone to risk Anglo Saxons are.

As the previous backdrop, it is important to add that in England, United States or Australia, there is a lot of programs, doctorates and Research-department associated to risk perception or terrorism, while the same does not apply for Spanish-related nations as Argentina, or Mexico. Therefore, the importance of risk perception in English speaking countries not only is an interesting point of convergence which was never checked, but also reveals the intersection of technology to prevent surrounding threats. This suggests likely Weber was in the correct side when confirmed that Protestants developed a frightening culture resulted from the rise and consolidation of predestination doctrine. If the Mediterranean cultures, coined the Latino-archetype, were based on piety and sacrifice, the Anglo-countries have deposited on the future much expectative. Equally important is to discuss why some countries are risk-oriented, while others are not, as well as the role played by some cultures as leaders of technology while others are excluded. As explained, Weber`s legacy inscribed into a gap between Protestant and Catholic cosmologies. The protestant logic would be determined by the sense of predestination (Weber, 1964; 1995; 1958). The roots of capitalism are based on the stimulation of competence among citizens, who devote their resources to show themselves they are special (Fromm, 2005). At a first glance, As Coleman puts it, the lack of certainness re-

Table 1. Entries on Google

Language	Spanish	English
Disaster / Desastre	19.500.000	151.000.000
Risk / Riesgo	83.200.000	527.000.000

specting to salvation led Protestantism, unlike other religious waves to develop a strong attachment to politics (Coleman 2013).

Since Protestantism has developed a negative image of the external world, its hopes of salvation where in predestination. The war for the redemption of souls became in one of the concerns of evangelicals once arrived to the US. This sentiment, which combined fear and love, formed the American character and a particular way of producing politics (Greven 1988). Recently, Korstanje (2012) suggested that Weber did not take the wrong way in diagnosing capitalism derived from Reform, but he opted to an easy explanation in lieu of delving into the Norse Mythology. Unlike others form of mythologies, in Norse Culture Walkyrias, were sent by Odin to embrace all fallen warriors. They know beforehand who will fall, even before the involving victim. This paved the ways for the advent of the predestination principle, Luther and Calvino will develop in their theologies. Unlike other cultures as Greece, where events resulted from a previous human decision making process, in Norse Mythology, human beings are minimized to the extent of accepting a closed-view of future. If Agamemnon embraces the sacrifice of his daughter to defeat Troy, he is aware of his freedom to opt for one or another way. For the ancient Norse culture, fate is given to mortals to be accepted and followed as it has been revealed. Untouchable for Protestants, the future never can be changed. As explained, the sense of predestination is older than Weber´s presupposed. By the way, E. Dodds (1997), sees in the Greek world the point of indetermination. Parents are fitted against sons are at odds but in all circumstances, the future is drawn by the practices and things that people accomplish in this world. Gods are only divine advisers.

In middle Ages, Catholic Church monopolized not only the production of knowledge but served as a platform of consult to gain understanding of the world. This was radically changed whenever a strong quake hit Lisbon in 1775. This disaster not only was the epicenter of secularization but the birthday of the Science. Its effects evidenced serious problems in the existent-already Faith to forecast similarly events in the future. Philosophers of the caliber of Voltaire or Rousseau questioned seriously the dogmatism of Catholicism in presenting a God that may protect people in all conditions. This quake dissipated the idea God as a benefactor questioning that God not necessarily should be the good-doer we have learned (Nigg, 1995; 1996). Although many scholars have seen Lisbon´s earthquake as the launching platform of modern Science, less attention was conferred to the fact that this new embarrassment surfaced in German countries; of course, whose mythology emphasized on the pre-destination's principle.

The Nature of Risk

Time permits the classification of experiences, to serve as moral guidelines in times of uncertainty. Whenever events are not framed, or cannot be framed, risk surfaces. First of all, risk connotes a social narrative enrooted in a situation of danger. But risk is not real, this represents a future condition. Whenever the danger takes places in forms of disasters, or emergencies risk disappears (Douglas & Wildavski, 1983; Douglas, 1992). This is exactly the way insurance policies works in modern world. Lay-people are promoted by experts to buy insurance based only in a precautionary doctrine. Any insurance-policy cannot be bought once the risk has been transformed in a real danger. As debated in earlier chapters, U. Beck acknowledges that the modern society is linked to an on-going risk because the grounding social institutions that regulated the social life are in decline. While religion, education and economy not only has changed but also are in process of disappearance, other new types of mediators emerged. Risk connotes an interesting mediator (like money) to connect people otherwise would not be familiar. Without risk, society would disintegrate in question of years (Beck, 2006). Thus, the concept of future seems to be inextricably intertwined to risk.

In his valuable book, *the Transformation of Intimacy,* A. Giddens (1997) sheds light on the ways both sexes, males and females, men or women have re-developed novel tactics of negotiation and self-identity. Giddens openly argues that the sex, as any other human practices, is structured in the economical forces. Whenever the marriage was driven to fertility issues, women have a defined role, subordinated but protected by men. Although in these times the mother-death was a common-placed thing, their ontological security was based on a much wider reproduction-stricken force. At some point, the modernity extended the expectance of life, modifying not only the role played by women but their security in births. As a result, women, emancipated to patriarchal order, were equalled to men. The fertility, as a mediator between the family and the production of goods, sets the pace to new forms of sex-induced conflicts. The classical sense of marriage, as we known, was based on a covenant among two defined genres. This pact signalled to cultural values linked to tradition, lore and past. The introduction of modernity and self-reflexibility not only gives to lay people further information about sex, but also makes from it a science. When women were liberated of their reproduction-oriented role, the sexuality has transformed in a plastic-sexuality, based on the pleasure as a primary concern. Paradoxically, this objectified the women, in a commodity, to be sold in pornography moulded and framed under male´s desires. What today is important to debate is the radical shifts are lovers facing. Unlike passion, the romantic love resulted from the advent of future-oriented view that accompanied the surface of modernity. The romantic love may be narrated as a fiction, which is based on a future. Under some circumstances, this process would generate addictive behaviour because the lover lives its life in illusory spheres, but in others it inflicts panic if the loved-other is absent. Addictive behaviour alludes to sex as a form of enhancement and escapement, reinforcing the equality but in material terms. Now, women struggle in similar conditions with men, but unfortunately deriving in results in the upsurge of domestic violence.

One of the pioneers in studying the prominent benefits of technology has been Daniel Bell. His optimism on the promises of technology respecting to the optimization of work (Bell, 1974), was not supported by other colleagues who saw in the same process a big problem (Webster & Robins, 1986; Schiller, 1981; Rifkin, 1998). Detractors of digital age proposed to state to regulate the information trade by the following goals:

1. Avoid the systematization of work that leads to unemployment.
2. Prevent act of terrorism because of excessive information-dependency of society.
3. The excess of technology results in a decline of trust.
4. Human beings can play to be gods by the manipulation of bio-technology.

In this vein, William Dutton (1998) explains technology exerts today considerable influence on the life of people. However, it is almost impossible to determine if technology determines social agents or vice-versa. What the specialized literature suggests is that people coactively may select those ITC as their discretion. This process is understood by the previous cognitive framework of user. Therefore, in the digital era we can say technology is not good or bad, it hinges on how and what it is employed for. In the contemporary society, what would be interesting to discuss is the psychological dependency of information. This point begs a more than interesting question. Is risk and technology resulted from the predestination?

In the digital times, events occur first on the screen and afterward in reality. The boundaries between fiction and reality have been blurred. The Baudrillard legacy situates capitalism in an enigmatic scenario, where the future plays a vital role. Unlike Max Weber or Werner Sombart, who envisaged an "iron-cage" foreclosed to the advance of logic, Baudrillard opted to see the world from a poetic-view. Based on the premise that theory may be equaled to fiction, his books focused on a fresh literary interpretation to transcend the limitations of science. Mistakenly, some scholars have precluded that Baudrillard was the philosopher of non-sense, obscuring the meaning of events to the extent of claiming polemic assertions as "the Gulf war did not take place". It is important not to loose the sight the concept of reversibility. Originally coined by Greek philosophers, reversibility connotes the possibility to weaken the foundations of system by means of its own functioning.

Empires expanded their hegemony by the imposition of force over the colonies, but at the same time a counter-force is gradually undermining their hegemony towards the final collapse.

In other terms, Coulter writes "reversibility is (ironically) a strong antidote to determinism and linear theories of progress. In our time (the time since Barthesian structuralism became pregnant with post-structuralism), reversibility has replaced dialectics" (p. 7). Poets emerge from the ambiguity that comes after success. One of the troubling aspects of post-structuralism is that it destabilizes the horizons of certainty and firm knowledge. Both scholars, Coulter urges, contributed to the understanding of post-structural emptiness in ways that the whole portion of philosophers avoid. In sharp opposition to philosophy that proclaimed the world is predestined by knowledge, Baudrillard acknowledged that the spirit of ongoing uncertainness was rather determined by language. Since any interpretation is open and subject to meaning, any word signifies something different depending on the reader. Nonetheless, both philosophers share commonalities but substantial discrepancies respecting to what "writing" means.

While Barthes saw writing as a limited position to the being simply because I exist beyond what I can write, Baudrillard preferred to take the road of the poetic. Even he argued that intellectuals, in a pejorative way, devoted efforts to legitimate the "empire of meaning". Condemned to deconstruct its own aura, any text disturbs the conscience but intellectuals are not enthroned to speak in the name of other (Baudrillard, 1994; 1993; 2002).

In his life, Baudrillard struggled to develop an all-encompassing conceptual framework that helped people understanding the acceleration of history that recycled the present history. By comparing the war on terror declared by Bush, minority report, where "precogs" facilitate to police the arrests before the crime to be committed. This notion of legal jurisprudence defies the roman meaning of crime. While Barthes employs the term terror, Baudrillard contends that irony is based on the terrorism of meaning because it makes the system working against itself, as an autoimmune virus.

Recognizing that "nothing can be said about the world," Baudrillard is convinced that appearances have replaced the meaning, in a type of vertigo of interpretations. The violence of interpretation lies in the impossibility to separate the reasons from effects in the events as the media portrays. News about natural disasters produced by global warming effects is equaled to terrorism and quakes without any kind of distinction. Not only the system of meaning has been altered by capitalism and the media, but upended.

Baudrillard said that West is debating itself in a quandary between real and its double, simulation. However, we live in a moment where the real is framed by the principle of simulation. Since the real cannot override the copy, the hyper-real substitutes the real, at the same time the acceleration of images emptied the pastime. What today we may learn from disasters comes from cinema and the film-industry. In view of this argument, disasters as are often commoditized and sold by the media are not real events, but pseudo-events. The visual technology in the digital times has not only changed the perceptual horizons accelerating the time and space but created the end of resiliency as known today. This striking point represents a fertile ground in potential research in the fields of disasters and risk-management (Baudrillard, 1983; 1993; 2002; 2003). The legacy of Baudrillard reminds us that risk serves, as invention (construal) enrooted in the future, that serves to mould the daily behaviour today. What policy-makers plan, may not occurred as they imagined. At what extent technology accelerated the present and future, is the second interesting point of discussion, wherein Paul Virilio based his last work, *the University of Disaster.*

His theory is aimed at exploring the role of "dromology", a neologism coined to denote the study of velocity, in late modernity. The advances of technologies and mobilities not only have created new forms of displacements, but also blurred the relation between time and space. As a result of this, Virilio adds, people have fully access to any geographical point of this globe in hours. The time of waiting has changed forever. Travelers now are moved by the indifference and visual consumption, there is not genuine contact in the visited lands. At some extent, The University of Disaster not only synthesizes years of investigation, but represents the corollary of a critical thought respecting to mobility and mass media.

The events in past formed the history as a continuance of ordered facts, but the real-time makes people any longer would be able to synchronize watches. Citizens has transformed in consumers. Certainly, the history has been emptied in a fragmentation of events, dispersed elsewhere and broadcasted once and once again. The knowledge that characterized the labor of University has been polarized to delocalize territories. Based on an ongoing future that never takes room the presentiment of disaster announces the eschatology of neurosis. In other terms, Virilio argues that everything happens at the same time in the hyper-reality without a logical sequence. The world stage is being represented outside the planet, in an exo-earth. The days of Science, as an all-encompassed instrument based on rational understanding, has changed. Transformed in an exo-science that promotes the simultaneous globalization of fear, biology and astronomy are eclipsed by the "eternal present". As the previous argument given, the "University of Disaster" reminds the "mea culpa" of science for its failure in creating an ethic of life. Based on the belief that the global warming is not reversible in the short-run terms, Virilio strongly believes in the importance to analyze the sense of homeland safety and security.

To be protected, the big corporations, banks and capital elite call climatologists and geographers (experts) to design the necessary Catastrophe simulation software that provides some information where the next disaster will take hit. In this vein, a new profession is uprising, the "economic disaster- modeling-geek". This expert seems to be more interested in finding and eliminating the risks of businesses than in protecting the environment. The philosophy of the science is today determined by the logicism of digital screens. The simulation of future that characterizes the digital world has replaced the daily life. Of course, this begs an interesting question, what is the role of experts in this process?.

The advance of science moved to snail´s pace by prioritizing the quality of knowledge. Its objectivity lies in the observation of facts enrooted in reality. However, things have changed a lot. The digital world has blurred the time, prompting the Science to study thousand of simultaneous events, which do not lead to any coherent logic. The reality is not any-longer the object of scientific research. Virilio´s work inspires a new reflection about the design of disasters in the early modern times. His critique view gives a conceptual model to understand the current "show of catastrophe" televised 24 hours day to a wider globalized audience. Similar in the argument to Baudrillard or Augé, Virilio is convinced that humankind has to come back to an ethic of the science whose concerns have been aimed at protecting the integrity of human beings (Virilio, 2010). The technology expanded the limits of cities towards the border of planet. The contours are drawn while the deep space is situated as the only line of horizon that defines human habitat. Virilio insists on the belief that "the technical consciousness is what you put on the screen". What is important to discuss here is not the "Techno-phobia", but "techno-philia" which prioritizes the measure than meaning. The growth of simulation software has been adopted in the domain of education, administration and sports. Basically, these types of tools intend to lead people to the most efficient decision to optimize their performances. It is important not to loose the sight that knowledge only may be understood in accordance to a specific time and space. Without places and present, the information is circulating through the lens of televisions. The problem seems to be that the existent flow of information created by technology, which accompanied the transport in other times, is now the transport itself. Proponents of ecology, climatologists, geographers and other scientists preoccupied in Global warming have no clue in what is going to happen, but they are recruited by insurance companies and corporations to design the next protection-related products in the market (Virilio, 2010). This tendency, which is only facilitated by the concept of predestination, leads societies to develop instruments capable to perform an accurate prediction on the future events. Although predestination is still a key factor in English speaking countries there are others elements which should be further discussed. In next, we will place the books authored by professor Geoffrey Skoll, emeritus from SUNY at Buffalo, under the lens of scrutiny.

The Culture of Fear, Genesis, and Expansion

This section is oriented to explain, through two seminal books, the genesis and expansion of America, as well as its predisposition to fear. The idea that capitalist system reproduces by the growth of inequality and poverty is not new since it has been discussed widely by Marxism and post-Marxism in the four corners of the world. However, the intersection of terrorism and capitalism seems be a heated discussion which today retains considerable concern from scholarship. We are told that working is the best project in what we can spend our time whereas terrorism is an act of hatred-filled hearts or maniacs. But to what extent it is true?

To what extent, the sentiment of fear is manipulated for governments to impose their interests to workers, is an ethical dilemma widely discussed within the circles of philosophy. From its inception, the founding parents of democracy in America concerned to the possibilities few hands use the fear to concentrate further power leading the Republic into a Dictatorship. Their efforts in constructing a genuine Republic which was based on the autonomy of branches not only set the pace to more reified forms of panic, which are triggered by the essence of terrorism, but has been placed on the tapestry after US-led Invasions to Iraq. This point of entry adjoined to the doctrine of lesser evil, is present in the book *Social Theory of Fear where* Skoll dissects Wallerstein`s prelude on the collapse of capitalism.

Pushed to a dichotomy between two contrasting alternatives, capitalism, like the Roman Empire, not only has been exhausted but also is globalized to the extent of running a serious risk of collapsing. From this vein, Skoll writes:

... catastrophes occur when systemic regulators no longer contain the conflict through various institutional responses. Such a crises always hold the potential for bifurcation of the system. Bifurcation occurs at a tipping point where the system stops organizing itself and enter in a chaotic state. (p. 28)

The elite's reaction in times of crisis seems not to be consistent with their own interests. Aristocrats do not try to save the system and its well-functioning, which paradoxically facilitated the collapse. Egocentrism coupled with individual interests and a wider sentiment of fear, are at a first stage, a big problem for involving societies and a sentence of death. Comparable with the Roman Empire's and Feudal disintegration, late-capitalism's crisis seems to recur to military-machine expansion to keep the control and trade, but unlike these past-empires, in our modern nation-state they are inextricably interconnected. As a result of this, the collapse may be very well being further apocalyptic than at other times. The reliance on fear works as a conduit for the status-quo and elites to maintain their privileges.

As the previous argument gives, Skoll introduces readers in a fascinating text, not only very well-written but also illustrative precisely in a theme that now always remains clear. Throughout the existent economic crisis, Skoll adds, 9/11 can be equaled to the Reichstag Fire in 1933 where leaders imposed the need to a preventive attack (against communists) and even beyond the boundaries of Germany. At some extent, the U.S. took advantage of the tragedy that represented the World Trade Center.

Regardless of questions about the origins, causes, or perpetrators of the attacks, 9/11 marked a turning point in the history of the world, because the political and economic leadership took advantage of the opportunity Bush had identified. At this point, and maybe forever, their precise thoughts, plans and strategies remain hidden. (p. 35)

Following this reasoning, the consequent interventions in Middle East follow significant lines based on "military action" to achieve dominance in strategic zones, mass-support in metropolitan towns, and extraction of local resources or surplus from conquered countries. Undoubtedly, for Skoll, the policy of fear conducted by the U.S. and its war on terror are a result of a planned-campaign conducted by ruling classes in times of uncertainty. Of course, whenever these classes feel under attack or in danger, fear allows two important things. The most important is the internal indoctrination that gives sense to a shared-territory and culture, but secondly fear revitalizes consumerism by trivializing the critique.

The diverse chapters that structure this book are suggested but the third can be surely considered the stepping stone of Skoll´s development. Throughout the reading of this book a interesting question arises: Why the Anglo-world supported the U.S. preventive war on terror? A coincidence? Skoll brilliantly said Canada, Australia and the U.K. not only share similar cultural background with the U.S., but also they represent an exclusive club of rich countries that resist slumping down. This sentiment of terror paves the ways for the advent of dictatorship simply because it gradually modifies the rights of people, constitution, and the legal-framework that assures the principle of civility. Skoll´s deep gaze examines slippery matters as torture, fear, local-crime aversion, modernity, evolution and state and the end of liberty in a fluid and all-encompassed way. With this book, one starts to understand the events that characterized the U.S. international strategies in the last years.

Discussing this in depth, the function of the state is to maintain the equilibrium by exerting power and violence over populations. In times of low-conflict, the legitimacy of the state rests on the market, which confers to the system a certain stability. Nonetheless, in contexts of chaos and disorder the state recurs to violence to refashion the lost order. Similarly, the market mediates among human beings, imposing a state of gratification in lieu of constraints, but in the moment the control weakens, fear seems to be the last attempt to recover legitimacy. Skoll goes on to acknowledge that "fascism exercises social control through deprivation, identification with a powerful leader, and aggression against internal and external enemies. Liberal capitalism replaces denial with indulgement consumerism and lifestyles replace status identifies such a race" (p. 45). Afterwards the attacks in New York in 2001, these subtle forms described in the earlier paragraph set the pace to the logic of fascism. As previously explained, Skoll argues that fascisms (and Stalinism) were certainly constructed under the logic of emergency. These policies permit to the taking of control, suppressing elementary rights.

The hegemony of the liberal mind is associated with consensus and not violence. If hegemonic control is based on two forms, alliances with neighbors and violence, then the U.S. took the monopoly of capital, expanding loans to other developing countries by means of international organisms as World Bank and IMF. The economic dependence of peripheral countries was indeed counter-productive for the U.S. for many reasons. The first attempt to articulate a systematic expansion was the issuance based on interest that created a huge debt, and secondly, a growing sentiment of resentment of the Third World since these loans aggravated their situations. The globalized capital engendered for the liberal state a special form of obedience that in contrast with other authoritarian republics shaped a specific consciousness centered on calculation, cultural consumption and rationale. The policies of terror are the necessary result for U.S. expansion that led involuntarily to the current crisis. While authoritarian regimes suppress the dissent, liberal hegemonies marginalize it. That way, one of the main contributions of Skoll to political fields is linked to the need of reconsidering the roots of democracy and dictatorship, and re-thinks under what context a supposed democratic country can trespass the boundaries of autocracy (the power of only one) and of course the pervasive role of fear in such a process. Trauma can alter the reality of self, and 09/11 changed forever U.S. policy with respect to international affairs (Skoll, 2010)

In a recent publication, *Globalization of American Fear Culture,* Geoff Skoll continues a much deeper discussion instilled by former work, *Social theory of Fear.* At a closer look, the act of governing through fear is used by US to create a culture of mistrust. From the onset, US was based on the belief that the world as a dangerous place to live. Over recent decades in XXth century, fear was used to enhance legitimacy of elite, sometimes oriented for the workforce to accept policies otherwise would be rejected. At the time, capital and American Empire expanded to colonize new worlds, a much deeper sentiment of fear accompanied the politics. Therefore, it is safe to say we live in a world characterized by a "global fear" which is functional to a particular way of making politics. One of the aspects facilitated the expansion of capitalism was the idea that citizens live in the best of the possible worlds. Beyond the boundaries of consuming society, of course, any change represents a threat for popular parlance, a barrier to overcome. In the first chapter, Skoll traces the historical roots of the culture of fear in America. In retrospect, the capitalist system and theories of economy showed widely that accumulation only is feasible if we introduce exploitation as a key factor to produce and distribute wealth in few hands. The monopolization of surplus value, as Marx puts it, resulted not only from human creativity but by the means of elite to commoditize labor into exchangeable goods. The number of rank-and-file workers involved in a process of production affects directly the profits of capital owners. From that moment on, the capital reproduction seems to be always in detriment of workforce. In this mayhem, whenever the conflict rises, fear undermines the possibilities of claimers and protesters to impose their views. Two major instruments were used by privileged classes to keep the control, ideology and repression. While the latter appealed to surveillance to exert violence against the pathological agents, the latter one was enrooted in a process of fear-mongering that limited the negotiation of worker unions. After 9/11, total forms of control were established in private life subordinating individual rights to the collective well-being, which means a more secure society. Leisure industries were witness of obstructive methods of surveillance over lay citizens. It was unfortunate that this trend makes from US a fascist state. This is a very interesting introductory section where author combines his erudition pitting historical cases where governments used fear in their favor against US and its Anglo-allies in the war on terror. In the second chapter, Skoll discusses to what extent elite in America devoted its resources to forge a culture of fear which passed from communism towards terrorism. The organization of labor conjoined to profit maximization is two key factors behind the manipulation of fear. Though actors changed, the dynamic are the same. Doubtless, this is one of the merits place Geoffrey Skoll now as a must-read author in terrorism issues. Chapter 3 explores the model of four wars as it was formulated by Skoll where he confirms a polemic thesis. From its inception, US was always an imperialist power which struggled in four major events, *Philippine War, Korean War, Cold War and Vietnam War.* Though the two total world wars involved US as well, no less true is that in these events US played the role of an empire inspiring a model that situates United States as the "administrator" of capitalism. The management of exploitation centers in a genocidal campaign by disciplining communists. The period 1968-1973 not only encouraged a liberalization of human relationship, it induced substantial changes in economies worldwide. During this age the spin doctors of capitalism precaritized the power of workers paving the pathways for the rise of neoliberalism during 90s decade. As this backdrop, the expansion of US as the unique imperial power was possible after the collapse of Soviet Union, but without the legacy of UK in financial leadership, it would have never taken place. One mantle passed from one power to other as US enthralled as the center of manufacture and trade. The globalization of fear is assessed in chapter six, seven and eight. The direct intervention or full-scare led wars are ideologically legitimized by the needs of bringing the ideals of American democracy, liberty, freedom and mobility. However, at the bottom, this globalized culture of fear hidden dark interests associated to exploitation. Paradoxically, these types of interventions suggest that terrorism needs the use of force, but in so doing, impotence and deprivation surface. To set an example, Skoll adds, whilst US supported by financial investors, IMF or ONG arrives to Middle East to take the local politics on its hands, a state of chaos and political instability dominates the environment.

CONCLUSION

Undoubtedly, Anglo and Latin worlds have created, according to their cultural matrices, diverse tactics to adapt to environment, as the form of understanding the future. While Anglo-countries developed a fascinating attraction to risk and future, the Latin-speaking countries based their idiosyncrasy on poverty and sacrifice. If Catholics opted for the safety of dogma, Anglo Saxons launched to domesticate death. In this process the concept of predestination plays a crucial role. As previous argument given, the Anglo-culture goes to a high sentiment of anxiety because the future is unreachable, any attempts to regulate risk in a future only engenders new worse risks, not contemplated up to date. The sense of predestination alludes to what today has not occurred yet. Technology only helps to mitigate the temporal effects of uncertainty triggered by the orientation to future. This in part obliged to many Anglo countries to be on the top of technological and economic pyramid, as Weber put it, but they ran a big risk. This means the risk of reversibility or self-destruction. Last but not least, while Catholics deposited their expectances in religion and gods, Anglo hates their gods, because precisely they do not know who would be saved or condemned in the book-life. The hereafter is closed to forget the axiom of human death. This is the reason behind the bio-technology made serious efforts in extending life. In this chapter, we discussed the fear of terrorism as the touchstone of American culture in contemporary society to understand some of Weberian hypotheses. The allegory of war on terror leads governments not to tackle off real dangers produced by capitalism as pollution or global warming. Unless otherwise resolved, the question whether US and its domains pays attention to terrorism as the main threat of West covering other most terrible risks, one dangles that the probabilities our civilization collapses are higher than thought. Anyway, citizens are prone to develop sustainable practices of consumption that encourages a real democracy from the bottom to the top.

REFERENCES

Baudrillard, J. (1983). *Simulations*. New York: Semiotexte.

Baudrillard, J. (1993). *The Transparency of Evil: essays of extreme phenomena*. New York: Verso.

Baudrillard, J. (1994). *Simulacra & Simulation*. The University of Michigan Press.

Baudrillard, J. (2002). *Cool Memories IV*. New York: Verso.

Baudrillard, J. (2003). *The Spirit of Terrorism*. New York: Verso.

Beck, U. (1998). *La Invención de lo Social*. FCE.

Beck, U. (2006). *La Sociedad del riesgo, hacia una nueva modernidad*. Buenos Aires: Paidos.

Bell, D. (1974). *The Coming of postindustrial society. A venture in social forecasting*. London: Heinemann.

Coleman, S. (2013). *Actors of History? Religion, Politics, and Reality within the Protestant Right in America. In Religion, Politics & Globalization* (pp. 171–188). New York: Berghan Book.

Dodds, E. R. (1997). *Los Griegos y lo Irracional*. Madrid: Alianza.

Douglas, M. (1992). *Risk and Blame*. Routledge.

Douglas, M., & Wildawsky, A. (1983). *Risk and Culture: an essay on the selection of technological and environmental Dangers*. Los Angeles, CA: University of California Press.

Dutton, W. (1998). *Society on the Line: information politics in the digital Age*. Oxford, UK: Oxford University Press.

Fromm, E. (2005). *El Miedo a la libertad*. Buenos Aires: Paidos.

Giddens, A. (1991). *Modernity and Self-Identity: Self and Society in the Late Modern Age*. Stanford University Press.

Giddens, A. (1997). *The Transformation of Intimacy*. Oxford, UK: Polity Press.

Greven, P. (1988). *The Protestant Temperament. Pattersn of child-rearing religious experience, and the self in Early America*. Chicago University Press.

Korstanje, M. (2012). Aportaciones de la Mitología Nórdica para comprender la modernidad, revisión de Max Weber. *Historia Actual.*, *28*(1), 175–190.

Nigg, J. (1995). *Disaster Recovery as a social process.* Preliminary Paper #219. Disaster Research Center, Universidad of Delaware.

Nigg, J. (1996). *Policy Issues for post disaster mitigation the need for a process.* Preliminary Paper # 244. Disaster Research Center, Universidad of Delaware.

Rifkin, J. (1998). *The biotech Century: a coming age of genetic commerce.* London: Victor Gollancz.

Schiller, H. (1981). Who knows: Information in the age of the fortune 500. Norwood.

Skoll, G. (2010). *Social Theory of Fear: Terror, Torture, and Death in a Post-Capitalist World.* New York: Palgrave-MacMillan.

Skoll, G. (2016). *Globalization of American Fear Culture: the empire in the Twenty-First Century.* New York: Palgrave-Macmillan.

Virilio, P. (2010). *The University of Disaster.* Oxford, UK: Polity Press.

Weber, M. (1958). *Essays in Sociology.* New York: Oxford University Press.

Weber, M. (1964). *Economía y sociedad: esbozo de sociología comprensiva.* FCE.

Weber, M. (1995). *La Ética Protestante y el Espíritu del Capitalismo.* Barcelona: Península.

Webster, F., & Robins, K. (1986). *Information Technology: a luddite Analysis.* Norwood.

ADDITIONAL READING

Bustamante, A. M. (2013). The Impact of Post-9/11 US Policy on the California–Baja California Border Region. *Journal of Borderland Studies, 28*(3), 307–319. do i:10.1080/08865655.2012.751729

Caraway, T. L. (2008). Explaining the dominance of legacy unions in new democracies comparative insights from Indonesia. *Comparative Political Studies, 41*(10), 1371–1397. doi:10.1177/0010414007303094

Caraway, T. L., Cook, M. L., & Crowley, S. (Eds.). (2015). *Working Through the Past: Labor and Authoritarian Legacies in Comparative Perspective.* Ithaca: Cornell University Press.

Centeno, M. A. (2002). *Blood & Debt. War and the Nation-State in Latin America.* Pennsylvania: The Pennsylvania State University Press.

Clara, L., Crespo, H., & Yankelevich, P. (2008). *Argentina 1976: estudios en torno al golpe de Estado*. Buenos Aires: FCE.

Collier, D., & Norden, D. L. (1992). Strategic choice models of political change in Latin America. *Comparative Politics*, *24*(2), 229–243. doi:10.2307/422280

Cropf, R., & Bagwell, T. (2016). *Ethical Issues and Citizen rights in the era of digital Surveilance*. Hershey: IGI Global. doi:10.4018/978-1-4666-9905-2

Davis, D. W., & Silver, B. D. (2004). Civil liberties vs. security: Public opinion in the context of the terrorist attacks on America. *American Journal of Political Science*, *48*(1), 28–46. doi:10.1111/j.0092-5853.2004.00054.x

de Imaz, J. L. (1984). *Sobre la identidad iberoamericana*. Buenos Aires: Editorial Sudamericana.

Dinges, J. (2005). *The Condor Years, how Pinochet and his allies brought terrorism to three Continentes*. New York: The New Press.

Eid, M. (2014). *Terroredia: Exchanging Terrorism Oxygen. Exchanging Terrorism Oxygen for Media Airwaves: The Age of Terroredia*. Hershey: IGI Global. doi:10.4018/978-1-4666-5776-2

Entel, A. (2007). *La Ciudad y sus Miedos: la pasión restauradora* [The city and its fears]. Buenos Aires: La Crujía Ediciones.

Feierstein, D. (2006). Political violence in Argentina and its genocidal characteristics. *Journal of Genocide Research*, *8*(2), 149–168. doi:10.1080/14623520600703024

Feierstein, D. (2014). *Genocide as Social Practice: Reorganizing Society Under the Nazis and Argentina's Military Juntas*. Washington, DC: Rutgers University Press.

Foucault, M. (2000). *Defender la sociedad*. Buenos Aires: Fondo De Cultura Económica.

Franco, M. (2012). *Un enemigo para la nación. Orden interno, violencia y subversión, 1973-1976*. Buenos Aires: FCE.

Fukuyama, F. (2006). *The end of history and the last man*. New York: Simon and Schuster.

Gadarian, S. K. (2010). The politics of threat: How terrorism news shapes foreign policy attitudes. *The Journal of Politics*, *72*(02), 469–483. doi:10.1017/S0022381609990910

Grossman, M. (2014). *Artists of the possible. Governing networks and American policy change since 1945*. Oxford: Oxford University Press. doi:10.1093/acprof:oso/9780199967834.001.0001

Guest, I. (1990). *Behind the disappearances: Argentina's dirty war against human rights and the United Nations*. Pennsylvania: University of Pennsylvania Press.

Guidotti-Hernández, N. (2011). *Unspeakable Violence: Remapping U.S. and. Mexican National Imaginaries*. Durham, NC: Duke University Press. doi:10.1215/9780822394495

Haggard, S., & Kaufman, R. K. (1995). *The Political Economy of Democratic Transitions*. Princeton: Princeton University Press.

Hobsbawm, E. (1984). *The Age of Extremes. A history of the world 1914-1991*. New York: Vintage.

Howell, J., & Lind, J. (2009). *Counter-terrorism, aid and civil society: Before and after the war on terror*. New York, Palgrave-Macmillan.

Howie, L. (2007). The terrorism threat and managing workplaces. *Disaster Prevention and Management: An International Journal*, *16*(1), 70–78. doi:10.1108/09653560710729820

Huntington, S. P. (1993). If not civilizations, what? Samuel Huntington responds to his critics. *Foreign Affairs*, *72*(4), 186–194. doi:10.2307/20045880

Ignatieff, M. (2013). *The lesser evil: Political ethics in an age of terror*. Princeton: Princeton University Press.

Jelin, E. (2003). *State repression and the labors of memory* (Vol. 18). Minneapolis: University of Minnesota Press.

Keck, M., & Sikkink, K. (1998). *Activist Beyond borders*. Ithaca: Cornell University Press.

Kekes, J. (2005). *The roots of evil*. Ithaca: Cornell University Press.

Keohane, R. O. (2002). The globalization of informal violence, theories of world politics, and the "liberalism of fear". *Dialogue IO*, *1*(01), 29–43.

Knudson, J. W. (1997). Veil of Silence: The Argentine Press and the Dirty War, 19761983. *Latin American Perspectives*, *24*(6), 93–112. doi:10.1177/0094582X9702400605

Korstanje, M. (2012) *Estado, Política y Religión: reflexiones para comprender la Argentina contemporánea*. Lap Lambert Academic Publishing, Saarsbrucken.

Korstanje, M. E. (2006). Identidad y Cultura: un aporte para comprender la Conquista de América. *Iberia: Revista de la Antigüedad*, (9), 191-212.

Korstanje, M. E. (2010). El miedo político en C. Robin y M. Foucault. *Revista de Antropología Experimental, 10*(6), 111–132.

Korstanje, M. E. (2015). *A Difficult World, examining the roots of Capitalism*. New York: Nova Science.

Korstanje, M. E. (2015). *A Difficult World: Examining the roots of Capitalism*. New York: Nova Science Publishers.

Korstanje, M. E. (2016) "A Review on Working through the past". *Review of Books*. London School of Economics, Available at http://blogs.lse.ac.uk/lsereviewofbooks/2016/04/14/book-review-working-through-the-past-labor-and-authoritarian-legacies-in-comparative-perspective-edited-by-teri-l-caraway-maria-lorena-cook-and-stephen-crowley. Extracted May 02, 2016

Korstanje, M. E., & George, B. P. (2012). Falklands/Malvinas: A re-examination of the relationship between sacralisation and tourism development. *Current Issues in Tourism, 15*(3), 153–165. doi:10.1080/13683500.2011.569535

Korstanje, M. E, Skoll G, & Timmermann F. (2014) "Terrorism, Tourism and Worker unions, the disciplinary boundaries of fear". *International Journal of Religious tourism and Pilgrimage.* 2 (2): 1-12

McCauley, C., & Moskalenko, S. (2008). Mechanisms of political radicalization: Pathways toward terrorism. *Terrorism and Political Violence, 20*(3), 415–433. doi:10.1080/09546550802073367

Murillo, S. (2008). *Colonizar el dolor: la interpelación ideológica del Banco Mundial en América Latina: el caso argentino desde Blumberg a Cromañón*. Buenos Aires: Clacso.

Norden, D. L. (1996). *Military Rebellion in Argentina: Between Coups and Consolidation*. Lincoln: University of Nebraska Press.

Norden, D. L. (1998). Party relations and democracy in Latin America. *Party Politics, 4*(4), 423–443. doi:10.1177/1354068898004004002

Palacios, M. (2012). *Violencia Publica en Colombia, 1958-2010*. Bogotá: FCE.

Piazza, J. A. (2006). Rooted in poverty?: Terrorism, poor economic development, and social cleavages 1. *Terrorism and Political Violence, 18*(1), 159–177. doi:10.1080/095465590944578

KEY TERMS AND DEFINITIONS

Culture of Fear: Climate fostered by State to impose policies otherwise would be neglected by citizenry.

Democracy: Form of government where a majority of citizens are involved to make decisions.

Fear: One of the basic emotions which serves for the organism to survive.

Globalization: Interconnection between economies and cultures to form as an all-encompassing entity.

Risk: The determination of necessary assessment to evaluate what are potential threats for a person or a group.

Terrorism: Understood as any act which is associated to illegal violence.

Chapter 6 Trolls Just Want to Have Fun:
Electronic Aggression within the Context of E-Participation and Other Online Political Behaviour in the United Kingdom

Shefali Virkar
University of Oxford, UK

ABSTRACT

Over the last two decades, public confidence and trust in Government has declined visibly in several liberal democracies, giving way instead to disillusionment with current political institutions, actors, and practices; rendering obsolete or inappropriate much of traditional democratic politics. Simultaneously, digital technologies have created huge opportunities for public bodies and agencies. In analysing the No. 10 Downing Street ePetitions Initiative based in the United Kingdom, this chapter engages with issues related to the innovative use of digital network technology by Government to involve citizens in policy processes and to buffer national security within existing democratic frameworks. The work examines whether the application of new digital platforms to participatory democracy in the Government 2.0 era leads eventually to radical transformations in government functioning and the body politic, or merely to modest, unspectacular political reform and to the emergence of technology-based obsessive-compulsive pathologies and trolling behaviours amongst individuals in society.

DOI: 10.4018/978-1-5225-1938-6.ch006

INTRODUCTION

During the last two decades, there has been increased questioning of traditional democratic politics in Western liberal democracies, largely due to a decline in and a lack of opportunity for public participation in these processes. Such concerns are largely thought to be embodied in (amongst other phenomena) low voter turnout during elections; a trend particularly noticeable amongst young people where only half of those eligible to vote actually do so (The Electoral Commission Report, 2005). This is especially worrying and problematic for governments, as it speaks of growing political apathy and a broader, more general disillusionment with current political institutions, actors and practices. Whilst it is impossible to comprehensively untangle all the reasons for the decline in civic participation in these countries, there is little doubt that many citizens feel distanced from any sense of political relevance or power, often under the impression that not only will their votes and individual voices be drowned out in the clamour of the crowd, but that the rules which govern their daily lives are drawn up by politicians and bureaucrats whom they will never meet and who are usually extremely difficult to contact (Eggers, 2005).

Leading commentators have described the political processes and institutions integral to Western democracies as undergoing what has been variously described as 'a crisis of legitimacy', a 'credibility crisis' or a 'crisis of democracy' (cf. Habermas, 1985, Archibugi & Held, 1995), and are fast reaching agreement that the fundamental flaw lies in traditional decision-making practices which are, in their current form, often democratically inadequate as they fail to provide extensive and relatively equal opportunities for citizens, communities and groups to contribute towards the shaping of decision-making agendas (Sclove 1995). The focus of discourse and scholarly activity, both in academic and policy circles, has thus gradually shifted away from a more centralised, top-down conception of 'government' – those formal institutions and processes which operate at the level of the nation state to maintain public order and facilitate collective action (Stoker, 1998) – towards the notions of 'governance', an idea which, while traditionally a synonym for government, has been captured in recent theoretical work as signifying 'a change in the meaning of government referring to a new process of governing; or a changed condition of ordered rule; or the new method by which society is governed' (Rhodes, 1996: 652).

Governance is thus seen to be ultimately concerned with crafting the conditions for ordered rule and collective action, or 'the creation of a structure or an order which cannot be externally imposed, but which is the result of the interaction of a multiplicity of governing and each other influencing actors' (Kooiman & Van Vliet, 1993: 64), and is thus a conceptual way of capturing shifts in the character of political rule which has been stretched to encompass a range of different transformations including an emphasis on drawing citizens and communities into the process of collaborative participation in political processes and the creation of new forms of governable subjects (Newman, 2005).

This chapter aims to assess, through the use of a relevant exploratory case study, whether the innovative use of network technology (in particular the Internet and its associated applications) in the Government 2.0 era would, in attempting to involve citizens in policy processes within existing democratic frameworks, eventually lead to radical transformations in government functioning, policymaking, and the *body politic*, or merely to modest political reform and to the emergence of obsessive-compulsive technology-based pathologies and Internet-based addictions amongst individuals in society. In doing so, it seeks to explore the factors, particularly those rooted in a country's legal and institutional foundations, that might hinder or enable the successful implementation of e-consultation projects at different levels of government and develop a set of recommendations for overcoming any barriers encountered. The main idea behind Government 2.0 is participation by citizens, through the use of Web 2.0 Technologies, a term referring to the collection of social media through which individuals are active participants in creating, organizing, editing, combining, sharing, commenting, and rating Web content as well as forming a social network through interacting and linking to each other.

GOVERNANCE, ICTs, AND PUBLIC ENGAGEMENT IN THE EUROPEAN UNION

Whilst early speculations about the Internet and democracy emphasised the potential for direct, unmediated debate and discussion and stressed the radically transformative nature of the process of public engagement in a manner that led some scholars to go as far as to predict the end of the nation-state in favour of direct democracy (Margolis, 2007), this chapter follows the view of those who believe that whilst e-democracy in its purest sense may be altogether incompatible with a political culture of elitism (often unavoidable within the framework of a representative democracy), it is in practice sometimes neither feasible nor indeed desirable to replace what has evolved so far. Both theorists and practitioners talk of creating a 'civic commons in cyberspace' that would elicit and coordinate citizen comments and reactions to problems facing public institutions in order 'to create a link between e-government and e-democracy – to transcend the one-way model of service delivery and exploit for democratic purposes the feedback paths that are inherent to digital media' (Coleman, 2011).

The emergence of the European Union (E.U.), and its evolution as a policy-making and legislative body, particularly over the last 20 years, has, raised several questions regarding the extent to which Europe constitutes a transnational public sphere in which citizens can debate and participate (Scharpf, 1999), particularly as the process of European integration over the last decade and a half has initiated a profound restructuring of the region's 'public space'; within member-state polities, at the European level and in a complex interplay between European, national, regional and local levels (Koopmans & Statham, 2000). This intricate arrangement is clearly illustrated by the fact that whilst policy decisions in Europe are increasingly taken at the supranational and international levels, the nation-state has remained the primary focus for collective identities; with public debates and citizen participation in policy processes both still primarily situated at the national level and generally directed by national or local authorities.

Trust in European institutions and support for the integration process has steadily declined since the early 1990s (Thomassen & Schmitt, 1999), and many scholars point to a democratic deficit at the core of existing legitimacy problems within the E.U. which they attribute mainly to, on the one hand, a dichotomy resulting from the development of Europe's institutions and their increasing competencies and influence on European citizens' conditions of life (Bicking & Wimmer, 2010) and, on the other, inherent flaws present in those same institutions resulting in a limited 'Europeanisation' of public discourse across the European Body Politic in comparison with the emphasis placed on the national political space as an arena for public debates and a source for collective identification and notions of citizenship (Peters et. al., 2005). It is fast being recognised that in order to promote greater inclusiveness, efforts must be made to better connect citizens to the people and institutions that represent them, and better embed these links within the networks of government (Kies & Wojcik, 2010).

RESEARCH METHODOLOGY

Case study research consists of a detailed investigation of phenomena within a given context, often with data being collected over a period of time, the aim of which is to provide an analysis of the surrounding environment and processes to throw light on the theoretical issues being investigated (Eisenhardt, 1989). The phenomenon under examination is thus not isolated from its context, rather it is of interest precisely because the aim is to observe and understand actor behaviour and/or organisational processes and their interplay with the surrounding environment. The use of a case study itself is therefore not as much a method as it is a research strategy, where the context is deliberately included as part of the overall design. Today, case studies are widely used in organisational research across the social sciences, indicating growing confidence in the approach as a rigorous research strategy in its own right (Hartley, 2005). As research done by adopting this strategy is typically done in the field, the presence of too many observations and uncontrollable 'variables' makes the application of standard experimental or survey approaches infeasible.

Further, information tends to be scattered and generally cannot be picked up using one single method. Case studies thus typically combine a number of data collection methods such as participant observation, direct observation, interviews, focus groups, ethnography, document analysis, questionnaires etc., where evidence may be quantitative or qualitative depending on the research issues at hand. The approach is consequently flexible, allowing for new methods to be incorporated as new sources of data and new actors present themselves. The case study approach may thus be and has been used for various purposes – to provide a descriptive narrative, to generate new theory, or to test existing theory through the triangulation of data (Virkar, 2011).

From the above discussion, it follows that the use of a case study for this chapter is particularly apt for two reasons. Firstly, the approach is particularly useful for examining research issues that require a detailed understanding of socio-political, economic, or organisational processes through the collection and analysis of rich data. Secondly, as discussed above, case study research design is also more flexible than other frameworks such as laboratory-based or survey-based approaches, in that it is able to reconcile different research methods and harness the evidence gathered to generate novel theory from any creative insights that might ensue from the juxtaposition of data at various points in the analysis.

The choice of case study aims to fill the gaps in the existing literature on ICTs and public participation in the European Union, firstly by focusing on the factors arising from the political dynamics and organisational culture within the context of online political participation on the Prime Minister's e-Petitions Gateway, and examining how these in turn influenced élite and non-élite political actor perspectives as circumscribed by the relationships between actor perspectives, organisational reform, and institutional change. It was felt that the case study method would be best suited to this exploratory study given the multiplicity of sources of evidence available, and the reliability and validity of conclusions drawn would benefit from a convergence or 'triangulation' of data collected from these sources.

The empirical study presented in this chapter has one key limitation, which is that the discussed findings and their implications are obtained from a single case study that examined a particular mode of e-participation using an experimental technological platform targeting a specific user group. Thus, although the findings throw light on some of the non-technical factors influencing the outcome of e-consultation projects, indeed e-participation initiatives, worldwide, caution needs to be taken when generalizing these findings and discussion to other technologies, groups, or governments. That said, despite this threat to the validity of the case at hand, the use of mixed method data triangulation has ensured that study fully captures the key elements of and challenges to the development of innovative e-participation technologies based on old forms of citizen engagement.

DIS(CONNECTED) CITIZENSHIP? AN EXPLORATION OF KEY CONCEPTS

The idea of governance, and by extension e-Governance, may therefore be said to comprise of two distinct but complementary elements: that of e-Government – which encompasses all the formal institutional and legal structures of a country, and **e-**Democracy – which can be said to refer to the participative and deliberative processes which operate within those structures (Virkar, 2007). Broadly speaking, on the one hand, e-Government itself may be divided into two distinct areas:

1. e-Administration, which refers to the improvement of government processes and to the streamlining of the internal workings of the public sector often using ICT-based information systems, and
2. e-Services, which refers to the improved delivery of public services to citizens through multiple electronic platforms (Virkar, 2011).

On the other, the concept of e-Democracy may be further subdivided into two distinct areas: e-Engagement (or e-Participation), which emphasises opportunities for greater consultation and dialogue between government and citizens, and e-Voting, the expression of fundamental democratic rights and duties online (Virkar, 2007).

e-Engagement as a policy, if defined by an express intent to increase the participation of citizens in decision-making through the use of digital media, would consequently involve the institutionalised provision of resources to facilitate the responsible and collaborative decision making involved ultimately in institutional and social change. Whilst the earliest speculations about the Internet and Democracy emphasised the potential for direct, unmediated participation (OECD 2001) and the transformative nature of the process of public engagement, this chapter follows the view of scholars such as Coleman and Gotze (2001) that whilst e-Democracy is incompatible with a political culture of élitism, it is not about replacing what has evolved so far but instead, rather than seeking to radically transform governance along any particular ideological line, it aims to complement the institutions and processes of representative democracy.

In this view, facilitating the involvement of different sections of society in the process of government is now seen as a democratic prerequisite in many advanced liberal democracies, with some commentators such as Fishkin (1995) highlighting the need for 'mass deliberation', and emphasising the need for people and their representatives to be brought together to collaborate on issues of mutual interest. The recent exponential growth in access to new Information and Communication Technologies (ICTs), and the expansion of a newly-created digital environment wherein people shop, talk, and otherwise spend large parts of their lives in online spaces, has opened up a plethora of new opportunities for interaction between power elites and the various constituent elements of civil society. At the same time, their rapid proliferation has raised important questions and triggered debates as to who is able to participate and to what extent they may do so, as well as the types of participation such technologies make possible at different levels of government and their impact on different government institutions and democratic processes (Virkar, 2011).

Technologies used include blogs, wikis, social networking hubs, (such as Facebook and MySpace), Web- based communication modes like chatting and online chat groups, photo-sharing tools such as Flickr and Picasa, video casting and sharing platforms like YouTube, audio-sharing media such as Podcasts, mashups, widgets, virtual worlds, microblogs like Twitter, and the social annotation and bookmarking of Web sites (Ferguson, 2006). Through these social media individuals act as active agents in creating, organizing, combining and sharing information. The emphasis is on an outside-in wisdom of crowds approach, where data and information are created by a network of users outside of an organizational boundary in a collaborative manner. This is different from the inside-out authoritative know-all approach typical to the Web 1.0 era, where an organization or apex body is the key creator and organizer of the content and the people are considered passive mere consumers of information (Chun et.al, 2010).

THE INTERNET, POLITICS, AND THE PUBLIC: UNDERSTANDING POLITICAL PARTICIPATION ONLINE

The design and implementation of complex computer systems, such as those that support e-consultation platforms, requires a better understanding in practitioner circles of the users of such networks and the settings in which they work. Part of the problem resides in the implicit treatment of ordinary people as unskilled, non-specialist users of technology and their networks comprising of elementary processes or factors that can be studied in isolation in a field laboratory setting (Bannon, 1991). Although psychology has a long tradition of contributing to computer systems design and implementation, it has been a neglected discipline in scholarly circles and key issues such as those relating to the underlying values of the people involved in large-scale system design and their motivation in the work setting have been missed out in recent computer science-based scholarly analysis (Salvendy, 2012). Conceptualising and understanding people as actors in situations, on the other hand, with a set of skills and shared practices based on work experiences with others, requires a reorientation in the way in which the relationship between key elements of computer system design, namely people, technology, work requirements, and organisational constraints in work settings, is negotiated (Kuutti, 1996).

The use of the terms 'human factors' and 'human actors' give us a clue as to how people in system design clusters are approached (Virkar, 2011). More particularly, the terms highlight difference in how people and their contributions are perceived, the former connotating a passive, fragmented, depersonalised, somewhat automatic human contribution to the systems environment; the latter an active, controlling, involved one (Carayon et. al, 2012). More precisely, within the human factor approach, the human element is more often than not reduced to being another system component with certain characteristics that need to be factored into the design equation for the overall human-machine system (Czaja & Nair, 2012). In doing so, the approach de-emphasises certain important elements of work design: the goals, values, and beliefs which technologists and system-users hold about life and work (Jacko et. al., 2012). By using the term human actor, emphasis is placed on users and developers as autonomous agents possessing the capacity to control, regulate, and coordinate their behaviour, rather than them being on par and analysed as mere information processing automatons (Proctor & Vu, 2012).

The study of actor interactions is key to e-consultation initiatives, as it is important determine the impact that actor motivations have on the consultative process and subsequently on policy outcomes. One approach to understanding behaviour is to look at the composition of individual actors, rather than the system as a whole. This is largely because political actors are driven by a combination of organisational and institutional roles and duties and calculated self-interest, with political interaction being organised around the construction and interpretation of meaning as well as the making of choices (Virkar, 2013). The main actors in electronic consultation process may be placed into two groups:

1. **Internal Actors:** Comprise chiefly of those institutional actors responsible for the maintenance, upkeep and running of a project, including officers of the assembly who are responsible for the operation of the system such as IT specialists and forum moderators, and elected representatives (and their support staff) who respond to petitions individually and collectively.
2. **External Actors:** Comprise of two distinct categories including participants or the person (or group) who initiates an online interaction after identifying an issue and follows its progress through from submission to final feedback and outcome and citizens: those individuals who may or may not be entitled to participate but who will invariably impact the outcome of a policy process through their ability to shape public opinion.

The central issue that needs to be understood whilst studying the development of ICT platforms and their implementation in public sector organisations through an analysis of actor interactions is thus: Why do people do what they do? One approach to understanding behaviour is to look at the rationality of individual actors, rather than at human factors or at the computer system network as a whole. This is largely because human actors are driven by a combination of organisational and institutional roles and duties and calculated self-interest, with political, social, and economic interactions being organised around the construction and interpretation of meaning as well as the making of choices.

One approach to the study of political e-participation begins by defining and examining the motives and goals that prompt actors to interact and participate in decision- and policy making processes online. All behaviour is motivated in some way and individuals will engage in a particular behaviour in order to achieve a desired end (Atkinson & Birch, 1970). Political actors, in particular, have a complex set of goals including power, income, prestige, security, convenience, loyalty (to an idea, an institution or the nation), pride in work well done, and a desire to serve the public interest (as the individual actor conceives it). Added to this, individuals and private citizens tend to participate in politics for altruistic or conformist reasons, to boost their self-esteem, to self-enhance, and to achieve self-efficacy (Cruickshank, et. al., 2010). Actors range from being purely self-interested 'climbers' or 'conservers' motivated entirely by goals which benefit themselves and their status quo rather than their organizations or the society at large, to having mixed motives as 'zealots', 'advocates' and 'statesmen' motivated by goals which combine self interest and altruistic loyalty with larger values (Downs, 1964).

For citizens and users of the e-participation application, the motivation to use the system may be either intrinsic or extrinsic (Cruickshank et. al., 2010). Intrinsic motives include the desire to feel competent and self-determining, to show altruism, or to seek to increase the welfare of others. On the other hand, extrinsic motives are usually associated with some sort of external reward in the social, economic, or political sphere. Both these manifest themselves in conditional co-operation, social pressure, thresholds and the bandwagon effect (Margetts et. al., 2012). Different motives and goals may underlie the same surface behavior, with the social and psychological consequences of participation may be different for different users (i.e. some participate to gain information or support, others to communicate), resulting in a set of nested, interrelated interactions with the framework of a large meta-game or playing field (Virkar, 2011). Consequently, the motivations and goals for using the online resources will determine how they will they be used, by whom, and when.

An in-depth analysis of the ICT for development literature by this researcher identified five actor groups involved in games relating to the implementation of e-government projects:

1. **Politicians:** The first group identified comprises of elected representatives of various hues, guided and influenced chiefly by electoral imperatives and a need to maintain their public image, and are therefore concerned with directing both key economic policy issues as well as issues of public service delivery.
2. **Administrators/Civil Servants:** This group of actors is guided by their perceptions of existing institutional 'culture' and practices and their positive (or negative) attitudes towards internal bureaucratic reforms such as concerns about the down-sizing of administrative services to promote 'efficiency' and a sense of being policed by elected government through the introduction of ICTs.
3. **Organisations Dealing with Technical Designing of IT and ICT Systems:** The approach private IT suppliers take to e-government might be considerably different to what the adopting government agency actually needs or wants from a system.
4. **Citizens:** This is another particularly interesting group of actors as one is never quite sure what their reaction to the implementation of e-government will be. Whilst in theory citizens should welcome the introduction of a system that simplifies administrative processes, in practice it is equally possible that some citizens might not be very happy if a more efficient system was put into place.
5. **International Donors:** This final actor group controls the purse-strings and oftentimes comes to the table with 'higher' ideals coloured by ideas prevalent in international politics (such as the desire to see a particular brand of 'good governance' in the developing world).

The empirical study of the No. 10 Downing Street ePetitions Initiative presented in this chapter has one key limitation, which is that the discussed findings and their implications are obtained from a single case study that examined a particular mode of e-government using an experimental technological platform targeting a specific user group to solicit electronic feedback on key issues of politics and policy. Therefore, although the findings throw light on most of the key factors influencing the behaviour of actors and actor groups involved with the development of e-consultation platforms, indeed e-governance initiatives, worldwide, care needs to be taken to avoid over-generalizing these findings and conclusions whilst examining other technologies, groups, or governments. That said, despite this potential threat to the validity of the case at hand, the use of mixed method data triangulation as a research strategy ensures that the study fully captures the key elements of and challenges to and the motives and machinations prevalent behind the development and use of innovative e-government technologies based on old forms of democracy, citizen participation, and engagement.

REFLECTIONS ON THE INSIDER THREAT: CYBER SECURITY AND THE ARCHITECTURE OF SYSTEMIC CONTROL

Today the Internet is much more than a collection of esoteric discussion forums. It is a mass communications medium that been instrumental in defining the current age we live in and who we are to ourselves and to others. As humankind's emotional investment in the Internet has grown, the stakes for trolling and the indulgence in aggressive behaviour online have risen. Trolling has evolved from ironic solo skit to vicious group hunt (Schwartz, 2008). Technology, apparently, does more than just harness the wisdom of the crowd. It has intensified its hatred as well.

The converse viewpoint can be considered as the threat posed by insiders to a given architecture of systemic control. This perspective, in contrast to the phenomenon of trolling itself, is not new (Bellovin, 2008). Despite a distinct lack of non-insider access to military-grade technology during the pre-Internet era of the 1970s, the period witnessed a rise in the number of scholarly and practitioner canons documenting the problems associated with the exercise of internal discretion within the organisation globally (Pfleeger, 2008), particularly with relation to technological systems and their disabuse (Fink et.al., 2013). Parker (1978), for instance, in his treatise *Crime by Computer*, estimated that 95% of all computer attacks were committed solely by authorised users of a system (Parker, 1978).

What once held true for organisations and agencies before the advent of the Internet and its associated platforms and technologies, remains equally pertinent today (Virkar, 2011). More precisely, although criminality and other forms of illegal behaviour have existed from time immemorial, it has taken the combined power of the personal computer, the computational speed of the micro-chip processor, and the sheer scale and nature of the Internet to worsen the dimensions and the magnitude of electronic aggression and cybercrime in the modern era (Virkar, 2014).

At its most fundamental, the insider threat to systems architecture comprises of three basic forms of attack (Bellovin, 2008):

1. **Misuse of Access:** Refers to the misuse or disabuse of legitimate access privileges and is probably the hardest type of attack to detect or counter,
2. **Defence Bypass:** Refers to targeted localised attacks facilitated by the insider advantage leveraged unfairly by authorised personnel, and
3. **Access Control Failure:** Involves attacks by primarily technical authorised personnel on pre-set, functioning, and accepted standard operating system configurations.

To counter threats to digital systems architecture and from insider attacks, two base strategies are prescribed and usually deployed (Bellovin, 2008). The first approach is colloquially termed *to defend against an attack*; whereby in-built systemic defences are initially considered schematically, the depth of all resources available gauged, and the requisite parameters then adequately and effectively buffered against attacker penetration (Fink et. al., 2013). The second approach employed is to detect an attack after the fact; wherein security tools and procedures are deployed in a hindsight capacity immediately following an attack (Carroll et. al., 2012).

Whilst strengthening existing system defences is generally regarded as being more preferable (Fink et. al., 2013), even a brief examination of the taxonomy of insider attack renders this claim to superiority inherently incomplete (Bellovin, 2008). By their very definition, authorised personnel possess greater insider access to a protected or an encrypted system, and are consequently far more difficult to detect (Pfleeger, 2008). Technical rules and procedures put in place to detect intrusions, particularly online, break down comprehensively in the face of aggressive human behaviour and negative cognitive frameworks (Virkar, 2011).

TECHNOLOGY- AND INTERNET-BASED ADDICTIONS: KEY DEFINITIONS AND CONCEPTS

Extensive discussions in scholarly and practitioner circles, of the new technologies and the frequency and the manner of their use in society, reveal that an excessive, poorly controlled use of the Internet can be pathological or otherwise addictive, leading to what is generically referred to as *technology addiction* or *technological addiction* or, more specifically, when applied to a compulsive dependence on the World Wide Web and its associated platforms and hardware, *Internet addiction* (Griffiths, 1998). In this context, Technological addiction may be operationally defined as a non-chemical, behavioural addiction that involves human-machine interaction (Marks, 1990). The behaviours manifest in this sort of disorder can either be passive, as in television watching, or active, as in non-stop computer game playing, and usually contain both inducing and reinforcing features which may contribute to the promotion of addictive tendencies; including *salience*, *mood-modification*, *tolerance*, *withdrawal*, *conflict*, and *relapse* (Griffiths, 1995).

Definitionally, purists might argue that the term "addiction" be applied only to cases involving the ingestion of a drug or of a similar stimulant substance (Rachlin, 1990). However, in more closely delineating the concept of a *compulsive* or a *pathological addiction* or a *dependence*, the diagnosis of such impulse-control disorders has moved beyond this narrow conceptualisation to include a number of clinically abnormal behaviours that do not directly involve an intoxicant, such as *compulsive gambling* (Mehroof & Griffiths, 2010), *video game playing* (Keepers, 1990), *overeating* (Lesieur & Blume, 1993), *overexercising* (Morgan, 1979), *unhealthy love relationships* (Peele & Brodsky, 1979), and *inveterate television viewing* (Winn, 1977).

The notion of both *Technological Addiction* and *Internet Addiction*, and their acceptance as legitimate medical disorders is a fairly recent phenomenon where, over the past decade, a growing body of peer-reviewed scholarly literature has developed, adapted, and popularised the key terms and diagnoses of the condition (Hansen, 2002). Key derivatives of technological addiction related to the pathological use of the Internet, either as the primary source of obsessive impulse-control behaviour or, in its excessive compulsive usage, as the medium to fuel other addictions, include Internet Addiction Disorder, Pathological Internet Use, Excessive Internet Use, and Compulsive Internet Use (Widyanto & Griffiths, 2007). Broadly speaking, all the derivatives of Internet Addiction cover a wide variety of behaviours and impulse-control problems that might be categorised into five basic subtypes (Young, 1999):

1. **Information Overload:** Or compulsive web surfing, downloading, uploading, or electronic database searching;
2. **Computer Addiction:** Or continuous small-screen movie-watching, remote close circuit television addiction, or obsessive computer-game playing;
3. **Cybersexual Addiction:** Or the compulsive use of adult websites for cybersex and for accessing digital pornography;
4. **Cyber-Relationship Addiction:** Or the overinvolvement in online relationships, particularly with strangers, to the detriment of face-to-face, familiar, familial, and interpersonal contact; and
5. **Other Internet-Based Compulsions:** Including obsessive online gambling, frequent online shopping or online day-trading, and excessive participation in online political activities via Web-based platforms or digital media.

Unlike other addictive substances, particularly known chemicals and stimulants that have the propensity to create severe dependencies in people, technology – both hardware and software – offers several direct benefits to people and societies in the form of technological advancement and organisational streamlining, and is therefore not an artefact to be criticized as being either inherently "addictive" or "isolating". (Levy, 1996).

On the contrary, however, the ways in which the benefits that accrue from the development, integration, and the use of technology in everyday life manifest in society range from innovations in work-flow processes to the practical application of digital hardware and software to old routines; from the way politics is conducted, business transactions are performed, libraries are accessed, and research is co-ordinated, to the ways in which interpersonal communication is structured; pre-empting severe societal disruption and personal psychological damage. This observation holds especially true for the Internet and its associated platforms and software, and the psychological and functional benefits that these digital technologies bring to our daily living; where, by comparison, addictive chemical substances and stimulants are not an integral part of our day-to-day professional or private lives, nor does their abuse or any dependence on them offer either society or the individual any direct benefit (Turkle, 1995).

TECHNOLOGY ADDICTION DISORDERS, DIGITISED ENVIRONMENTS, AND PUBLIC ENGAGEMENT: CYBER-BULLYING, LURKING, AND TROLLING ONLINE

In general, the Internet being a highly-promoted technological tool makes the detection and the diagnosis of a technology-based addiction imperative, together with the need for the condition at hand to be relatively simple to comprehend. More specifically, it is essential to understand the criteria that differentiate normal Internet usage from Internet- and technology-based pathological behaviours, the diagnosis of which is often complicated by the fact that there is currently no accepted descriptive yard-stick or set of principles available from even recent medical and psychiatric sources, against which the extent and nature of an affliction might be judged.

For the purposes of this article, therefore, Internet addiction is characterised by four key criteria or symptoms:

1. **Excessive Use:** Often associated with a loss of sense of time or a neglect of basic drives,
2. **Withdrawal:** Including feelings of anger, tension, and or depression when computer-related technology becomes inaccessible,
3. **Tolerance:** Including the constant need for better computer equipment, software, or more hours of use, and
4. **Other Negative Repercussions:** Including arguments, lying, poor achievement, social isolation, and fatigue (Block, 2008).

Similar to other forms of detrimental substance abuse, an individual may be said to be addicted to the Internet, to digital appliances, or to other manifestations of digital environments when the technology is found to cause extreme harm or even significant problems resulting in disruption to a user's day-to-day routine (Shah et. al., 2002). Similar to an alcoholic who needs to consume greater levels of alcohol to achieve satisfaction, addicts not only routinely spend significant amounts of their time online, but also go to great lengths to mask the nature and direction of their online activities; primarily driven by the need to conceal the extent of the pathology behind their routine, everyday behaviour. As an Internet-based addiction deepens, an addict's use of computer technology becomes less about using it as a tool for information-seeking and for communication, and more about finding a compulsive psychological escape or *fix* to cope with boredom and with other problems of living and working (Morahan-Martin, 2007).

As corporations and government agencies rely on more-and-more on Management Information Systems and on the new Information and Communication Technologies to run almost every facet of their business and routine functioning, employee and civil servant technology- and Internet-based misuse and abuse and its potential for being a detrimental, lifelong addiction has become a potential workplace epidemic, in both the private and public sectors of the global economy (Turkle, 2011). Several studies show, for instance, how employee abuse of the Internet during work hours results in lost productivity, negative publicity, and increased legal liability (Young & Case, 2004). Firstly, although time is not a direct function used in the diagnosis of Internet addiction, the online usage behaviour of addicts is generally excessive; characterised by prolonged periods of indoor Internet use without recourse to proper exercise, disrupted sleep patterns, and bouts of sleep deprivation, all resulting in, for the civil servant user, impaired occupational performance and a deterioration in the quality of everyday life caused by acute loss of focus, excessive fatigue, and a chronic vulnerability to disease.

Underlying compulsive citizen technology usage, on the other hand, and a similar increased propensity for an addiction to ICTs and their overuse, is the relative anonymity afforded to the participants of the online, electronic transactions comprising the virtual context that circumscribes a subjective escape from emotional difficulties such as stress, depression, or anxiety; from problematic situations caused by other communication disorders; or from personal hardships resulting from boredom, job burnout, academic troubles, sudden unemployment, or from marital discord. When online addicts are forced to go without the technology, in most cases the Internet or access to forms of mobile computing, they experience a sense of disempowerment and withdrawal often associated with diagnoses of the sort of impulse-control disorder that are accompanied by increasingly painful states of tension and agitation, relieved only through their reconnection with the so-called "outside" electronic world.

The proliferation of social grassroots database technologies and microblogging websites in the early 2000s led to the widespread use of the Internet and its associated platforms and applications to build and develop a broad spectrum of online discussion groups, participatory forums, and sub-culture focused online communities.

Centered especially around the use of social networking services, the popularity of this phenomenon has grown, but so has the amount of Internet abuse and anonymous bullying that carries on over these digitised platforms. These technologies have made possible the instantaneous expression of and access to information, news, and opinion, creating what is commonly known as The Public Square (Neuhaus, 1989); in other words, making instantaneously available to all those people proficient at using the Internet and with ready access to digital technologies and online social networking services the ability to publish and to control editorial policy.

The study of online communities in recent years, from the standpoint of a variety of established disciplines and practitioner fields, has led to the emergence and the development of a number of colourful expressions to identify and to describe computer-mediated electronic aggression and the manifestation of intrinsic maladjustive behaviours online. Key words include *lurking*, *trolling*, *flaming*, *spamming*, and *flooding*, developed in order to describe those obsessive-compulsive and aggressive behaviours that benefit the perpetrators through the disruption, annoyance, and aggravation of other people (Lampe & Resnick, 2004; Pyzalski, 2012).

The most significant problem stalling the growth of an online community is the lack of participation of its members in posting and in interacting with the group platform, its population, and its available site content. As with even the most appropriate fit of technology to a given purpose, documented evidence attests that there still exists often a large number of non-participants, passive actors, or *lurkers* who do not involve themselves fully online (Bishop, 2007b). Lurkers may be concisely defined as members of an online community who clandestinely visit and use designated virtual spaces and environments but who do not interact either actively or visibly with other participants and community actors. Unlike active posters, communicators, or socializers, the behaviour of lurkers does very little to overtly build or enhance a particular community in any way, especially when considered within the context of give-and-take relationships and direct social interaction (Beike & Wirth-Beaumont, 2005).

From a psychological standpoint, it has been demonstrated that the average lurker is often less enthusiastic about the benefits accrued from community membership than their average counterparts, indeed having themselves become socially isolated, disengaged, or withdrawn from their peer group beforehand (Howard, 2010; Chen et. al., 2009). Lurking, however, may nonetheless be considered as normal behaviour amongst individuals participant in an online group, forum, or community; indeed altogether benign if undertaken only to a moderate degree as standard preferential practice or to protect and preserve individual privacy in a digital environment (Efimova, 2009).

THE CHANGING FACE OF AGGRESSIVE DISORDERS: FROM THE SCHOOL PLAYGROUND TO THE VIRTUAL SANDPIT

Bullying is an age-old societal problem, beginning in the schoolyard and often progressing to the boardroom (McCarthy et. al., 2001). It may be defined as an aggressive, intentionally harmful, maladaptive behaviour occurring without provocation over time that involves the abusive treatment of a person by means of force or coercion (Campbell, 2005; Peterson, 2001). Bullying may be either *physical*, including behaviours such as hitting, punching, biting, and spitting; or *verbal*, involving the use of sound or language that equates with verbal assault, that is at once browbeating, teasing, scapegoating, laced with ridicule and/or sarcasm (DiGuilio, 2001; Slee & Rigby, 1993). In recent years, a new form of bullying – *cyber-bullying* – has emerged which makes use of and is transmitted by the diverse range of technology currently available (Campbell, 2005). The new digital technologies, and in particular The Internet, provide users with an anonymous environment wherein it is easy to seek out and to explore one's niche, however idiosyncratic. In consequence, antisocial and poorly adjusted individuals have greater opportunities to connect with similar others and to pursue their own personal brand of self-expression than they did in the past.

In this respect, the frequency of activity is an important correlate of antisocial uses of technology. For instance, cyber-bullying is often perpetrated by heavy Internet users (Juvonen & Gross, 2008), and disagreeable persons who use mobile technologies more than others; again not for socializing, but for the sake of personal entertainment, low conscientiousness, high extraversion and the attainment of individual satisfaction (Andreassen et. al., 2013; Phillips & Butt, 2006). Similarly, gamers who profess non-social motivations such as competition, revenge, and personal achievement for their continued involvement in online gaming, demonstrate lower levels of sociability and conscientiousness than others (Graham & Gosling, 2013). These patterns parallel gender differences in online behaviour and socialization: men are prone overall to greater Internet use (Joiner et. al., 2005) and rate higher in antisocial behavior online (Zweig et. al., 2013).

As cyber bullying and other forms of Internet-based aggression increase in prominence with advances in technology, it has become important from the point of view of scholarship to determine exactly *what* this classificatory type of aggressive disorder is, as well as *how* and *why* it manifests (Campbell, 2005; Hinduja & Patchin, 2010; Ybarra & Mitchell, 2004). In this respect, the variant of the construct of bullying and/or aggression that manifests itself and occurs online has yet to be properly defined. Whilst some studies consider and examine online maladaptive behaviours as a single definitional entity within the broader scholarly discipline of passive-aggressive disorders (Campbell, 2005; Law et. al., 2012; Slonje et. al., 2013), cyber bullying may be identified, diagnosed, and investigated against a number of different traditional and unorthodox axes. More minutely, the many aspects of bullying online are thus seen to vary greatly according to the specific type of passive-aggressive behaviour encountered in virtual spaces and in virtual arenas (Law et. al., 2012).

Indeed, there is little consensus across scholarship about what to even call it, with terms such as *cyber bullying*, *online aggression*, *Internet harassment*, and *electronic aggression* used regularly in the literature (Dooley et. al., 2009; Kowalski et. al., 2008; Smith, 2009). One approach to this lack of clarity has been to assume that bullying online functions in a manner similar to more traditional forms, in that, whilst the essential nature of the disorder remains markedly recognizable, the triggers, catalysts, and arenas of action are circumstantially unique and often Internet-based or virtual (Dooley et al., 2009). When considered in this context, the previously established lack of a clear definition prevents a full understanding of the given construct and how it relates to developmental outcomes; a task that must be undertaken in order to accurately comprehend and to address the social, psychological, and emotional consequences associated with the phenomenon (Law et. al., 2012; Slonje et. al., 2013).

Whilst not exhaustive, lists of types of cyber bullying and electronic or online-based aggression may be used to thoroughly investigate the emergence and spread of new forms of electronic maladjustive behaviours that evolve as new technology develops. Within the bounds of the discipline, therefore, some studies place the analysis of electronic aggression and cyber bullying inside the context of a base framework that involves two fundamental communications media:

1. The Internet, and
2. Mobile phone technology (e.g., Ortega et. al., 2009).

However in recent years, the advent of smart phone technology has blurred this distinction and has made the boundaries circumscribing the definition of online activity, indeed bullying, more problematic to discern; facilitating both the dispatch and the reception of emails and text-rich messages via either a personal computer or a mobile phone handset, as well as the use of these and other similar technologies to access the Internet and the World Wide Web more broadly.

Scholars such as Huang and Chou (2010), on the other hand, investigated and defined types of cyber bullying behaviour across three different subject or role groups: *victims*, *perpetrators*, and *bystanders*. The most frequent behaviour reported by both victims and perpetrators of electronic aggression, online maladjustive activity, or cyber bullying was *threatening* or *harassment*, followed by *making spiteful jokes about* or *making malicious fun of a victim*, and lastly *vindictive rumour-spreading* (Olweus, 1993; Olweus & Limber, 1999). For bystanders the order of frequency was found to be different, with *making spiteful jokes about or malicious fun a chosen victim* of being most frequent, followed by *threatening* or *harassment*, and then by *vindictive rumour spreading* (Huang & Chou, 2010).

Yet other scholars have investigated online aggressive activity and cyber bullying in terms of the range of media used by perpetrators and by oppressors to aggress both their victims and bystanders. Smith et al. (2008), for instance, in their seminal study of secondary school playground and online bullying, used seven main media listed by secondary school pupils - *mobile phone calls*, *text messages*, *multimedia messages*, *picture and video clips*, *e-mails*, *chatrooms*, *instant messaging*, and *websites* - to illustrate and then to describe the role of Information and Communication Technology in online bullying. Hinduja and Patchin (2010), likewise, used a 9-item cyber-victimization scale covering a similar media spectrum to explore related issues and arguments within a similar demographic. More recently, Wachs and Wolf (2011) used a 5-item scale, again covering an almost-identical range of communications media, to investigate the nature and influence of electronic aggression across a group of diverse subjects, grouping both behaviour and medium types together into predefined, now-seminal variable clusters.

An alternative experimental control to looking at the informational and communications media used is to consider either the type of action constituting electronic aggression or online maladjustive behaviour, or to investigate its content framed in terms of its overall sphere of influence. Willard (2006), for instance, described seven categories of behaviour to diagnose and to describe technology- and Internet-based maladaptive disorders that are to some extent independent of the media used: *flaming*, *online harassment*, *cyberstalking*, *denigration (put-downs)*, *masquerade*, *outing*, and *exclusion*. Similarly, Rivers and Noret (2010) developed ten main categories to describe and analyse the content and the repercussions of abusive text messages and e-mails in an English sample: *threat of physical violence*, *abusive or hate-related name calling* (including homophobia), *death threats*, *ending of platonic relationship(s)*, *homophobia*, *sexual acts*, *demands and instructions*, *threats to damage existing relationships*, *threats to home and to family*, and *menacing chain messages*.

THE PHENOMENON THAT IS ELECTRONIC AGGRESSION: FROM CYBER BULLYING TO ONLINE TROLLING

There are two types of cyber-bullying: *online bullying* and *trolling*. The two clinical disorders are similar terms but each with slight differences. On the one hand, online bullying involves both the overt and the clandestine use of technology to deliberately and repeatedly to cause psychological and physical harm and distress. Online bullying is also generally a chronic condition, where the bullies concerned usually know the person they are attacking and keep the assault ongoing (Law et. al., 2012). To the contrary, online trolling is the practice of behaving in a deceptive, destructive, or disruptive manner within a social setting on the Internet with no apparent instrumental purpose (Buckels et. al., 2014). Trolling is the act of deliberately trying to distress someone online through frequently inflammatory and abusive behaviour, but usually just to disrupt without direction and to often do so anonymously. More specifically, trolling may be considered as a favourite Internet pastime of the bored, the insecure, and the antisocial; usually trying to bait members of the virtual community or environment into descending to their level (Bishop, 2014). Simply described, it is a more nuanced, sophisticated form of bullying.

Despite public awareness of the phenomenon, there is little empirical research available on trolling (Buckels et. al., 2014). Existing literatures are scattered and multidisciplinary in nature (Hardaker, 2010; Herring et. al., 2002; McCosker, in press; Shachaf & Hara, 2010). Reliable case study evidence and examples of best practice are rare and equally difficult to come across (Law et.al., 2012). Shachaf and Hara (2010), for instance, conducted interviews of Wikipedia trolls and identified themes of boredom, attention-seeking, revenge, pleasure, and a desire to cause damage to the community amongst their expressed motivations for trolling. In other research, Hardaker (2010) conducted a content analysis of Usenet posts that identified four primary characteristics of trolling behaviour:

1. Aggression,
2. Deception,
3. Disruption, and
4. Unrestrained success.

The clandestine, the deceptive and the so-called pointlessly disruptive aspects of online interactions may in addition be used to distinguish trolling from other forms of online antisociality such as cyber-bullying, wherein perpetrator identities are usually clear (Lenhardt, 2013) and the intent is more straightforward (Campbell, 2005).

In suggesting that dark personalities leave large digital footprints, the 'noxious personality' variables known as the *Dark Tetrad of Personality* – narcissism, Machiavellianism, psychopathy, and sadistic personality disorder – are yet to be investigated in the trolling literature (Buckels et. al., 2013; Furnham et. al., 2013). Their importance and relevance to the discussion advanced in this chapter is further suggested by research linking these traits to bullying in both adolescents (Fanti & Kimonis, 2013) and adults (Baughman et. al., 2012; Jones & Paulhus, 2010; Linton & Power, 2013). Also equally suggestive is research showing that narcissists (Ljepava et. al., 2013) and those with antisocial personality disorder (Rosen et. al., 2013) use Facebook and other social networking and grassroots community media more frequently than others (Buckels et. al., 2014). Described in lay terms, trolls are poorly adjusted individuals who operate as agents of chaos on the Internet, exploiting 'hot-button issues' to make users appear overly emotional or foolish in some manner, preferably in public (Binns, 2012). If an unfortunate person falls into a laid trap, trolling intensifies for further, for merciless amusement of the trolling community (Bishop, 2012).

TROLLING AS ASYCHRONOUS COMPUTER-MEDIATED AGGRESSION: PROFILING TROLLS AND TROLLING BEHAVIOURS

The term *Internet Trolling* may be considered one of the most popular and fastest-spreading pieces of computer-related behavioural jargon of the 21st century, used to describe the collection of theories, concepts, and ideas focused around key, clearly defined forms of Internet- and digital technology-based abuse and misuse. Perhaps the first recorded attempt to define and characterise online trolls and Internet-based trolling activities was made in the mid-1990s, with the release of the book *Netlingo* (Jansen & James, 2002); wherein, despite its purported origins in Scadanavian linguistics and literature, the word '*troll*' when used to refer to persons who try to provoke others by way of a hobby or an almost-fulltime leisure activity, is reputed to have originated in the US military as a reference to aerial dog-fighting during the 1960s, just prior to the conception and realisation of the Internet by DARPA as a medium for military and popular mass communication (Wilcox, 1998; Bishop, 2014, Virkar, 2014).

By this extension, individuals within subversive, transgressive, and dissident factions and groups have long called themselves trolls in order to legitimise their abusive behavior, particularly online; with the term now originally referring to people who attempt to entertain their kind through the sustained provocation of others or via the propagation of misinformation and other forms of propaganda in order to attain a certain measure of dominance and satisfaction. Abusive behaviour such as this often poses a real problem for website owners and platform moderators, systems operators, and owners of computer-mediated environments (Binns, 2012; Phillips, 2013; Bishop, 2014a).

Over the years, the term '*trolling*' has likewise morphed significantly, transmuting the express connotations of the vocabulary from implying the mere provocation of people online for mutual sub-culture and group enjoyment to meaning the definitive and sustained abuse of others for an individual's own personal satisfaction, material gain, and entertainment (Hardaker, 2010; Bishop, 2014b). Trolling is consequently known as a form of sub-culture that amplifies types and forms of social exclusion, being essentially a form of online baiting and pseudo-martial patrolling or policing involving the movements and actions of a '*troller*', an identifiable entity that actively seeks out people not conforming to a particular reality or opinion with the aim of provoking them into a response or a corresponding situation (Hardaker, 2010; Buckels, 2014).

Trolling behaviour, at its most basic, may be described against four major categorizations (Hardaker, 2010; Bishop, 2014a): *Classical Trolling*, or maladaptive Internet-based behaviours indulged in by individuals and groups for a community's consensual entertainment and to build bonds between users; *Anonymous Trolling*, or those behaviours partaken in at the expense of an outsider to a particular group or community for the attainment of revenge or for satisfaction; *Flame Trolling*, or the initiation online or via digital media of deliberately harmful behaviour not intended to be humorous; and *Kudos Trolling*, consisting primarily of the expression of excessive and hence increasingly maladaptive attention-seeking behaviours online.

Trolling may be further investigated along two principal axes (Phillips, 2011; Buckels et. al., 2014): *Lolz*, or *trolling for the lolz*, or for laughing-out-loud behaviour generated by trolling sub-culture humour, however subversive or sadistic; and *Lulz*, or *trolling for the lulz*, or troll-like or troll-ish or sadistic behaviour undertaken solely to destroy, to harm significantly or to abuse. The former is often conducted in online communities over platforms such as message boards or forums frequented by like-minded people, and the latter in places or environments requiring sophisticated computer usage where there are vulnerable people or potential targets engaging who might not be as skilled.

DON'T FEED THE TROLLS!: UNDERSTANDING THE PSYCHOLOGY BEHIND HUMAN-COMPUTER INTERACTIONS

As academic scholarship and public dialogue surrounding the decline of social capital, and the role of the new Information and Communication Technologies in this dynamic, has expanded over the last couple of decades, a new medium – the Internet – has become prominent as the latest catalyst for the emergence of new socio-psychological environments globally. In this regard, some theorists and practitioners are extremely sanguine about the Internet and its impact on the psychology of individuals and communities within a society; many even assert that being "wired" has the potential to tighten existing social associations and to encourage community building (Coleman, 2005). In particular, individuals who use the Internet to explore interests, gather data, communicate with colleagues and friends, and send e-mail appear to be *more*, not less, socially and politically engaged (Chiu, 2006).

These activities are amongst the most common, most stable Internet-based behaviours, and their continued popularity over and above other niche interests online raises the possibility that Internet use, on the whole, has positive consequences for society at large. That is, the frequent import and application of the Internet and its associated technologies into mainstream work practices and into daily living may promote healthy levels of social interaction and civic engagement, as they allow for users to, regardless of other circumscribing societal factors, reinforce social bonds, forge new ties, gain new knowledge, and to coordinate their actions to address joint concerns (Huysman & Wulf, 2004).

Given the potential importance of the different forms of community involvement for political participation, it has become increasingly important to clarify the effects of Internet usage on them (Rheingold, 1993). More particularly, the flip side of scholarship optimistic about the impact that technology has on society is the contention that the digitisation of socio-political and economic processes and the increased use of the Internet and its associated technologies in day-to-day routines will hasten societal civic disengagement, considering that even infrequent Internet use subtracts from interpersonal contact, weakens real-world ties, and reduces community involvement; as individuals who invest significant amounts of time and energy into online activities consequently have less of both to invest in similar interactions offline (Turkle, 2011).

Examined within the context of pathological Internet usage or, more clinically-speaking, addictive technology- and Internet-related behaviours online, intangible losses to a polity, community, or to an economy from problems related to technological misuse and abuse constitute a real risk to society, even though they may not be either immediately quantifiable or wholly psychological. In this context, Sclove (1995) argues that "…an overwired world would be one in which virtual life and online activities increasingly crowd out traditional forms of social engagement, leisure activities or time spent in the natural world, thus creating a pathological self-reinforcing dynamic".

THE TROLLS AMONG US: PRIVACY, TRUST, AND FREEDOM OF EXPRESSION ONLINE

The global diffusion of the Internet, together with the ubiquitous nature of the technology and the rapid proliferation of its associated platforms and applications, has raised in recent years significant questions regarding the nature, scope, and direction of trust, privacy, and freedom of expression online within the express ambit of cyber security. Most particularly, has such technological ubiquity undermined or has it reinforced the values that once characterized and have come to signify the

Internet's early vitality and generativity? Further, do values such as these reflect in users' online interactions and current attitudes?

The Internet has allowed individuals to express themselves freely, as well as given them the opportunity to reach out, communicate, and to interact with a wider audience (Dutta et. al., 2011). The pervasive nature of Internet technology today has, therefore, at once both fostered the unconstrained, unrestrainable sharing of innovative ideas and interests, and has made online participants more self-conscious and guarded in their actions and their words, and more controlled and self-regulated in their chosen mode of expression; both online and offline.

Similarly, privacy and the protection of personal information, issues of express import ever since the incorporation of personal computing into the public sector (Virkar, 2011), have also become matters of appreciable concern following the advent of commercial complex computer systems and the rise of the Internet as a popular communications medium in the mid-1990s (Cranor et. al., 1999). Fear of and interest in the dimensions of these fundamental civil liberties manifest online, together with the manner in which they are breached or encroached upon, have been heightened by the widespread diffusion of the Internet in recent years, and as a direct consequence of the use of the medium in and its application to a growing number of areas salient to everyday living (Cavoukian & Jonas, 2012); including within commerce, medical care, and in online social networking (Awad & Krishnan, 2006).

Urban et al. (2009) point to online trust as a key factor pivotal to the Internet's overall success, failure, and/or popularity as a global communications technology, platform for unfettered personal expression, and social playground. People are most often guarded about their individual privacy when they lack trust in others (Dutton and Sheppard, 2006). In this respect, it has been noted that nearly all scholarly and practitioner working definitions of trust in existence to date involve the relationship between a minimum of two agents: the one who *must trust* and the one who *is trusted* (Dutta et. al., 2011). The adoption of new technology or digital platform, and its eventual diffusion across an international demographic could not, therefore, occur without a minimum level of trust in both the device and in the human-actor agents that use, maintain, and operate it (Virkar, 2011).

Does the defence of free speech and similar civil liberties online then tend to move toward the truth or away from it? When does scholarship evolve into a better collective understanding and thence to grassroots social direct action, both on- and offline? When do digitally-based global social movements collapse into the Babel of trolling; of *flaming*, *online harassment*, *cyberstalking*, *denigration (put-downs)*, *masquerade*, *outing*, and of *exclusion?* Is the effort to control what is said online always a form of censorship, or might certain rules be compatible with traditional and popular notions of fundamental rights and civil liberties?

Trolls Just Want to Have Fun

In the late 1980s, the first Internet users adopted and transmuted the word "troll" to denote a network user as someone who intentionally disrupts online communities (Jansen & James, 2002). Early trolling was relatively innocuous, taking place inside of small, single-topic Usenet groups (Binns, 2012); employing what in ethnographic studies is known as a *pseudo-naïve tactic*, involving the asking of stupid questions in order to determine who would rise to the bait (Donath, 1999). The fundamental aim of this "game" on the path to trolldom was thus to find out who would see through this stereotypical newbie behaviour, who would fall for it, and who would be harmed by it (Schwartz, 2008).

THE MIRROR HAS TWO FACES: e-CONSULTATION AS TWO-WAY PARTICIPATORY GOVERNANCE IN THE DIGITAL AGE?

The shift in thinking about e-governance[1] from being a technology-driven process to one that is more citizen-focused, and the recognition of the need to promote greater inclusiveness in decision-making processes, may be clearly seen in the emergence of concrete policies for e-democracy (in addition to those dealing with e-government) in many European countries over the last decade. This is most notably so in the United Kingdom, where several e-democracy projects are publicly funded, and both Parliament and the Government have over the years outlined sets of policy principles on the subject. One such set of guidelines, a consultation paper issued by the UK government in July 2002 (HM Government, 2002), which sets out a clear policy agenda for e-democracy and contains a detailed but useful division of the concept into two distinct areas:

- E-Voting (or the use of technology in elections and associated structures and processes), and
- E-Engagement (or e-participation, which emphasises opportunities for greater consultation and dialogue between government and citizens).

A long-established way of engaging citizens in dialogue with policy makers is that of Consultation, where citizens are given the opportunity to provide feedback to government on matters of public importance and participate in the shaping of issues relevant to them (OECD, 2001). Whilst there is a need for dialogue at several different stages during the policy process, the process of consultation has traditionally involved discussion based around a pre-determined policy issue defined by the government during its initial formative stages, on which citizen's views and opinions are then sought (Rosen, 2001). The government sets the questions and manages the process, often laying down the parameters within which the consultation is to take place. Only in rare cases are citizens are invited to suggest issues for discussion that they, as private individuals, might consider particularly important. At the heart of the consultative process, therefore, lies the provision of information and the establishment and maintenance of channels of communication between government and its citizens.

The use of ICTs in consultative processes is catching on as their potential for allowing policymakers to interact directly with the users of the services, to target the opinions of those at whom a policy is aimed, and to seek general citizen input on matters of national importance is gradually being recognised (Virkar, 2013b). The speed and immediacy of ICT networks allow people to communicate, give feedback, ask questions, complain, exchange information effectively and build relationships with their representatives; and governments too may benefit from any information they obtain by using it to enhance the quality of policymaking and general administrative functions. Broad guidelines for 'conventional' written consultation by more traditional means are already in place in most of the Western world, and these are now being used as a basis for e-consultation (OECD, 2003), with this type of e-democracy thus encompassing what may be referred to as 'a continuum of consultation', ranging from low level information gathering and aggregation towards a fuller quasi-deliberative level of interaction.

Despite contrary claims from scholars such as Rash (1997) and Bimber (1998), the consultative model is not without its problems as it is sometimes presented as facilitating direct, unmediated access to government for special interest groups of a sort that may distort opinion on particular issues. Information gathered from the consultative process is usually regarded as a passive resource, largely due to the fact that communication by direct question-asking is based on the need to generate quantifiable and comparable responses to particular policy innovations. The result is that the consultative model may only allow for inputs that fit within parameters already set down by policy makers, with a marked danger that opinions which question the necessity or legitimacy of a policy or otherwise be outside the ambit of pre-defined issues, are deliberately marginalised or excluded altogether, particularly if discussions are 'moderated' (Whyte & Macintosh, 2003). In order to further the analysis of issues affecting the impact of ICTs on existing democratic frameworks, this chapter sets out a three-fold categorisation of noteworthy cases across Europe along different axes depending on their duration, the level of the participating government organisation, and from whom feedback is sought. Three categories of e-consultation initiatives in Europe may be derived from this author's research:

Duration

Projects when classified according to duration may be examined under three sub-categories – long term, short term, and one-off consultations – depending on the length of time that they lasted for.

- **Long Term Consultations:** Include those e-consultation initiatives which are either specifically set up as long-lasting initiatives or which become permanent initiatives after an initial trial period. Projects such as these tend to provide permanent platforms to solicit continual citizen participation and feedback on highly topical issues as and when they arise, and are not bound around a particular political event or related to a particular occurrence.
- **Short Term Consultations:** Include those initiatives that seek to obtain citizen opinions around specific political events or during a designated fixed period of time. These initiatives are intentionally short-term, are focused on getting citizen input for a specific purpose, and come to a close once the event or time period is at an end.
- **One-Off Consultations:** Highly specialised issue-based e-consultations, generally held on an ad hoc basis for a fixed period of time. Such consultations tend to be used by government as a means of gathering information from a well-defined target group on a specific, often on a pressing issue and, of the three types of consultation discussed so far, are the most likely to have a visible impact on government policy.

Level of Government

Case studies may also be classified according to the level of government at which they are implemented; more specifically as projects implemented by local government agencies, at the level of national government and at the inter-governmental or pan-European level.

- **Local Government Consultations:** Refer to e-consultation projects of note initiated at the level of local government within the E.U.
- **National Government Level Consultations:** Include all e-consultation initiatives begun within national government ministries and associated institutions.
- **Inter-Governmental or Pan-European Initiatives:** Those projects initiated at the regional and pan-European level, generally as supranational agreements between different European government.

Nature of the Target Audience

Projects may also be categorised and discussed according to their target audience or in terms of the section of the population from whom feedback is sought. Whilst most initiatives are generally concerned with obtaining feedback from the general public on a variety of issues, a small number seek to obtain specific information from a carefully targeted, often specially selected group.

In building on the categorisation discussed above, and attempting to compile a rich and informative inventory of issues that are or can become significant to the implementation of e-consultation projects, this chapter will look briefly at the No. 10 Downing Street e-Petitions Website as a pioneering effort by a national government department to harness digital technology to recast and reinvigorate democratic processes within its respective spheres of influence.

PUBLIC ADMINISTRATION AND ICT POLICY IN THE UNITED KINGDOM

e-Governance policy in the United Kingdom in its most recent form began to take shape in the late-1990s, with the central overarching issue for almost all government-based ICT projects initiated during that period and after being to attempt to improve the democratic character of public decision-making and delivering greater legitimacy by involving the public directly (Harrison & Mort, 1998). Whilst there have been huge changes in public administration, particularly over the last couple of decades with the State seeking greater administrative efficiency through the computerisation and the digital storage of data, scholars are divided as to how effective these policy changes have really been in furthering public participation (Ward, Gibson, & Lusoli, 2003). To date much of what has been achieved in terms of practical e-democracy in the UK appears to centre more around information provision about the government and its activities and in ensuring that government services are available on the Web (i.e. electronic service provision), rather than on the other, more interactive side of e-democracy whereby feedback from the public is actively sought on issues relevant to their everyday lives (Dutton & di Gennaro, 2006).

In the UK, the actual mechanisms by which the public can communicate with government appear to have remained essentially unchanged; with some studies, for example, showing that interaction with elected representatives has not really moved on from where it was a few decades ago compared to the pace of change in other Western liberal democracies such as the United States, largely as a consequence of European integration (Pleace, 2007). In this vein, whilst local, national, and central government agencies within the UK today publish a vast amount of material online, their websites should not be taken simply as mere repositories of information, but instead fulfil a dual role of both informing and connecting with the public (Pleace, 2007). A good example of this two-fold role played by government websites is The No.10 Downing Street Website, gateway to the Office of the Prime Minister and host to the Prime Minister's e-Petitions Gateway, which on the one hand serves to inform the public about the role and functions of the Prime Minister, and on the other is uniquely placed to provide the crucial link for interaction between the Prime Minister and the public.

THE NO.10 DOWNING STREET e-PETITIONS INITIATIVE

One method by which members of the public have traditionally placed their views before government is by the submission of petitions, formal issue-based requests to a higher authority such as a Head of State or Parliament, which are signed by one or a number of citizens (Macintosh, Malina & Farrell, 2002). In the United Kingdom, the practice of petitioning is a long established one – reputed to be even older than that of voting – wherein paper-based petitions consisting of a bundle of sheets bearing the names and signatures of the petitioners were (and still are) usually presented to the Prime Minister at the door of his official residence of No.10 Downing Street.

In November 2006, the humble petition received a 21st Century makeover when the Prime Minister's Office in partnership with mySociety, a project associated with the charity UK Citizens Online Democracy[2], launched an electronic petitioning service on its website to provide citizens with a modern, more convenient parallel: the electronic or e-Petition. The No.10 Downing Street ePetitions System was brought firmly into the spotlight soon after its launch by a petition submitted to the website against the Government's proposed Vehicle Tracking and Road Pricing Policy which, between the time it began collecting signatures in December 2006 and its closure in February 2007, was widely publicised and attracted over 1.8 million signatories – a protest seen by some commentators as the 'biggest protest against government policy since the anti-war demonstrations of 2003' (Wheeler, 2007).

Simply defined, an e-Petition is a form of petition posted on a website (Pleace, 2007). e-Petitions may be created easily by an individual or group, posted on the host website for anyone to read and, by adding their details such as an email address to verify their authenticity, may be 'signed' by any visitor to the site. Whilst in theory there is little difference between a paper petition and an electronic petition, except for the way in which signatures are collected and delivered and for the fact that the system has built into it the opportunity for partial communication between government and petitioner, in practice, the introduction of the e-petitions section on the Prime Minister's website has opened up the possibility of a new channel by which groups and individual citizens may put across their viewpoint to the Prime Minister's Office directly without going through traditional media conduits and thereby potentially being able to influence the direction or nature of current political debate.

In combining traditional access to politics with technology, e-petitions are a powerful way of making politicians and decision-makers aware of those hidden issues which are currently important to a small group of people but which might in the future become important issues pertinent to larger sections of society. The key questions are, however, what in the long-term would be the impact of the No.10 e-Petitions Initiative on democratic processes within the UK and how might it contribute towards re-engaging the public?

RECONFIGURING GOVERNMENT-CITIZEN RELATIONS THROUGH e-PETITIONS

Any analysis must keep in mind that to begin with, the e-Petitions system is an almost-direct replica of the traditional petitions model, whereby citizens may petition the government electronically and receive partial feedback on their opinions (Virkar, 2007). The system does not, however, engage citizens further than this nor use their comments within a full-scale e-consultation forum. In addition, it must be remembered that e-petitions (much like their traditional counterparts) are not meant to be representative of a country's opinion as, say, an opinion poll might be. Instead, they simply indicate what one group of people think on a subject, with accepted wisdom stating that only the most vocal citizens who feel most strongly about an issue and who are comfortable with using the Internet as a medium of communication will sign up to them. For those citizens who are either usually satisfied with most proposals or for whom political participation (particularly online) is generally not a top-of-mind consideration, there is little compulsion to participate.

Whilst e-petitions are thus, in one sense, not a radically new democratic tool whereby the voice of the majority would translate directly into a legal change; there do remain a number of advantages to e-petitioning (Lindner & Riehm, 2009). In the first instance, citizens may obtain background information on an issue, make comments, sign petitions online, and receive feedback on the progress of their petition: a process which, at least in theory, would make for a more informed petitioner and an overall better quality of participation, as citizens may take their time to research the finer points of a particular issue before they decide whether or not to support it. From a government perspective, popular petitions and their associated comments may be used during the policy process, with extremely high-profile petitions serving to highlight issues and uncover underlying discontents which have not as yet been picked up by the mainstream media (Virkar, *forthcoming* 2013c).

There is, however, no formal obligation for the Government to actively pursue views raised once the initial communication has been established with the citizen, and there is no guarantee that this new form of voter- or community-led engagement would necessarily lead to concrete changes in or gains from policy. In this respect, questions may be asked as to how seriously both citizens and government officials would take this form of interaction should there be no popular or extremely topical issue to keep the mainstream media interested (Geoghegan, 2007). Further, when it comes to designing and deriving official directives and public policy, the Government is faced with the challenge of balancing and sustaining the overarching democratic requirements of openness, accessibility, and participation with the need for adequate gatekeeping and platform moderation to guard against the ravages of trolling and other malicious online behaviours, to protect participants and to buffer national security, and to stay within legally stipulated Data Protection standards. Similarly, given how easy it is to start and sign up to a petition – over 3,313 people, for example, supported a call for Tony Blair to stand on his head and juggle ice-cream in early 1999 – there is a good chance that more genuine petitions and worrying national mental health trends may get mixed up or overlooked if a fair and unbiased process is not put in place to identify and sift out more 'pertinent' views from 'less important' ones.

Keeping these considerations in mind it is, first and foremost, necessary for those administering a system to ensure that the more mundane but equally pressing issues that do not have either the media hype or numbers to support them get noticed and dealt with. There is also a further need for the Government to recognise and take into account the presence of a silent, but possibly far larger proportion of the population who may hold moderate but equally pertinent views on an issue. One way in which both concerns may be dealt with would be to frame a comprehensive set of general guidelines from existing policy surrounding electronic interactions which would allow views gathered from the website to feed into a much wider policymaking context such as an interactive consultation, whereby they may be discussed and deliberated upon by a wider section of the Body Politic representing different sides of the debate.

DESIGNING PUBLIC POLICY FOR e-DEMOCRACY IN THE UK

The case of the Prime Minister's e-Petitions illustrates that devising public policy for digital democracy may sound contradictory and in many cases an unnecessary waste of government resources. This is because, on the one hand, traditional notions of democracy intrinsically imply a limitation of State power in favour of its citizens through a bottom-up process where individuals initiate action and individual freedoms are protected. On the other, the implementation and development of the information superhighway and its associated infrastructure has been a market- and not state-led process which is said to attenuate the power of the State in favour of other forces in local, national, and world politics.

Particularly affected are those ideas dealing with access to public information, through the introduction into the political system of new ways of conceptualising 'old' values such as transparency and new styles of public administration which draw on a mixture of corporate management principles and Weberian models of bureaucratic ordering and functioning. With the current generation of citizens being not only information-consumers, but increasingly acting as *producers* of web content, the innovative use of Information and Communication Technologies offers up the possibility of new consultation spaces and the potential to increase the breadth and depth of citizen participation in the public sphere (Ferguson, 2006). Representative democracy in its most traditional form is thus being increasingly challenged as the Internet and its associated technologies make the logistical distribution of public information much easier and less costly, with citizens demanding greater participation in public affairs and the adoption of new forms of accountability and control of government

The increasing sophistication of technology often runs contrary to legalities relating privacy, anonymity, identity management, and privacy-enhancing tools if, through increasingly complex technology, the government is able to control and monitor citizens and in doing so encroach on basic fundamental rights (Dutton & Peltu, 2007). Such security concerns are generally manifested in discussions of risks posed by technology to citizens' anonymity and privacy, and the collection and storage of sensitive data, as the development of online government services opens up new possibilities of intrusion by third parties into government databanks and exchanges of information between different departments (Hiller & Bélanger, 2001). In addition, questions arise as to whether the responsibility for the accuracy and security of personal data should reside with government, given that it alone has the manpower and financial resources capable of handling such a task, or whether personal data should – as in the case of the Downing Street ePetitions system – be entrusted to an independent third-party (Irani, Elliman, & Jackson, 2007).

Government has a privileged *sui generis* position in the fabric of society, especially in terms of access to information. In recognising this, however, little attention has so far been paid to the *kind* of information and content necessary to promote digital citizenship and engender trust. It has been argued that ideally, digital democracy should provide both government and citizen with a diversity of content and quality information that has a high cultural value, notions of which are not easy to define (Bishop & Anderson, 2004). In this respect, Freedom of Information acts and ordinances which set out to define the ways in which 'information' and 'data' are perceived and dealt with within a country's legal framework often come in useful, as they set boundaries on the degree to which government may store and use citizen data that it has collected for specific, often publicly-stated purposes (Hiller & Bélanger, 2001). At the same time, the potential of Web 2.0 technologies to empower modalities necessary to functionalise the relationship between government and individual members of the public are also further enhance.

There also exists in the literature a well-documented record of cases in the UK where the reality of online public engagement or electronic service provision initiatives has often fallen short of initial – both government and citizen – expectations (Pleace, 2007). This often happens when those using the technology anticipate that its introduction will result in an increased simplification of government processes, and are disappointed when the complexity of those processes remains at best unchanged, or becomes even more complex (Virkar, 2011). Government officials and civil servants also are often apprehensive about using unfamiliar or innovative technology and are cautious about participating in initiatives that might either bind them to a course of action which in the long term could both damage the government and prove detrimental to their own public image or which, in the short term, could open them up to attack from the citizenry without there being adequate legal protection.

CONCLUSION

Scholars and pundits of the new technology and its associated impacts on society, economy, and polity have long looked to the new media for clues regarding the production and destruction of social capital and psycho-social well-being. Computers may impair thinking one fundamental way. Multimedia, instant messaging, and remote desktop capabilities, together with in-built three-dimensional display and virtual reality capability, promise society a future of an increasingly full sensory experience to the individual user. However, engaging the senses totally may also encourage the passive reception of a data feed, thereby both discouraging imaginative cognition and highly symbolic thought, and raising questions about the developmental value of workplace digital environments and highly digitised virtual selves. A seminal case in point has proven to be the digitisation of petitioning in the United Kingdom, where, as illustrated in this chapter, a once-serious form of political grassroots participation has been, in several cases, turned into nothing more than an interactive electronic novelty or idiot box, with several of its regular participants today straddling the verge of being diagnosed with pathological addictions to technology and related mental health disorders.

From a discussion of the themes and issues brought out in this chapter, the key to successful e-participation projects appears to be the presence of clearly spelled-out goals, adequate planning and preparation of the initiative, the presence of a disaster management plan in case a contingency should arise, and the innovative combination of technology and policy to reach out to as many citizens as possible: in short, the ability of government to combine not only the creative use of new technologies with a balanced understanding of the Internet and of what actually works online, but also to successfully understand both citizens and government servants and their motivations and be able to anticipate and tackle the (often unrealistic) expectations of technology held by them.

Currently practiced *laissez-faire* regulation is, however, usually based on the assumption that not only are users aware that their information is being monitored and being used in particular ways by those collecting it, but that there do exist ways and means by which they might guard against unnecessary intrusions and fraud. Issues surrounding user authentication and information security thus have the potential to impact the ability of a process to generate trust and increase participation by determining access to information, the protection of citizen identity, and the prevention of abuse of government systems.

ICT-based public engagement may not, in the long-term, wholly replace conventional, more personalised methods of government-citizen interaction, but can definitely be used to complement them so as to overcome their shortcomings and provide government with new methods and innovative ways in which it might reach out, communicate, and interact with its citizens. The relationship between technology and trust thus needs to be explored and dealt with thoroughly if a meaningful and sustained two-way interaction between external and internal actors is to be developed and maintained whilst simultaneously balancing existing notions of privacy, data protection, and grappling with new and emerging variants of digital fundamental rights and duties.

In conclusion, this author would suggest that, for any government which chooses to use ICTs to increase citizen participation in decision-making processes, the ultimate idea would be to neither use online public engagement to completely supplant the offline decision-making processes of elected representatives (as techno-enthusiasts might argue) nor altogether reject online public engagement (as Net-sceptics would prefer), but instead use the Internet and its associated technologies and applications to help elected representatives strengthen their democratic mandate and develop more informed, publicly-supported policy proposals to further strengthen State institutions and processes.

REFERENCES

Awad, N. F., & Krishnan, M. S. (2006). The Personalization Privacy Paradox: An Empirical Evaluation of Information Transparency and the Willingness to Be Profiled Online for Personalization. *Management Information Systems Quarterly*, *30*(1), 13–28.

Bannon, L. J. (1991). From Human Factors to Human Actors: The Role of Psychology and Human Computer Interaction Studies in System Design. In J. Greenbaum & M. Kyng (Eds.), *Design At Work: Cooperative Design of Computer Systems* (pp. 25–44). Lawrence Erlbaum Associates, Inc. Publishers.

Barber, B. R. (1984). *Strong Democracy: Participatory Politics for a New Age*. Berkeley, CA: University of California Press.

Baughman, H. M., Dearing, S., Giammarco, E., & Vernon, P. A. (2012). Relationships between Bullying Behaviours and the Dark Triad: A Study with Adults. *Personality and Individual Differences*, *52*(5), 571–575. doi:10.1016/j.paid.2011.11.020

Beike, D., & Wirth-Beaumont, E. (2005). Psychological Closure as a Memory Phenomenon. *Memory (Hove, England)*, *13*(6), 574–593. doi:10.1080/09658210444000241 PMID:16076673

Bellamy, C., & Taylor, J. A. (1998). *Governing in the Information Age*. Maidenhead, UK: Open University Press.

Bellovin, S. M. (2008). The Insider Attack Problem Nature and Scope. In S. J. Stolfo, S. M. Bellovin, S. Hershkop, A. Keromytis, S. Sinclair, & S. W. Smith (Eds.), *Insider Attack and Cyber Security: Beyond the Hacker* (pp. 1–4). New York, N.Y.: Springer Link. doi:10.1007/978-0-387-77322-3_1

Bicking, M., & Wimmer, M. A. 2010. Tools and Technologies in eParticipation: Insights from Project Evaluation. In F. De Cindio, A. Machintosh, & C. Peraboni (Eds.), *Online Deliberation:* Proceedings of the Fourth International Conference (pp. 75 – 86).

Bimber, B. (1998). The Internet and Political Transformation: Populism, Community and Accelerated Pluralism. *Polity*, *31*(1), 133–160.

Binns, A. (2012). Dont Feed the Trolls! *Journalism Practice*, *6*(4), 547–562. doi:10.1080/17512786.2011.648988

Bishop, J. (2012). The Psychology of Trolling and Lurking: The Role of Defriending and Gamification for Increasing Participation in Online Communities using Seductive Narratives. In H. Li (Ed.), *Virtual Community Participation and Motivation: Cross- Disciplinary Theories* (pp. 160–176). Hershey, PA: IGI Global. doi:10.4018/978-1-4666-0312-7.ch010

Bishop, J. (2014a). My Click is My Bond: The Role of Contracts, Social Proof, and Gamification for Sysops to Reduce Pseudo-Activism and Internet Trolling. In J. Bishop (Ed.), *Gamification for Human Factors Integration: Social, Education and Psychological Issues* (pp. 1–16). Hershey, PA: IGI Global. doi:10.4018/978-1-4666-5071-8.ch001

Bishop, J. (2014b). Representations of Trolls in Mass Media Communication: A Review of Media-Texts and Moral Panics Relating to Internet Trolling. *International Journal of Web Based Communities*, *10*(1), 7–24. doi:10.1504/IJWBC.2014.058384

Bishop, P., & Anderson, L. (2004). *E-Government to E-Democracy: 'High-Tech' Solutions to 'No Tech' Problems.* Paper presented to the Australian Electronic Governance Conference, Melbourne, Australia.

Block, J. (2008). Issues for DSM-V: Internet addiction. *The American Journal of Psychiatry*, *165*(3), 306–307. doi:10.1176/appi.ajp.2007.07101556 PMID:18316427

Blumler, J. C., & Coleman, S. (2001). Realising Democracy Online: A Civic Commons in Cyberspace. IPPR Citizens Online Research Publication No.2.

Buckels, E. E., Trapnell, P. D., & Paulhus, D. L. (2014). Trolls Just Want to Have Fun. *Personality and Individual Differences*, *67*, 97–102. doi:10.1016/j.paid.2014.01.016

Campbell, M. A. (2005). Cyber Bullying: An Old Problem in a New Guise? *Australian Journal of Guidance & Counselling*, *15*(1), 68–76. doi:10.1375/ajgc.15.1.68

Carayon, P., Hoonakker, P., & Smith, M. J. (2012). Human Factors in Organizational Design and Management. In G. Salvendy (Ed.), *Handbook of Human Factors and Ergonomics* – (4th ed.; pp. 534–552). John Wiley and Sons. doi:10.1002/9781118131350.ch18

Carroll, T., Edgar, T., Manz, D., & Greitzer, F. L. (2012). *Realizing Scientific Methods for Cyber Security.* Research paper presented at LASER '12.

Catinat, M., & Vedel, T. (2000). Public Policies for Digital Democracy. In K. L. Hacker & J. van Dijk (Eds.), *Digital Democracy: Issues of Theory and Practice* (pp. 184–208). London: Sage Publications. doi:10.4135/9781446218891.n11

Cavoukian, A., & Jonas, J. (2012). *Privacy by Design in the Age of Big Data*. Ontario: Information and Privacy Commissioner of Ontario, Canada.

Chen, Y., Harper, F. M., Konstan, J., & Li, S. X. (2009). Group Identity and Social Preferences. *The American Economic Review*, *99*(1), 431–457. doi:10.1257/aer.99.1.431

Chiu, C. M., Hsu, M. H., & Wang, E. T. (2006). Understanding Knowledge Sharing in Virtual Communities: An Integration of Social Capital and Social Cognitive Theories. *Decision Support Systems*, *42*(3), 1872–1888.

Chun, S., Shulman, S., Sandoval, R., & Hovy, E. (2010). Government 2.0: Making Connections Between Citizens, Data and Government. *Information Polity Journal*, *15*(1-2), 1–9.

Coleman, S. (2005). 'Just How Risky is Online Voting? *Information Polity*, *10*(1), 95–104.

Coleman, S., & Blumler, J. G. (2011). The Wisdom of Which Crowd? On the Pathology of a Listening Government. *The Political Quarterly*, *82*(3), 355–364. doi:10.1111/j.1467-923X.2011.02210.x

Coleman, S., & Gotze, J. (2001). *Bowling Together: Online Public Engagement in Policy Deliberation*. London: Hansard Society.

Cranor, L.F., Reagle, J., & Ackerman, M.S. (1999, August 17). Beyond Concern: Understanding Net Users' Attitudes About Online Privacy. *Privacy Study*.

Cruickshank, P., Edelmann, N., & Smith, C. (2010). Signing an e-Petition as a Transition from Lurking to Participation. In H. J. Scholl & M. Janssen (Eds.), *Electronic Government and Electronic Participation: Joint Proceedings of Ongoing Research and Projects of IFIP EGOV and IFIP ePart 2010* (pp. 275–282). Linz, Austria: Trauner Verlag.

Czaja, S. J., & Nair, S. N. (2012). Human Factors Engineering and Systems Design. In G. Salvendy (Ed.), *Handbook of Human Factors and Ergonomics* – (4th ed.; pp. 38–56). John Wiley and Sons. doi:10.1002/9781118131350.ch2

di Gennaro, C., & Dutton, W. H. (2006). The Internet and the Public: Online and Offline Political Participation in the UK. *Parliamentary Affairs*, *59*(2), 299–313. doi:10.1093/pa/gsl004

Donath, J. S. (1999). Identity and Deception in the Virtual Community. In M. A. Smith & P. Kollock (Eds.), Communities in Cyberspace (pp. 29-59). London: Routledge Press.

Dutta, S., Dutton, W. H., & Law, G. (2011). *The New Internet World: A Global Perspective on Freedom of Expression, Privacy, Trust and Security Online*. INSEAD Working Paper No. 2011/89/TOM.

Dutton, W. H., & Peltu, M. (2007). *Reconfiguring Government Public Engagements: Enhancing the Communicative Power of Citizens*. Oxford Internet Institute Forum Discussion Paper No. 9.

Dutton, W. H., & Shepherd, A. (2006). Trust in the Internet as an Experience Technology. *Information Communication and Society*, *9*(4), 433–451. doi:10.1080/13691180600858606

Eggers, W. D. (2005). *Government 2.0: Using Technology to Improve Education, Cut Red Tape, Reduce Gridlock and Enhance Democracy*. Lanham, MD: Rowman and Littlefield Publishers Inc.

Ferguson, R. (2006). *Digital Dialogues: Interim Report, December 2005 – August 2006*. London: The Hansard Society.

Fieldhouse, E., Tranmer, M., & Russel, A. (2007). Something about Young People or Something about Elections? Electoral Participation of Young People in Europe: Evidence from a Multilevel Analysis of the European Social Survey. *European Journal of Political Research*, *46*(6), 797–822. doi:10.1111/j.1475-6765.2007.00713.x

Fink, G., Best, D., Manz, D., Popovsky, V., & Endicott-Popovsky, B. (2013). Gamification for Measuring Cyber Security Situational Awareness. In D. D. Schmorrow & C. M. Fidopiastis (Eds.), *Foundations of Augmented Cognition: Advancing Human Performance and Decision-Making Through Adaptive Systems* (pp. 656–665). Berlin: Springer-Verlag. doi:10.1007/978-3-642-39454-6_70

Fishkin, J. S. (1995). *The Voice of the People: Public Opinion and Democracy*. New York: Yale University Press.

Geoghegan, T. (2007). The Petition, the 'Prat' and the Political Ideal. *BBC News – Magazine*. Available at: http://news.bbc.co.uk/1/hi/magazine/6354735.stm

Goodhart, M. (2007). Europes Democratic Deficits through the Looking Glass: The European Union as a Challenge for Democracy. *Perspectives on Politics*, *5*(3), 567–584. doi:10.1017/S1537592707071551

Griffiths, M. D. (1995). Technological Addictions. *Clinical Psychology Forum*, *76*, 14-19.

Griffiths, M. D. (1998). Internet Addiction: Does It Really Exist? In J. Gackenbach (Ed.), *Psychology and the Internet: Intrapersonal, Interpersonal, and Transpersonal Implications* (pp. 61–75). New York, NY: Academic Press.

Hansen, S. (2002). Excessive Internet Usage or Internet Addiction? The Implications of Diagnostic Categories for Student Users. *Journal of Computer Assisted Learning*, *18*(2), 232–236. doi:10.1046/j.1365-2729.2002.t01-2-00230.x

Hardaker, C. (2010). Trolling in Asynchronous Computer-Mediated Communication: From User Discussions to Academic Definitions. *Journal of Politeness Research*, *6*(2), 215–242. doi:10.1515/jplr.2010.011

Hoff, J. (2006). The Shaping of Digital Political Communication – Creating e-Democracy in a Danish Municipality: Intentions and Realities. In H. K. Hansen & J. Hoff (Eds.), *Digital Governance: Networked Societies. Creating Authority, Community and Identity in a Globalized World* (pp. 261–299). Copenhagen: Samfundslitteratur Press/NORDICOM.

Huysman, M., & Wulf, V. (2004). *Social Capital and Information Technology*. Hong Kong: M.I.T. Press.

Jacko, J. A., Yi, J. S., Sainfort, F., & McClellan, M. (2012). Human Factors and Ergonomic Methods. In G. Salvendy (Ed.), *Handbook of Human Factors and Ergonomics* (4th ed.; pp. 289–329). John Wiley and Sons. doi:10.1002/9781118131350.ch10

Jansen, E., & James, V. (2002). *NetLingo: The Internet Dictionary*. Oxnard, CA: Netlingo Inc.

Janssen, D., & Kies, R. (2004). *Online Forums and Deliberative Democracy: Hypotheses, Variables and Methodologies*. e-Democracy Centre e-Working Papers No. 1.

Kamis, B., & Thiel, T. (2013). *The Original Battle Trolls: How States Represent the Internet as a Violent Place*. Paper presented at the ECPR General Conference, Bordeaux, France.

Karwowski, W. (2012). The Discipline of Human Factors and Ergonomics. In G. Salvendy (Ed.), *Handbook of Human Factors and Ergonomics* (4th ed.; pp. 3–37). John Wiley and Sons. doi:10.1002/9781118131350.ch1

Keepers, G. A. (1990). Pathological Preoccupation with Video Games. *Journal of the American Academy of Child and Adolescent Psychiatry*, *29*(1), 49–50. doi:10.1097/00004583-199001000-00009 PMID:2295578

Kies, R., & Wojcik, S. (2010). European Web-Deliberation: Lessons from the European Citizens Consultation. In F. De Cindio, A. Machintosh, & C. Peraboni (Eds.), *Online Deliberation: Proceedings of the Fourth International Conference - OD2010* (pp. 198-211).

Kuutti, K. (1996). Activity Theory as a Potential Framework for Human Computer Interaction Research. In B. A. Nardi (Ed.), Context and Consciousness: Activity Theory and Human Computer Interaction (pp. 17-44). Boston, MA: M.I.T. Press.

Law, D. M., Shapka, J. D., Hymel, S., Olson, B. F., & Waterhouse, T. (2012). The Changing Face of Bullying: An Empirical Comparison Between Traditional and Internet Bullying and Victimization. *Computers in Human Behavior*, *28*(1), 226–232. doi:10.1016/j.chb.2011.09.004

Lesieur, H. R., & Blume, S. B. (1993). Pathological Gambling, Eating Disorders, and the Psychoactive Substance Use Disorders. *Journal of Addictive Diseases*, *12*(3), 89–102. doi:10.1300/J069v12n03_08 PMID:8251548

Levy, S. (1996). Breathing is Also Addictive. *Newsweek*, *128*, 52-53.

Loukis, E., & Wimmer, M. A. (2010). Analysing Different Models of Structured Electronic Consultation on Legislation Under Formation. In F. De Cindio, A. Macintosh & C. Peraboni (Eds.), *Online Deliberation: Proceedings of the Fourth International Conference – OD2010* (pp. 14 – 26).

Macintosh, A., Malina, A., & Farrell, S. (2002). Digital Democracy through Electronic Petitioning. In W. J. McIver Jr & A. K. Elmagarmid (Eds.), *Advances in Digital Government: Technology, Human Factors and Policy* (pp. 137–162). Dordrecht: Kluwer Academic Publishing.

Marks, I. (1990). Behavioural (non-chemical) Addictions. *British Journal of Addiction*, *85*(11), 1389–1394. doi:10.1111/j.1360-0443.1990.tb01618.x PMID:2285832

McCarthy, P., Rylance, J., Bennett, R., & Zimmermann, H. (2001). *Bullying: From Backyard to Boardroom* (2nd ed.). Sydney, Australia: The Federation Press.

Mehroof, M., & Griffiths, M. D. (2010). Online Gaming Addiction: The Role of Sensation Seeking, Self-control, Neuroticism, Aggression, State Anxiety, and Trait Anxiety. *Cyberpsychology, Behavior, and Social Networking*, *13*(3), 313–316. doi:10.1089/cyber.2009.0229 PMID:20557251

Morahan-Martin, J. (2007). Internet Use and Abuse and Psychological Problems. In *Oxford Handbook of Internet Psychology* (pp. 331–345). Oxford, UK: Oxford University Press.

Morgan, W. (1979). Negative Addiction in Runners. *The Physician and Sportsmedicine*, *7*, 56–69.

Neuhaus, R. J. (1989). The Naked Public Square: Religion and Democracy in America, Second Edition. Grand Rapids, MI: William B. Eedermans Publishing.

Olweus, D. (1993). *Bullying at School: What We Know and What We Can Do*. London: John Wiley.

Parker, D. B. (1976). *Crime by Computer*. New York, NY: Scribner's.

Peele, S., & Brodsky, A. (1979). *Love and Addiction*. Scarborough, Ontario: New American Library of Canada.

Pfleeger, C. P. (2008). Reflections on the Insider Threat. In S. J. Stolfo, S. M. Bellovin, S. Hershkop, A. Keromytis, S. Sinclair, & S. W. Smith (Eds.), *Insider Attack and Cyber Security: Beyond the Hacker* (pp. 6–15). New York, NY: Springer Link. doi:10.1007/978-0-387-77322-3_2

Phillips, W. (2011). LOLing At Tragedy: Facebook Trolls, Memorial Pages, and Resistance to Grief Online. *First Monday*, *16*(12), 1–13. doi:10.5210/fm.v16i12.3168

Phillips, W. (2013). The House That Fox Built: Anonymous, Spectacle, and Cycles of Amplification. *Television & New Media*, *14*(6), 494–509. doi:10.1177/1527476412452799

Proctor, R. W., & Vu, K.-P. L. (2012). Selection and Control of Action. In G. Salvendy (Ed.), *Handbook of Human Factors and Ergonomics* (4th ed.; pp. 95–116). John Wiley and Sons. doi:10.1002/9781118131350.ch4

Pyzalski, J. (2011). Electronic Aggression Among Adolescents: An Old House with a New Facade (or Even a Number of Houses). In C. Hällgren, E. Dunkels, & G.-M. Frånberg (Eds.), *Youth Culture and Net Culture: Online Social Practices* (pp. 278–295). Hershey, PA: IGI Global. doi:10.4018/978-1-60960-209-3.ch016

Rachlin, H. (1990). Why Do People Gamble and Keep Gambling Despite Heavy Losses? *Psychological Science*, *1*(5), 294–297. doi:10.1111/j.1467-9280.1990.tb00220.x

Rheingold, H. (1993). *The Virtual Community: Homesteading on the Electronic Frontier*. Reading, MA: Addison-Wesley.

Schwartz, M. (2008, August 3). The Trolls Among Us. *The New York Times*.

Sclove, R. E. (1995). *Democracy and Technology*. London: Guilford Press.

Shah, D. V., Kwak, N., & Holbert, R. L. (2002). Connecting" and" Disconnecting" with Civic Life: Patterns of Internet Use and the Production of Social Capital. *Political Communication*, *18*(2), 141–162.

Thomassen, J., & Schmitt, H. (1999). Introduction: Political Representation and Legitimacy in the European Union. In H. Schmitt & J. Thomassen (Eds.), *Political Representation and Legitimacy in the European Union* (pp. 3–21). Oxford, UK: Oxford University Press. doi:10.1093/0198296614.003.0001

Turkle, S. (1995). *Life behind the Screen: Identity in the Age of the Internet*. New York: Simon & Schuster.

Turkle, S. (2011). *Life on the Screen: Identity in the Age of the Internet*. New York, NY: Simon and Schuster.

Urban, G. L., Amyx, C., & Lorenzon, A. (2009). Online Trust: State of the Art, New Frontiers, and Research Potential. *Journal of Interactive Marketing*, *23*(2), 179–190. doi:10.1016/j.intmar.2009.03.001

Virkar, S. (2007). *(Dis) Connected Citizenship? Exploring Barriers to eConsultation in Europe. Report to the European Commission for the Breaking Barriers to eGovernment: Overcoming Obstacles to Improving European Public Services* Project.

Virkar, S. (2011). *The Politics of Implementing e-Government for Development: The Ecology of Games Shaping Property Tax Administration in Bangalore City* (Unpublished Doctoral Thesis). University of Oxford, Oxford, UK.

Virkar, S. (2014). Trolls Just Want To Have Fun: Electronic Aggression within the Context of e-Participation and Other Online Political Behaviour in the United Kingdom. *International Journal of E-Politics*, *5*(4), 21–51. doi:10.4018/ijep.2014100102

Wallace, P. (2001). *The Psychology of the Internet*. Cambridge, UK: Cambridge University Press.

Walters, T., Hourizi, R., Moncur, W., & Pitsillides, S. (2011). Does the Internet Change How We Die And Mourn? Overview and Analysis. *Omega*, *64*(4), 275–302. doi:10.2190/OM.64.4.a PMID:22530294

Widyanto, L., & Griffiths, M. (2011). Internet Addiction: Does It Really Exist? (Revisited). In J. Gackenbach (Ed.), *Psychology and the Internet: Intrapersonal, Interpersonal, and Transpersonal Implications* (pp. 142–166). London: Academic Press.

Winn, M. (1977). *The Plug-In Drug*. New York, NY: Viking Press.

Young, K. S. (1999). The Evaluation and Treatment of Internet Addiction. In L. VandeCreek & T. Jackson (Eds.), Innovations in Clinical Practice: A Source Book 17 (pp. 19-31). Saratosa, FL: Professional Resource Press.

Young, K. S. (2004). Internet Addiction a New Clinical Phenomenon and its Consequences. *The American Behavioral Scientist*, *48*(4), 402–415. doi:10.1177/0002764204270278

Young, K. S., & Case, C. J. (2004). Internet Abuse in the Workplace: New Trends in Risk Management. *Cyberpsychology, Behavior, and Social Networking*, *7*(1), 105–111. doi:10.1089/109493104322820174 PMID:15006175

ADDITIONAL READING

Andreassen, C. S., Griffiths, M. D., Gjertsen, S. R., Krossbakken, E., Kvam, S., & Pallesen, S. (2013). The Relationships between Behavioral Addictions and the Five-Factor Model of Personality. *Journal of Behavioral Addictions*, *2*(2), 90–99. doi:10.1556/JBA.2.2013.003 PMID:26165928

Buckels, E. E., Jones, D. N., & Paulhus, D. L. (2013). Behavioral Confirmation of Everyday Sadism. *Psychological Science*, *21*(11), 2201–2209. doi:10.1177/0956797613490749 PMID:24022650

DiGuilio, R. C. 2001. Educate, Medicate, or Litigate? What Teachers, Parents, and Administrators Must Do About Student Behavior. Thousand Oaks, C.A.: Corwin Press.

Dooley, J. J., Pyzalski, J., & Cross, D. (2009). Cyberbullying Versus Face-to-Face Bullying: A Theoretical and Conceptual Review. *Zeitschrift für Psychologie. The Journal of Psychology*, *217*(4), 182–188.

Efimova, L. 2009. Weblog as a Personal Thinking Space. In C. Cattuto, G. Ruffo, & F. Menczer (eds.) *Proceedings of the 20th ACM Conference on Hypertext and Hypermedia*, (pp. 289-298). New York, N.Y.: ACM Press. doi:10.1145/1557914.1557963

Eisenhardt, K. M. (1989). Building Theories from Case Study Research. *Academy of Management Review*, *14*(4), 532–550.

Fanti, K. A., & Kimonis, E. R. (2013). Dimensions of Juvenile Psychopathy Distinguish Bullies, Bully-Victims, and Victims. *Psychology of Violence*, *3*(4), 396–409. doi:10.1037/a0033951

Furnham, A., Richards, S. C., & Paulhus, D. L. (2013). The Dark Triad of Personality: A 10-year Review. *Social and Personality Compass*, *7*(3), 199–216. doi:10.1111/spc3.12018

Gackenbach, J. (2011). *Psychology and the Internet: Intrapersonal, Interpersonal, and Transpersonal Implications*. London: Academic Press.

Graham, L. T., & Gosling, S. D. (2013). Personality Profiles Associated with Different Motivations for Playing World of Warcraft. *Cyberpsychology, Behavior, and Social Networking*, *16*(3), 189–193. doi:10.1089/cyber.2012.0090 PMID:23438267

Hacker, K. L., & van Dijk, J. (2000). Introduction: What is Digital Democracy? In K. L. Hacker & J. van Dijk (Eds.), *Digital Democracy: Issues of Theory and Practice* (pp. 1–9). London: Sage Publications. doi:10.4135/9781446218891.n1

Hagen, M. (2000). Digital Democracy and Political Systems. In K. L. Hacker & J. van Dijk (Eds.), *Digital Democracy: Issues of Theory and Practice* (pp. 54–69). London: Sage Publications. doi:10.4135/9781446218891.n4

Hartley, J. (2005). Case Study Research. In C. Cassell & G. Symon (Eds.), *Essential Guide to Qualitative Methods in Organisational Research* (pp. 323–333). London: Sage Publications.

Heeks, R. (2000). The Approach of Senior Public Officials to Information-Technology Related Reform: Lessons from India. *Public Administration and Development*, *20*(3), 197–205. doi:10.1002/1099-162X(200008)20:3<197::AID-PAD109>3.0.CO;2-6

Heeks, R. 2003. Most eGovernment-for-Development Projects Fail: How Can the Risks be Reduced? *iGovernment Working Paper Series – Paper No. 14*, University of Manchester (UK).

Held, D. (1996). *Models of Democracy*. Cambridge: Blackwell Publishers.

Herring, S., Job-Sluder, K., Scheckler, R., & Barab, S. (2002). Searching for Safety Online: Managing‘ Trolling’ in a Feminist Forum. *The Information Society*, *18*(5), 371–384. doi:10.1080/01972240290108186

Hiller, J.S., & Bélanger, F. 2001. Privacy Strategies for Electronic Government. *The PriceWaterHouseCoopers E-Government Series*, January 2001.

Hinduja, S., & Patchin, J. W. (2010). Bullying, Cyberbullying, and Suicide. *Archives of Suicide Research*, *14*(3), 206–221. doi:10.1080/13811118.2010.494133 PMID:20658375

H.M.Government. (2002). *In the Service of Democracy – A Consultation Paper on a Policy for Electronic Democracy. Published by the Office of the e-Envoy.* London, United Kingdom: The Cabinet Office.

Howard, T. W. 2010. Design to Thrive: Creating Social Networks and Online Communities that Last. Burlington, M.A.: Morgan Kaufmann.

Huang, Y.-y., & Chou, C. (2010). An Analysis of Multiple Factors of Cyberbullying Among Junior High School Students in Taiwan. *Computers in Human Behavior*, *26*(6), 1581–1590. doi:10.1016/j.chb.2010.06.005

Irani, Z., Elliman, T., & Jackson, P. (2007). Electronic Transformation of Government in the UK: Research Agenda. *European Journal of Information Systems*, *16*(4), 327–335. doi:10.1057/palgrave.ejis.3000698

Jankowski, N., & van Selm, M. (2000). The Promise and Practice of Public Debate in Cyberspace. In K. L. Hacker & J. van Dijk (Eds.), *Digital Democracy: Issues of Theory and Practice* (pp. 149–165). London: Sage Publications. doi:10.4135/9781446218891.n9

Jansen, E., & James, V. 2002. *NetLingo: The Internet Dictionary*. Oxnard, C.A..: Netlingo Inc.

Jenkins, H., & Thorburn, D. 2004. Introduction: The Digital Revolution, The Informed Citizen and the Culture of Democracy. In H. Jenkins & D. Thorburn (eds.) Democracy and New Media (pp. 1-20). Cambridge, M.A.: M.I.T. Press.

Joiner, R., Gavin, J., Duffield, J., Brosnan, M., Crook, C., Durndell, A., & Lovatt, P. et al. (2005). Gender, Internet Identification, and Internet Anxiety: Correlates of Internet Use. *Cyberpsychology, Behavior, and Social Networking*, *8*(4), 371–378. doi:10.1089/cpb.2005.8.371 PMID:16092894

Jones, D. N., & Paulhus, D. L. (2010). Different Provocations Trigger Aggression in Narcissists and Psychopaths. *Social Psychological & Personality Science*, *1*(1), 12–18. doi:10.1177/1948550609347591

Juvonen, J., & Gross, E. F. (2008). Extending the School Grounds? Bullying Experiences in Cyberspace. *The Journal of School Health*, *78*(9), 496–505. doi:10.1111/j.1746-1561.2008.00335.x PMID:18786042

Kooiman, J., & van Vliet, M. (1993). Introduction: Governance and Public Management. In K. A. Eliassen & J. Kooiman (Eds.), *Managing Public Organisations: Lessons from Contemporary European Experience* (2nd ed., pp. 1–7). London: Sage Publications.

Koopmans, R., & Statham, P. 2002. The Transformation of Political Mobilisation and Communication in the European Public Spheres: A Research Outline. *Report to the European Commission dated 18th February 2002.*

Kowalski, R. M., Limber, S. P., & Agatston, P. W. 2008. Cyberbullying: Bullying in the Digital Age. Malden, M.A.: John Wiley.

Lampe, C., & Resnick, P. 2004. Slash (dot) and Burn: Distributed Moderation in a Large Online Conversation Space. In J.A. Konston, E. Chi, & K. Hook (eds.) *Proceedings of the SIGCHI Conference on Human Factors in Computing Systems*, (pp. 543-550). New York, N.Y.: ACM Press. doi:10.1145/985692.985761

Lenhardt, A. (2013). Teens, Smartphones & Texting. Washington, DC: Pew Research Center's Internet & American Life Project; Retrieved from http:// pewinternet.org/Reports/2012/Teens-and-smartphones.aspx

Lindner, R., & Riehm, U. 2009. Electronic Petitions and Institutional Modernisation. *eJournal of eDemocracy and Open Government*, 1(1), 1-11

Linton, D. K., & Power, J. L. (2013). The Personality Traits of Workplace Bullies are Often Shared by Their Victims: Is There a Dark Side to Victims? *Personality and Individual Differences*, *54*(6), 738–743. doi:10.1016/j.paid.2012.11.026

Ljepava, N., Orr, R. R., Locke, S., & Ross, C. (2013). Personality and Social Characteristics of Facebook Non-Users and Frequent Users. *Computers in Human Behavior*, *29*(4), 1602–1607. doi:10.1016/j.chb.2013.01.026

Lorenz, J., Rauhut, H., Schweitzer, F., & Helbing, D. (2011). How Social Influence Can Undermine the Wisdom of Crowd Effect. *Proceedings of the National Academy of Sciences of the United States of America*, *108*(22), 9020–9025. doi:10.1073/pnas.1008636108 PMID:21576485

Lowndes, V., Pratchett, L., & Stoker, G. (2001). Trends in Local Government: Part 1 – Local Government Perspectives. *Public Administration*, *79*(1), 205–222. doi:10.1111/1467-9299.00253

Margolis, M. (2007). E-Government and Democratic Politics. In P. G. Nixon & V. N. Koutrakou (Eds.), *E-Government in Europe: Rebooting the State* (pp. 1–18). London: Routledge Press.

McCosker, A. (2014). Trolling as Provocation: YouTubes Agonistic Publics. *Convergence (London)*, *20*(2), 201–217. doi:10.1177/1354856513501413

National Audit Office 2002. *Better Public Services through eGovernment: Academic Article in Support of Better Public Services through eGovernment*. Report by the Comptroller and Auditor General, April 2002.

Newman, J. (2005). *Remaking Governance: Peoples Politics and the Public Sphere*. Bristol: The Policy Press.

OECD. (2001). *Citizens as Partners: Information, Consultation and Public Participation in Policymaking*. OECD Press.

OECD. (2003). *Promise and Problems of E-Democracy: Challenges of Online Citizen Engagement*. Paris: OECD Press.

Olweus, D., & Limber, S. 1999. Bullying Prevention Program. In D. Elliott (Ed.), Blueprints for Violence Prevention. Boulder, C.O.: Institute of Behavioral Science, Regents of the University of Colorado.

Ortega, R., Elipe, P., Mora-Merchan, J. A., Calmaestra, J., & Vega, E. (2009). The Emotional Impact on Victims of Traditional Bullying and Cyberbullying: A Study of Spanish Adolescents. *Zeitschrift für Psychologie. The Journal of Psychology*, *217*(4), 197–204.

Peters, B., Sifft, S., Bruggmann, M., & Konigslow, K. (2005). National and Transnational Public Spheres: The Case of the EU. In S. Liebfried & M. Zurn (Eds.), *Transformations of the State?* (pp. 139–160). Cambridge: Cambridge University Press. doi:10.1017/CBO9780511752193.007

Petersen, L. 2001. Anti-Bullying Programs-Avoiding Bullying the Bullies. In *Promoting Wellbeing*, AGCA Conference Proceedings (pp. 51-60). Brisbane: Australian Academic Press.

Phillips, J. G., Butt, S., & Blaszczynski, A. (2006). Personality and Self-Reported Use of Mobile Phones for Games. *Cyberpsychology, Behavior, and Social Networking*, *9*(6), 753–758. doi:10.1089/cpb.2006.9.753 PMID:17201601

Pleace, N. (2007). E-Government in the United Kingdom. In P. G. Nixon & V. N. Koutrakou (Eds.), *E-Government in Europe: Rebooting the State* (pp. 61–74). London: Routledge Press.

Powazek, D. M. 2006. Design for Community: The Art of Connecting Real People in Virtual Places. Indianapolis, I.N.: New Riders Press.

Rhodes, R. A. W. (1996). The New Governance: Governing without Government. *Political Studies*, *44*(4), 652–667. doi:10.1111/j.1467-9248.1996.tb01747.x

Rivers, I., & Noret, N. (2010). I h8 u: Findings From a Five-Year Study of Text and Email Bullying. *British Educational Research Journal, 36*(4), 643–671. doi:10.1080/01411920903071918

Rosen, L. D., Whaling, K., Rab, S., Carrier, L. M., & Cheever, N. A. (2013). Is Facebook Creating iDisorders? The Link between Clinical Symptoms of Psychiatric Disorders and Technology Use, Attitudes and Anxiety. *Computers in Human Behavior, 29*(3), 1243–1254. doi:10.1016/j.chb.2012.11.012

Rosen, T. 2001. E-Democracy in Practice: Swedish Experience of New Political Tool. Available at www.svenkom.se/skvad/e-democracy-en.pdf

Scharpf, F. W. (1999). *Governing in Europe: Effective and Democratic?* Oxford: Oxford University Press. doi:10.1093/acprof:oso/9780198295457.001.0001

Sclove, R. (1995). *Democracy and Technology*. New York: Guilford Press.

Shachaf, P., & Hara, N. (2010). Beyond Vandalism: Wikipedia Trolls. *Journal of Information Science, 36*(3), 357–370. doi:10.1177/0165551510365390

Slee, P. T., & Rigby, K. (1993). Australian School Childrens Self-Appraisal of Interpersonal Relations: The Bullying Experience. *Child Psychiatry and Human Development, 23*(4), 273–281. doi:10.1007/BF00707680 PMID:8325135

Smith, P. K. (2009). Cyberbullying: Abusive Relationships in Cyberspace. *Zeitschrift für Psychologie. The Journal of Psychology, 217*(4), 180–181.

Smith, P. K., Mahdavi, J., Carvalho, M., Fisher, S., Russell, S., & Tippett, N. (2008). Cyberbullying: Its Nature and Impact in Secondary School Pupils. *Journal of Child Psychology and Psychiatry, and Allied Disciplines, 49*(4), 376–385. doi:10.1111/j.1469-7610.2007.01846.x PMID:18363945

Stoker, G. (1998). Governance as Theory: Five Propositions. *International Social Science Journal, 50*(155), 17–28. doi:10.1111/1468-2451.00106

The Electoral Commission. (2005). *Election 2005: Engaging the Public in Great Britain – an Analysis of Campaigns and Media Coverage*. London: The Electoral Commission.

The Royal Academy of Engineering. 2007. *Dilemmas of Privacy and Surveillance: Challenges of Technological Change*. Report available at: www.raeng,org.uk/policy/reports/default.htm

Virkar, S. 2014a. Designing and Implementing e-Government Projects: Actors, Influences, and Fields of Play. In S. Saeed & C. G. Reddick (eds.) Human-Centered Design for Electronic Government. Hershey, P.A.: IGI Global.

Virkar, S. 2014b. Re-engaging the Public in the Digital Age: e-Consultation Initiatives in the Government 2.0 Landscape. Encyclopedia of Information Science and Technology 3rd Edition (pp. 427-435). Hershey, P.A.: IGI Global.

Wachs, S., & Wolf, K. D. (2011). Correlates Between Bullying and Cyberbullying: First Results of a Self-Report Study. *Praxis der Kinderpsychologie und Kinderpsychiatrie*, *60*(9), 735–744. doi:10.13109/prkk.2011.60.9.735 PMID:22242255

West, D. M. (2004). E-Government and the Transformation of Services and Citizen Attitudes. *Public Administration Review*, *64*(1), 15–27. doi:10.1111/j.1540-6210.2004.00343.x

Whyte, A., & Macintosh, A. (2003). Analysis and Evaluation of E-Consultations. *e-Service Journal*, *2*(1), 9–34. doi:10.2979/ESJ.2002.2.1.9

Wilcox, R. K. (1998). *Wings of Fury: From Vietnam to the Gulf War-The Astonishing True Stories of America's Elite Fighter Pilots*. New York, N.Y.: Simon & Schuster.

Willard, N. E. 2006. Cyberbullying and Cyberthreats: Responding to the Challenge of Online Social Aggression, Threats, and Distress. Eugene, O.R.: Center for Safe and Responsible Internet Use Press.

Ybarra, M. L., & Mitchell, K. J. (2004). Online Aggressor/Targets, Aggressors, and Targets: A Comparison of Associated Youth Characteristics. *Journal of Child Psychology and Psychiatry, and Allied Disciplines*, *45*(7), 1308–1316. doi:10.1111/j.1469-7610.2004.00328.x PMID:15335350

Zweig, J. M., Dank, M., Yahner, J., & Lachman, P. (2013). The Rate of Cyber Dating Abuse Among Teens and How It Relates to Other Forms of Teen Dating Violence. *Journal of Youth and Adolescence*, *42*(7), 1063–1077. doi:10.1007/s10964-013-9922-8 PMID:23412689

KEY TERMS AND DEFINITIONS

Actor Goals and Motivations: The aims that key actors seek to attain and maintain from interacting with other players, encompassing both broader long-term achievements as well as more short- to medium-term rewards.

Actor Perceptions: Include the preferences and opinions of key institutional players that help determine the disjoint between project design and current ground realities, together with the nature and direction of organisational reform and institutional change.

Actor(s): The individuals, groups or other entities whose interactions shape the direction and nature of a particular game being considered.

Country Context Gap: Refers to the gap that arises when a system designed in theory for one country is transferred into the reality of another.

Design-Actuality Gap Model or Framework: A framework for project evaluation which contends that the major factor determining project outcome is the degree of mismatch between the current ground realities of a situation ('where are we now'), and the models, conceptions, and assumptions built into a project's design (the 'where the project wants to get us').

e-Administration: The improvement of government processes and to the streamlining of the internal workings of thc public sector often using ICT-based information systems.

e-Consultation: The process whereby citizens are given the opportunity to provide feedback to government online on matters of public importance and participate in the shaping of issues relevant to them via the new digital media.

e-Democracy: May be defined by the express intent to increase the participation of citizens in decision-making through the use of digital media and the application of Information and Communication Technologies to political processes. e-Democracy may be subdivided into e-Engagement (or e-Participation), e-Voting, e-Consultation.

e-Engagement (or e-Participation): The overall enhancement of opportunities for greater consultation and dialogue between government and its citizens through the encouragement of online citizen action and citizen participation in political processes electronically.

e-Governance: The use of ICTs by government, civil society, and political institutions to engage citizens in political processes and to the promote greater participation of citizens in the public sphere.

e-Government: The use of Information and Communication Technologies by government departments and agencies to improve internal functioning and public service provision. Broadly speaking, e-government may be divided into 2 distinct areas: e-Administration and e-Services.

e-Services: The improved delivery of public services to citizens through multiple electronic platforms

e-Voting: May be defined broadly as the expression and exercise of fundamental democratic rights and duties online through specially developed digital platforms.

Game(s): Arena(s) of competition and cooperation structured by a set of rules and assumptions about how to act in order for actors to achieve a particular set of objectives.

Hard-Soft Gap: The difference between the *actual, rational design* of a technology (hard) adopted within a project and the *actuality of the social context*, namely people, culture, politics, etc., within which the system operates (soft).

Internet Trolling: The practice of deliberately trying to aggress electronically or to distress participants online through frequently inflammatory and abusive behaviour; usually just to disrupt without direction and to often do so anonymously.

Managerial Variables: Those institutional variables relating to project management and other *soft* variables of project design and implementation, which include the efficiency and effectiveness of a supply chain, the characteristics of an agency's culture, and the capacity of an adopting agency to adapt to and to manage change.

Moves: Actions, decisions and other plays made by key actors taken to arrive at key goals, usually if not always based on their strategy of choice.

Partial Failure: An initiative is a situation in which major goals are unattained or where there are significant undesirable outcomes.

Political Variables: Those *soft* institutional variables relating to the perceptions and impressions that public servants have regarding potential labour cuts, administrative turnover, and changes in executive direction generated by the development of e-government.

Private-Public Gap: The mismatch that results when technology meant for private organisations is used in the public sector without being adapted to suit the role and aims of the adopting public organisation.

Project Outcome: The sum total of the interaction between organisational and institutional realities and the project design carried out within the constraints of the current organisational and institutional set-up.

Rules: The written or unwritten codes of conduct that shape actor moves and choices during a game.

Strategies: Include tactics, ruses, and ploys adopted by key actors during the course of a game to keep the balance of the engagement in their favour.

Success: An initiative is a situation in which most actor groups attain their major goals and do not experience significant undesirable outcomes.

Technological Addiction: Of which *Internet Addiction Disorder* is a subset, is the non-chemical, behavioural addiction that involves human-machine interaction; wherein the behaviours that manifest can be passive or active, and usually contain both inducing and reinforcing features which may contribute to the promotion of addictive tendencies.

Technological Variables: Those institutional variables relating to technology and other *hard* elements of project design and implementation, which include the ability of a user-population to access ICTs, the quality of the user population's Internet use, the availability of an internal technological infrastructure, and the provision of technical skills to the government workforce.

Total Failure: An initiative is a situation where a project is either never implemented or in which a new system is implemented but is almost immediately abandoned.

Troll(s): The person or set of people who attempt to entertain their kind through the sustained provocation of others or via the propagation of misinformation and other forms of propaganda in order to attain a certain measure of dominance and/or satisfaction; the term may be extended to include any individual within a subversive, transgressive, or dissident faction and/or group, particularly online.

Troller: An identifiable entity that actively seeks out people not conforming to a particular reality or opinion with the aim of provoking them into a response or a corresponding situation.

Trolling: An expression of electronic aggression or online maladjustive behaviour that ranges from implying the mere provocation of people online for mutual sub-culture and group enjoyment to meaning the definitive and sustained abuse of others for an individual's own personal satisfaction, material gain, and entertainment.

ENDNOTES

1 Otherwise called Electronic Governance, this term refers to the use of technology to enhance the structural (institutional and legal) frameworks and processes of governance. Based on discussions in earlier sections of this paper, the notion of eGovernance may be conceptually divided into eGovernment (which is concerned with electronic public service delivery) and eDemocracy (where technology is used to impact the quality and nature of democratic processes).

2 UK Citizens Online Democracy was a project which began life by setting up a website to discuss the Freedom of Information Act proposed by the Blair Administration in 1999/2000.

Chapter 7 The Risk, the Real, and the Impossibility of Life: Examining the Matrix

Maximiliano Emanuel Korstanje
University of Palermo, Argentina

ABSTRACT

The discourse of the Matrix not only wakes up philosophy from the slumber they are, but reflects the rise of fears which are proper of a culture, that manipulates anxieties to fix policies otherwise would be rejected. Basically, Neo is subject to an ethical dilemma which means that we are free to make the decision to live in a fake without suffering or in "the real of the dessert" as Morpheus put it. At this extent, if technology is used to protect humankind, Matrix evinces under some circumstances, it can oppress human beings when they renounce to their "free-will". This chapter not only explored the limitations of the society of risks within liberal thought but taking into consideration the role of authenticity serves ideologically to legitimate suffering. Indeed, accepting to live as Neo did, may lead to oppression or liberation.

DOI: 10.4018/978-1-5225-1938-6.ch007

INTRODUCTION

The film Matrix, a saga starred by Keanu Reeves, Carrie Anne Moss and Lawrence Fishburne, represented the philosophical dilemma between reality and fiction. The plot is as follows, Trinity who is a beauty hacker convinces Neo, jointly to Morpheus, the life he is living is a simple fake, produced by the matrix. To break with this Neo should take a pill and abandon the pseudo-reality where he was embedded with. Henceforth, Neo Trinity and Morpheus together other rebels start a rebellion against the oppression of matrix, a machine who subject humankind to live in a state of simulation where suffering and pain does not exist. Though philosophically speaking the Matrix did not add nothing new after the Cavern of Plato, it woke up much debate among philosophers and epistemologists, many of them gathered in the book *The Matrix and Philosophy* edited by Senior lecturer William Irwin. In this chapter we will discuss the ebbs and flows of this book, concentrating efforts in one of Morpheus`s phrases "welcome to the desert of the real" as well as the betrayal of Cypher, who opts for returning to Matrix to avoid the reality. The arrival of modernity not only altered the promise of well-farian state, but also proved to be a world where survival is not granted. If the world of our grand-parents was stable for them, postmodern cosmologies inscribes in an atmosphere of conflict and pain, where the basic resources for surviving are not ensured by state.

The Law within Liberal Tradition

One of the questions that concerned philosophers and social scientists was how society is united. Thomas Hobbes (1651) envisaged the social bond as based on two powerful but contrasting tendencies: the appetite for the property of others, and the needs of personal safety. Hobbes goes on to say that one might fight to gain further wealth competing with others, but sooner or later, the concept of protection prevails. To avoid the war of all against all, people do confer to a third party, Leviathan (state) the monopoly of force. Therefore, in Hobbesian theory, human beings are prone to develop a lasting peace. In contrast, Jean Jacques Rousseau (1762) argued that human beings corrupt themselves when they abandon the state of nature as given by God. The societal order gives to person an illusory view of reality. The division of labour, accompanied with all ideological mechanism of indoctrination leads the individual savage mind to the covenant. Certainly, the conflict we in our societies observe today, results from the frustration and not vice-versa. The Rousseau legacy gave an all-encompassing idea of how a group is formed, and it paves the ways for the contributions of other French scholar, Emile Durkheim. The fact is that Durkheim (1893, 1895, 1912) turned to Rousseau's concept of the *volonté générale* for a conception of social solidarity that did not depend on the atomistic individualism of liberal economics, today neoliberal economic, nor on the Hobbesian leviathan. Durkheim recognized that contractual relations among people, even those that ostensibly founded the state, depended on pre-existing norms and shared values. Here, Durkheim inserted the *conscience collectif* as a way to talk about how humans carry on their intercourse on a quotidian basis—that is, neither through formal contract nor coercion. People collectively create their normative worlds, which they then project onto various organizational and symbolic entities, hence arise churches, states, and religions.

While Durkheim sought an explanation for how a society is possible in collective explanations, the liberal alternative stays close to individualism. Instead of a coercive leviathan as in Hobbes, an apparently softer version was articulated by John Locke (1689). In the Lockean version of the social contract, the state acts as arbiter, whose main role is to settle disputes—a sort of general arbitration panel. John Rawls was one of the main exponents of this liberal tradition.

Based on his theory of justice, John Rawls identifies five forms of political organizations. In his 1999 *The Law of Peoples* he focuses on some questions linked to a utopia within the framework of a democratic constitutional society. Starting from this premise, Rawls explains not only why some nations fail in this process but also the idea that motivates the law of peoples in modern democracies. The first, he adds, is one of the evils of human history, unjust war and political oppression. The second refers to social policies to eradicate deep-seated injustices. Rawls' five subtypes of political organizations are:

1. Reasonable law,
2. Descent people,
3. Outlaw states,
4. Societies burdened by unfavourable conditions, and
5. Benevolent absolutism.

The theory of reasonable law rests on the belief that people sacrifice their appetite for war and ambitions to achieve wider forms of political, economics, and social cooperation. Therefore, trade and negotiations are of paramount importance to balance the international relationships. Of course, Rawls is criticized simply because after Auschwitz this idea would seem a simplistic utopia, an allegory. His response to these allegations is not convincing. Rawls echoes Kant's doctrine of international law considering that only the liberal society may lead human beings to a sustainable state of well-being. Reasonable and liberal people form a constitution that species institutions and a democracy as the primary form of government. Reasonable people seem to be united by three factors, constitution, common sympathy, and moral nature. Each view is counter-balanced by the view of a second, centrifugal force, which leads the society to pluralism. The law of people not only limits the sovereignty of state, and the problems this generates, but also limits states' propensity to declare war to other states. This idea corresponds with Rawls' concepts of justice and security.

A difference between liberal peoples and states is that just liberal peoples limit their basic interests as required by the reasonable. In contrast, the content of interests of states does not allow them to be stable for the right reasons: that is, from firmly accepting and acting upon a just Law of People. Liberal people do, however, have their fundamental interests as permitted by their conception of right and justice. They seek to protect their territory, to ensure the security and safety of their citizens, and to preserve their free political institutions and the liberties and free culture of their civil societies. (p. 29)

What is important to discuss in the Rawls's theory is to what an extent the liberality he proposes creates asymmetries or inequalities. He proposes the veil of ignorance in establishing the constitution in which the founders of a polity will not know where they will stand in the social and political-economic hierarchy. By this means, he argues, the constitution will ensure a measure of equality and equity, because no one wants to be at the bottom and if they are, wants the ability to rise.

Following this, Rawls sets forward a model to understand how conflict may be controlled. The veil of ignorance not only facilitates an egalitarian dialogue among liberal peoples, but also determines the lack of self-determination. People behave without knowing their own possibilities or the settings where they move.

As stated, Rawls introduce in his model of laws eight possible points.

1. People are free and independent.
2. People are prone to observe treaties.
3. People are part of a much broader covenant that binds them
4. People are to observe a duty of non-intervention
5. People reserve the right of self-defence
6. Human rights should be honoured by peoples.
7. War engenders serious restrictions to freedom and democracy.
8. Peoples must assist other people's living under unfavourable conditions that prevent their having a just descent political regime.

As a theory, the law of peoples should be revisited, but not for the reasons Rawls notes. Not surprisingly, he describes democratic society as decent. In so doing, he cuts the world in two: democracies which are based on the law and toleration, and burdened societies. It is plainly that the duty of assistance, formulated as the Truman Doctrine by the United States after the Second World War implies that all peoples wish to be not only democratic, but well-ordered. At a first glance, this belief overlooks the fact that order and wealth are inextricably intertwined. The neoliberal development in 1990s not only issued a lot of uncontrolled loans that indebted many poor countries, but also generated an iron chain of dependency between first and third world. Shown to be a resounding failure, the first social scientists who encouraged the theory of development—Milton Friedman possibly the most well-known, especially in his design of the Pinochet-led Chilean economy—replied that cultural asymmetries were the reason for the third world not to gain the benefits of financial assistance. That is, it was not the economy; it was the value system of Third World peoples that kept them in poverty. During years, liberal thought precluded that law-making and zum-zero society were two inexpungable realities. However, in recent times, some voices have questioned to what extent the concept of reality and politics are inextricably intertwined.

Risk in Liberal Zum Zero Society

Needless to say that Rawls' theory not only rests on shaky foundation, his shakiness lies in the ignorance of the history of democracy, which is only the history of western law. The concept of zum-zero society is changed whenever the catastrophe surfaces. Bottom days, exemplify not only our fear of chaos, but to what extent that chaos will bring out into reality.

In his reading of Baudrillard, G. Coulter acknowledges that liberal illusion rests on believing that the world is ordered by rationale. This view not only ignores how the reversibility works, but also how process of stability/instability alternates each other. In the digital times, events occur first on the screen and afterward in reality. The boundaries between fiction and reality have been blurred. When one delves into Baudrillard's legacy, it common to see a pioneer who alerted about "an enigmatic and intelligible world" that we know through the lens of the media. To unravel the concept of truth, Coulter says that since things are created they are condemned to be destroyed. Do not recognize this, is rejecting the nature of life. For that Baudrillard struggled to develop an all-encompassing conceptual framework that helps people understanding the acceleration of history, the transformation of past-time in present history. Therefore, he compared the war on terror declared by Bush or the ever increasing disasters as the movie, minority report, where "precogs" facilitate to police the arrests before the crime to be committed. While Barthes employs the term terror of meaning, Baudrillard admits that irony is based on the terrorism of meaning because it makes the system working against itself, as an autoimmune virus. Fiction, like theory gives the meaning to our mind to appreciate the world, fiction precedes every what we may perceive and feel. Therefore, the truth is not circumscribed to previous verification, but what results from the reversible side of thought. This means that if any system is explained by the construction of concepts (theory), its hidden side is nothing (the heart of the system). As this explanation is given, truth rests on the enigmatic nature of the world. Recognizing that "nothing can be said about the world," Baudrillard is convinced that appearances have replaced the meaning, in a type of vertigo of interpretations. The violence of interpretation lies in the impossibility to separate the reasons from effects in the events as the media portrays. News about natural disasters produced by global warming effects is equalled to terrorism and quakes without any kind of distinction. Not only the system of meaning has been altered by capitalism and the media, but upended. His destabilization of what truth is, leads us to think in the following axiom,

Western rationality has always been based, as regards discourse, upon the criteria of truth and falsehood. However, in consumer society, the neo-language of advertising lies beyond truth and falsehood where it increases in codes and models rather than actual reference or veracity. (Coulter, 2012: p. 26)

Following this, Coulter deciphers the dichotomy considering that if Baudrillard's axiom is right, the "absence of truth" reveals two things. On one hand, we are unable to find it, on another, the truth remains covert to the human beings because it is enrooted in the fiction, and beyond human access, fiction may not be empirically verified. To put this in bluntly, the truth is as the fish that bites its own tail. From irony to critique, his texts would go around the world defying previously-stereotyped beliefs and stable knowledge. If we have some certainness about how World Trade center happened, Baudrillard will destroy our perceptual frames, even admitting this event never existed in reality. Once again, as Coulter writes, he is not the philosopher of non-sense but his methods help us uncovering our "veil of ignorance", our prejudices enrooted in the mechanic turn of mind. To cut the long story short, something necessary in a book review, section one is dedicated to the complexities of the world, perception and the concept of reversibility. Coulter had special interests by analysing the question and limitations of Marx in respect to otherness. This issue requires attention because it reveals the influence of post Marxian legacy in Baudrillard as well as his view respecting to pseudo-reality and alienation. Baudrillard said that West is debating itself in a quandary between real and its double, simulation. However, we live in a moment where the real is framed by the principle of simulation. Since the real cannot override the copy, the hyper-real substitutes the real, at the same time the acceleration of images emptied the pastime. What today we may learn from disasters comes from cinema and the film-industry. In view of this argument, disasters as are often commoditized and sold by the media are not real events, but pseudo-events. The visual technology in the digital times has not only changed the perceptual horizons accelerating the time and space but created the end of resiliency as known today. If technology is used to make a safer life, following Baudrillard one might speculate how humankind may avoid a climate of oppression in hands of machines.

The Ideology of the Matrix

In the first chapter, editor William Irwin discusses to what extent The Matrix is not an attempt to retell ancient Greek philosophy. At some extent, Neo is a liberator, a Messiah who is in charge of liberating humankind from the illusion of Matrix. This begs the point on what life means? What is a good life? William adds.

Self Knowledge is the key and without it we can unlock no other knowledge worth having: This is theme important not just to Socrates and the Matrix but to other outstanding philosophical films. Fight Club poses the seemingly adolescent question; how much can you know about yourself if you have never been in fight? (Irwin, 10)

Undoubtedly, the quandary of Matrix signals to the intersection of technology and knowledge production. It echoes an old Marxist belief that the introduction of technology for labour enhancement is conducive towards alienation. Whatever the case may be, the Matrix symbolizes the cage (cavern) which makes from our life a better place, but unreal. The main discourse of the movie rests on the following axiom, while Neo horrifies from the grim present of his brothers and sisters, who are plugged into gooey pink cave-pods, he understands that after all not all should be liberated. Some of them like to be living a pseudo-reality with the end of avoiding the pain (William 2005).

Following this argument, David M Nixon (2005) calls the attention to the possibilities all our beliefs turn false, for what we think lacks of epistemological verifiability. If this is the case, to what extent we can really know anything?

Rather, James Lawler argues convincingly that no matter what people think or are feeling, the concept of reality induces pain and fear. Neo accepts the prophecy because he avoids the fear of death, but at the same time, the Matrix emulates the needs of humankind to sacrifice freedom by the rise of fear. It is preferably to live a life which is unreal than a real life fraught of threats and uncertainness.

The vitality of the physical body depends on the mind`s belief in the ultimate power of death. This is the basic rule that regulates the Matrix. Your power, your reality depends on your beliefs, and your beliefs are ultimately ruled by fear of death. (Lawler, 2005: 150)

After further discussion, the concept of The Matrix opens a hot-debate not only in regards to perception, or illusions, but to the problems of pluralism, as Gregory Bassham (2005) observed. As Jesus, Neo is entitled as "the messiah" who takes resemblances of Christ`s story documented in the bible. In fact, this movie discursively encompasses all elements of religion, such as the needs of salvation, and the questioning of ultimate reality. Although at a preliminary stage, religions expand in basis of a new pluralism which is defiance with status quo, with the passing of time, it encloses into the members of religion, who share the same faith (Bassham, 2005). Of course, religion plays a leading role not only in surveilling life, but trying to respond on the mystery of death.

The fact is that the problem of reality is not new for Hollywood, as Slavoj Zizek (2005) adhered, Truman Show situates as one of the first attempts to theorize on the power of discovery. Being or not happy does not depend on the materiality, but on what one is willing believe. In consonant with Baudrillard, Zizek reminds that one of the strongholds of capitalism are the production of a fictionalized reality (hyper-reality) dotted of substanceless.

In Late capitalism, consumerist society, real social life itself somehow acquires the features of a staged fake. With our neighbours behaving in real as stage actors and extras. The ultimate truth of the capitalist utilitarian despiritualized universe is the dematerialization of real life itself, its reversal into a spectral show. (Zizek 2005: 243)

This doctrine fits with the argument of Michel Maffesoli who understands modernism as "the necessary return of tragedy". As the tragedy evolves into an atmosphere of predestination, the future for us was diluted forever. This leads people to live the present with extreme hedonism and intensity. Similarly, to a concentration camp, where inmates embrace the cult of achievements, now workers are extremely overexcited by the cult of achievers. These spaces of death, which characterized the destiny of millions innocent civilians during WWII, were successfully accepted by capitalism today. Whether these camps exhibited a banner saying "die arbeit macht frei", which means "the work sets free", not surprisingly, one might confess that workers sacrifice their ideal freedom by a real slavery. The same applies for capitalist corporations nowadays.

Ideologically, Maffesoli understands capitalism as a symbolic trap, where always the immanent disaster is coming. Given this, all else will be immaterial if the tragedy will take place someday. Its nature not only is irreversible but irremediable tragic. It creates a space of encounters where consumption, hedonism and the aesthetics of individualism converge. We should not lose the sight that the culture of pleasure, the sentiment of disaster, the confronting of destiny, resulted from the adoption of a new ethics that takes the instant as the primary value.

As the previous argument given, the myth of tragedy is aimed not only to enthrall a mysticism of destructive creation, but a new continuity of life where past and future dilute. In the return of tragedy in postmodern societies the otherness serves as a commodity to be exploited by means of senses. We are prone to visit exotic landscapes, attending to musical parties as "love parade", taking the "Others" as a criterion of attraction. In a world of attractive appearances, where the future only brings destruction, ethics are subject to what experts thought as the best option. The decisions about our future were made beyond our capacities of understanding; the concept of closed-destiny plays a vital role in cementing our postmodern life. To put this in bluntly, the dramatic conception of the world in the modern times evinced the quest for happiness, in a paradise where the subject enjoy of its properties or merits in the life, capitalizing the individual pleasure and minimizing the costs. This sort of individualism was the hallmark of modernity. However, things have changed a lot. Postmodernism places the life as a collective property, where the "disaster" imposes over our individualism, by posing the destiny as something unaffordable for humankind. There is nothing in this world to do when disasters hit. Everything is there; predestined by external forces that has a design for us. However, unlike Weber, who envisaged a disenchantment of the world, Maffesoli formulates a new type of re-enchantment of hedonism, or an ethic of instant as it has been earlier explained in this book review.

Here, we hold the thesis nor Weber neither Maffesoli formulated the concept of predestination with accuracy. Although Weber, who was closer than Maffesoli in the description of capitalism, took the correct way at confirming the Reform introduced a new thinking associated to predestination, he ignored the influence of Norse Mythology in the configuration of predestination. It is clear how protestant embraced achievements to know if they were in the Book of Life, but there is evidence enough, earlier Christendom, for example in Norse Myths of persistence of predestination. In earlier studies, M. Korstanje explained that cultures are disciplined forms of battlegrounds and warfare. Not surprisingly, Anglo-American Empires built a cosmology according to their founding cultural values.

In this respect, Maffesoli is wrong at time of interpreting the meaning of *Moira* as a closed destiny, even in the battleground Greeks have a free choice to live or death. This happens simply because Gods (unlike for Germans) were only advisors. The fate of the warrior is never determined by the Gods, but what can be achieved now and here. Ancient Greece was never familiar with anything like predestination. Rather, readers of Norse Mythology acknowledge that Walkyrias (daughter of Odin/Wodan) were commissioned to pick up the bodies of fallen warriors. They not only knew *the destiny of warrior* beforehand, which is the basis of predestination, but also presented death as a form of redemption. Being selected by Odin to enter "Valhalla", was for Germans a great honor. Not surprisingly, Anglo-world from the outset was marked by disaster and predestination. This was the reason simply why detractors of Weber criticized he was not directed his attention to Catholic Holland, which was the epicenter of an early capitalism. Well, Holland was in conditions of fomenting capitalism because of his Norse cultural matrix. At the same time, the prone of Anglo-societies to advance in scientific knowledge is derived from its concern for future and risk. Unlike catholic societies which are based on charity and tradition, Protestantism is particularly concerned by the closed-fate, thus they devote all resources to colonize the future. To cut the long story short, the current postmodern cultures are future-oriented societies (Korstanje, 2014; Skoll & Korstanje 2013). We feel Maffesoli undoubtedly pose the horse before the cart. Philosophically speaking, risk reduction management is an eternal struggle with future.

FUTURE GUIDING DIRECTIONS: THE END OF RESILIENCY

The rise of risks and the globalized of fear were two contemporary phenomena that worked together in current times. While technology played a vital role by delineating the borders of more secure society, this does not mean it can be used as an alienatory force. The Matrix film represents an epistemological discussion which poses the problem of fear as the main reason, we opted to live a pseudo-reality in terms of Baudrillard. We are living through a new capitalism which not only ignores the alerts of disasters (as global warming or the nonrenewable resources), but profits from disasters. Naomi Klein dubbed these new trends as enmeshed into the shock doctrine, which was used by officials in order for citizens to accept policies otherwise would be neglected (Klein, 2007). For these critical voices, disasters and even terrorism produced a great shock in society which unless recycled in forms of spectacles or cultural consuming, would place the legitimacy of elite in jeopardy. In view of that, dark tourism places offered a good alternative for consumers to expand their current understanding of reasons why disasters happened, but as Korstanje adds, the message is based on a biased diagnosis on what the event happened. This leads not only to an ethnocentric discourse because survivors are over-valorized over witnesses, but society does not obtain a good lesson from disasters. It represents the end of resilience paving the ways for the next similar disasters appear (Korstanje, 2015).

One of the aspects that difficult the correct understanding of disasters, is the role played by survivors during the first facet of the event. After the landscape of obliteration or mass-destructions, survivors develop psychological mechanism of defence to accept the losses their suffered. As victims, the lives of friends, relatives or even offspring have gone. This exhibits a great suffering which may be reminded. However, after all, they feel not everything is lost, they have survived. This second stage paves the pathways for the creation of an *ethnocentric discourse*, where survivors think they have been protected by fate, destiny or Gods. A midst of mass destruction, they are enthralled as smarter, stronger and virtuous to give a message to the world. Though this facet is necessary in order for the group to enhance the self-esteem, it should be dully regulated and limited in timeframe; unless this occurs as noted, expressions of xenophobia, message of hate against Otherness and chauvinism may very well emerge. A sense of supremacy monopolizes the collective soul of affected community to the extent, their sense of reality is seriously affected. Unless otherwise resolved, survivors believe they are special and the only way of reaching happiness is by means of suffering. Expressions of this nature were documented by Korstanje during the Chile's earthquake in the Triathlon Chile Ayuda a Chile. Although interesting sums of investors devoted considerable financial resources in order to help Chilean cities to recover, less attention was given to bribes suspicions and proceedings against real-estate businessmen who sold common skyscrapers and buildings as anti-seismic. The spectacle of disasters sublimated the liabilities of some actors were covered. Following Klein, capitalism has been expanded in view of its efficacy to disorganize social bondage. This disorganization produces a gap which is fulfilled by a narcissist viewpoint of the world where the "Other" should be subordinated to my desires (Lasch, 1991). The culture of narcissism not only is useful to subordinate citizens to the market, as Z. Bauman noted, but also undermines the existent levels social trust (Bauman, 2013). The last stage consists in the construction of an allegory, which confers a symbolism to disaster. In this vein, it entices the visit of thousands of other peoples (tourists). However, for the site to be consolidated as an attraction, the political power should invest material and symbolic resources in order for message not to cause social fragmentation, or the lack of their legitimacy. Though necessary, it obscures the real reasons of tragedy, cementing the possibility to prevent a similar disaster takes place in a long run future. Capitalism has developed a great capacity to captivate audience even in context of disaster, but declining the process of resilience that helps learning of such experiences. This connotes a paradox which is stimulated by disaster cultural consumption, in which case, the society had problems to understand further on their disgraces. Otherwise, it presents a serious obstacle for the process of resiliency in the times of fear, where tourism plays a crucial role by engaging death and consumption. Anesthetization of pain is conducive towards pseudo-resilience where the disaster is a matter of time.

CONCLUSION

The discourse of the Matrix not only wakes up philosophy from the slumber they are, but reflects the rise of fears which are proper of a culture, that manipulates anxieties to fix policies otherwise would be rejected. Basically, Neo is subject to an ethical dilemma which means that we are free to make the decision to live in a fake without suffering or in "the real of the dessert" as Morpheus put it. At this extent, if technology is used to protect humankind, Matrix evinces under some circumstances, it can oppress human beings when they renounce to their "free-will". This chapter not only explored the limitations of *the society of risks within liberal thought* but taking into consideration the role of authenticity serves ideologically to legitimate suffering. Indeed, accepting to live as Neo did, may lead to oppression or liberation. Whatever the case may be, we should face the risk, which is determined by the contingency. While accepted to take the correct pill Neo launched to revere the situation of others, but very well he would have been killed in the intent. Last but not least, this chapter, is far from resolving the dilemma of authenticity vs. unauthenticity. Rather, the intention is given in another direction. Basically, it held the thesis that zero-sum game as well as risk pro precautionary society leads towards an alienatory platform that paves the ways for the rise of the Matrix.

REFERENCES

Bassham, G. (2005). *The Religion and the Matrix and the problem of pluralism*. Chicago: Open Court.

Baudrillard, J. (1989). *La Transparencias del Mal*. Barcelona: Anagrama.

Baudrillard, J. (1995a). *The systems of the objects*. Siglo XXI.

Baudrillard, J. (1995b). *The Gulf War Did Not Take Place*. Sydney: Power Publications.

Baudrillard, J. (2006). *Virtuality and Events: the hell of power* (Vol. 3). Baudrillard Studies.

Bauman, Z. (2013). *Consuming life*. New York: John Wiley & Sons.

Coulter, G. (n.d.). *Jean Baudrillard: from the Ocean to the desert or the Poetics of Radicality*. New Smyrna Beach, FL: Intertheory Press.

Durkheim, E. (1915). *The Elementary Forms of the Religious Life* (J. W. Swain, Trans.). New York, NY: The Free Press.

Durkheim, E. (1938). The Rules of the Sociological Method (S. A. Solovay & J. H. Mueller, Trans.; G. E. G. Catlin, Ed.). Glencoe, IL: The Free Press.

Durkheim, E. (1984). The Division of Labor in Society (W. D. Halls, Trans.). New York, NY: The Free Press.

Hobbes, T. (2004). Leviatán. Buenos Aires: Ediciones el Libertador.

Irwin, W. (2005). Computers, Caves and Oracles: neo and Socrates. In W. Irwin (Ed.), *The Matrix and Philosophy* (pp. 5–15). Chicago: Open Court.

Klein, N. (2007). *The shock doctrine: The rise of disaster capitalism*. New York: Macmillan.

Korstanje, M. E. (2014a). Why Risk-Research is More Prominent in English Speaking Countries in the Digital Society. *International Journal of Cyber Warfare & Terrorism*, *4*(1), 8–18. doi:10.4018/ijcwt.2014010102

Korstanje, M. E. (2014b). Chile helps Chile: Exploring the effects of earthquake Chile 2010. *International Journal of Disaster Resilience in the Built Environment*, *5*(4), 380–390. doi:10.1108/IJDRBE-01-2012-0002

Korstanje, M. E. (2015). *A Difficult World, examining the roots of Capitalism*. New York: Nova Science.

Lasch, C. (1991). *The culture of narcissism: American life in an age of diminishing expectations*. New York: WW Norton & Company.

Lawler, J. (2005). We are the one: Kant explains how to manipulate the Matrix. In W. Irwin (Ed.), *The Matrix and Philosophy* (pp. 138–152). Chicago: Open Court.

Maffessoli, M. (2003). *L'instant éternel. Le retour du tragique dans les sociétés postmodernes*. Paris: La Table Ronde.

Nixon, D. (2005). The Matrix Possibility. In W. Irwin (Ed.), *The Matrix and Philosophy* (pp. 28–40). Chicago: Open Court.

Rawls, J. (1999). *The Law of Peoples*. Cambridge, MA: Harvard University Press.

Rousseau, J. J. (1993). El contrato social. Barcelona: Editorial Altaya.

Skoll, G. R., & Korstanje, M. E. (2013). Constructing an American fear culture from red scares to terrorism. *International Journal of Human Rights and Constitutional Studies*, *1*(4), 341–364. doi:10.1504/IJHRCS.2013.057302

Weber, M. (1958). *Essays in Sociology*. New York: Oxford University Press.

Weber, M. (2002). *The Protestant Ethic and the Spirit of Capitalism: and other writings*. New York: Penguin.

Zizek, S. (2005). *The Matrix: or, the two sides of perversion*. Chicago: Open Court.

ADDITIONAL READING

Alderman, E., & Kennedy, C. (1997). *The right to privacy*. New York: Vintage.

Altheide, D. L. (2004). Consuming terrorism. *Symbolic Interaction, 27*(3), 289–308. doi:10.1525/si.2004.27.3.289

Altheide, D. L. (2006). Terrorism and the Politics of Fear. *Cultural studies critical methodologies, 6*(4), 415-439.

Altheide, D. L. (2009). *Terror post 9/11 and the media* (Vol. 4). New York: Peter Lang.

Bazan, E. B. (2002). *The Foreign Intelligence Surveillance Act*. Nova Publishers.

Colarik, A. M. (2006). *Cyber terrorism: Political and economic implications*. Hershey: IGI Global. doi:10.4018/978-1-59904-021-9

Furnell, S. M., & Warren, M. J. (1999). Computer hacking and cyber terrorism: The real threats in the new millennium? *Computers & Security, 18*(1), 28–34. doi:10.1016/S0167-4048(99)80006-6

Howie, L. (2007). The terrorism threat and managing workplaces. *Disaster Prevention and Management: An International Journal, 16*(1), 70–78. doi:10.1108/09653560710729820

Janczewski, L. (Ed.). (2007). *Cyber warfare and cyber terrorism*. Hershey: IGI Global. doi:10.4018/978-1-59140-991-5

Korstanje, M. E., & Clayton, A. (2012). Tourism and terrorism: Conflicts and commonalities. *Worldwide Hospitality and Tourism Themes, 4*(1), 8–25. doi:10.1108/17554211211198552

Kraft, E., & Wang, J. (2010). An exploratory study of the cyberbullying and cyberstalking experiences and factors related to victimization of students at a public liberal arts college.[IJT]. *International Journal of Technoethics, 1*(4), 74–91. doi:10.4018/jte.2010100106

Lyon, D. (2004). Globalizing Surveillance Comparative and Sociological Perspectives. *International Sociology, 19*(2), 135–149. doi:10.1177/0268580904042897

Lyon, D. (2007). *Surveillance studies: An overview*. Cambridge: Polity.

Lyon, D., & Zureik, E. (Eds.). (1996). *Computers, surveillance, and privacy*. Minneapolis: U of Minnesota Press.

OBrien, M., Tzanelli, R., Yar, M., & Penna, S. (2005). The Spectacle of Fearsome Acts: Crime in the Melting P (l) ot in Gangs of New York. *Critical Criminology*, *13*(1), 17–35. doi:10.1007/s10612-004-6111-9

Palanisamy, R., & Mukerji, B. (2012). Security and Privacy issues in e-Government. IGI Global, 236-248.

Rathmell, A. (1997). Cyber-terrorism: The shape of future conflict? *The RUSI Journal*, *142*(5), 40–45. doi:10.1080/03071849708446185

Reitz, J. C. (2006). E-government. *The American Journal of Comparative Law*, *54*, 733–754.

Shaw, L. E. (2004). USA Patriot Act of 2001, the Intelligence Reform and Terrorism Prevention Act of 2004, and the False Dichotomy between Protecting National Security and Preserving Grand Jury Secrecy, The. *Seton Hall Law Review*, *35*, 495.

Stohl, M. (2006). Cyber terrorism: A clear and present danger, the sum of all fears, breaking point or patriot games? *Crime, Law, and Social Change*, *46*(4-5), 223–238. doi:10.1007/s10611-007-9061-9

Wallace, W. L. (1988). *Toward a disciplinary matrix in sociology. Handbook of Sociology* (pp. 23–76). Newbury Park, CA: Sage.

KEY TERMS AND DEFINITIONS

Capitalism: An economic project which introduced the operation for profits.

Narcissism: A psychological pathology expressed by an exaggerated pursuit of self-gratification and admiration.

Protestant Reform: A religious movement which confronted to the hegemony of Catholic Church.

Pseudo-Reality: This term was originally coined by Jean Baudrillard to describe the advance of capitalism and globalization.

Risk: It signals to any potential or perceived harm over a person or their group.

Saga Matrix: A futurist Saga situated in a world where Machines took the control of human bodies, creating a pseudo-reality, where they are involved.

Section 3
CyberTerrorism and Civil Riots

Chapter 8
OSNs as Cyberterrorist Weapons against the General Public

Nicholas Ayres
De Montfort University, UK

Leandros Maglaras
De Montfort University, UK

Helge Janicke
De Montfort University, UK

ABSTRACT

Conventional terrorism has been around for hundreds of years and even though its tactics and the weapons of choice have evolved over time as well as the use and deployment of weapons may have changed the root definition of terrorism has remained relatively untouched. With the advent of mass computing, cybercrime has increased year on year. This chapter will look at three differing viewpoints of cyberterrorism and its ultimate effects on society. Many industry and academic experts warn that it is only a matter of time before conventional terrorist acts will migrate to the digital arena in the form of cyberterrorism. Current literature suggests that a countries critical national infrastructure will be the main focus of attack for the cyberterrorist but this chapter will address another possible target for the cyberterrorist using a different type of cyber weapon: a mimetic virus. This chapter also looks at how a mimetic virus could use social media to spread throughout the target audience using what is known as Internet memes.

DOI: 10.4018/978-1-5225-1938-6.ch008

INTRODUCTION

In 1990 the National Security Council envisaged that computers could in the future be used to not only facilitate crime but as the main tool for criminal acts; 'The modern thief can steal more with a computer than with a gun. Tomorrow's terrorist may be able to do more damage with a keyboard than with a bomb' (National Research Council, 1990). With the rise in global terrorism and the mass use of computers a new and potentially more destructive form of terrorism has come to the fore: cyberterrorism. Conventional terrorism has proved over time to be a controversial subject due to the lack of a unified global definition of what constitutes an actual terrorist act. Controversy also surrounds cyberterrorism but these centre around whether a cyberterrorist attack using digital weapons can in fact have the same profound impact as conventional terrorism using traditional kinetic weapons. This chapter looks at three different viewpoints when it comes to what is a cyberterrorist and what constitutes as a cyberterrorist attack. Current literature states that if there was a cyberterrorist attack the most likely target would be against a countries critical infrastructure (CI) or critical national infrastructure (CNI) as these systems underpin a countries society and if attacked could have nationwide effects not just on services but the population as a whole. This chapter will also look at why CI / CNI may indeed not necessarily be the main focus of an attack but will show how the public could be directly attacked using a much more focused and specific kind of digital weapon.

Conventional Terrorism vs. Cyberterrorism

In order to fully comprehend the term cyberterrorism; first introduced by Barry Collin in the 1980s (Gordon & Ford, 2002) we must begin by looking at the evolution of conventional terrorism. Terrorism has always been a controversial subject, which 'has been well studied and documented' (Collin, 1997:pp 15-18) but there is a lack of a globally accepted academic or legal agreement on its true definition (Schmid, 2011). The quote "One man's terrorist is another man's freedom fighter" (Seymour, 1975) has been cited many times in relation to Middle East conflicts and uprisings in Africa and Central America (National Research Council, 1990) to name but a few. This lack of a definitive global definition is largely because the act is a highly political and emotionally charged one and governments are reluctant to agree on a legally binding definition (Hoffman, 2013). That being said there is a general global consensus that terrorism is an act which is perpetrated by individuals, groups or governments to instil fear and terror in non-combatants through the use of violence, in order to pursue a political, ideological or religious goal. In essence terrorism is the use of or threat of violence or damage to life and property with the intention of influencing a government or society through a political, religious or ideological cause. Terrorists believe they are 'legitimate combatants' fighting for their beliefs using whatever means possible in order to 'attain their goals' (International Terrorism and Security Research, 2015). Conventional terrorism has had a huge global impact in modern history. For an act to be considered terrorism the target of an attack must be non-combatants; in essence the general public. During the 1972 Munich Olympics the Palestinian terrorist organisation Black September took hostage and ultimately killed 11 Israeli athletes. The 11 victims were the direct targets of the attack but the indirect or actual target was the estimated 1 billion people who were watching the televised event around the world who were 'introduced to fear - which is terrorisms ultimate goal' (Anon, 2015). The United States went to war with the ruling Taliban party in Afghanistan with the sole intention of dismantling the terrorist organisation al-Qaeda, which was using the country to train and indoctrinate fighters, import weapons and plot terrorist actions (Zelikow, 2011). The U.S. invasion of Afghanistan was the principle reaction to the 9/11 terrorist attacks on the World Trade Centre and Pentagon (Thruelsen, 2006) by al-Qaeda. Another motive for the 9/11 attacks may have been to coerce the U.S. into a war that would incite a pan-Islamist revolution (Duran, 2002, pp. 22-42). Whatever the ultimate motive was for the 9/11 attacks it was an unlawful use of violence, motivated by ideological beliefs, which instilled fear and coerced a government in pursuit of al-Qaeda's goal.

Cyberterrorism is seen by many to be a digital extension of conventional terrorism but the act of terrorism occurs solely in cyberspace. In 2000 Dorothy Denning's testimony before the Special Oversight Panel on Terrorism is one of the most cited papers on the issue of cyberterrorism. She stated that 'Cyberterrorism is the convergence of terrorism and cyberspace. Cyberterrorism is generally understood to mean unlawful attacks and threats of attack against computers, networks, and the information stored therein when done to intimidate or coerce a government or its people in furtherance of political or social objectives' (Denning, 2000). This statement generally falls in line with the accepted definition of terrorism but with the terrorist act occurring within solely in cyberspace with computers, networks or stored information being the initial direct targets of an attack. The main difference between the two types of terrorist acts is the deployment of a weapon and the ultimate effect on the target. Unlike the conventional terrorist the cyberterrorist no longer has to rely on kinetic weapons such as bullets or bombs nor do they require a specific geographical location; the cyberterrorist uses digital weapons to attack their victims anywhere in the world with their preferred choice of weapon the computer virus. The cyberterrorist has no need for the traditional weapons of the conventional terrorist they now have a whole new arsenal of digital armaments, which can potentially attack a target that is anywhere in the world and equally be deployed from anywhere. The postmodern cyberterrorist can deploy a digital weapon such as a virus that can be programed to 'explode' or activate at a specified time or if a specific condition or criteria is met. The cyberterrorist has a distinct advantage over their conventional comrades as once detonated the digital weapon could hide and re-emerge at a later date and time to wreak havoc once more. This is of course in stark contrast to the kinetic, non-digital weapon of the conventional terrorist such as a bomb, once it goes off, it goes off and cannot be used again and again. The weapon of the traditional terrorist is considered analogue because it only operates in a linear, sequential manner. There are many different interpretations as to what constitutes as a cyberterrorist attack it has been shown that any cyberterrorist attack must conform to the attributes of terrorism but include the use of a computer as the attacker and the victim. If these attributes of terrorism are not met and do not conform to the definitions of cyberterrorism then any potential attack by a cyberterrorist would be classified as another form of cyberattack. (Kenney, 2015) in his paper Cyberterrorism in a Post Stuxnet World compared a number of different cyber phenomenon to four distinct attributes to determine the type of cyberattack.

Attribute 1 - Cyber Vector: A computer attack targeting other computers, computer systems, or the information stored within. This first attribute can be applied to any of the four identified cyber phenomena. This attribute demonstrates that a cyber-attack must be instigated from a computer or computer system and subsequently directed towards a target computer or computer system or the information held on such a device.

Attribute 2 - Cyber Objective: The attack is in pursuit of a political, social or religious agenda or aim. This second attribute is shared with hacktivism, cyberwarfare and cyberterrorism. Each of these phenomena will fit into one of the objectives of an attacker, for example a hacktivist who is concerned with raising awareness in relation to a topic they feel is either unjust or wrong in their opinion and uses cyber technology to cause disruption, defacement or damage. In contrast to this cyberterrorism falls in line with either a religious or political aim.

Attribute 3 - Cyber Consequence: The attack threatens or produces physical violence against persons, property or critical infrastructure. This third attribute is the most controversial when defining a cyberattack to cyberterrorism. There are many people who believe that a cyber-attack is not achievable or in the same league as a conventional terrorist attack where bombs and bullets are used to attack a target. It could be argued though that the thought and threat of potential physical violence against a person is just as effective as the act of violence itself regardless of the method of attack. Kenney suggests that this attribute solely fits within the arena of cyberterrorism but as Stuxnet showed the world physical damage to property could be achieved using the computer as the weapon and the target with this in mind this third attribute can be applied to cyberwarfare.

Attribute 4 - Cyber Propagation: The attack causes widespread fear or physical intimidation in the target of the attack and beyond the immediate victims. This fourth and final attribute of the cyber phenomenon is in essence the main goal of conventional terrorism and as such must also be an attribute of cyberterrorism. A terrorist attack has the initial shock factor when executed but as with any news story will over time drop from mainstream media and social attention. The true goal of the terrorist is to affect not just the immediate victims of any attack but to spread its effects as far as possible. This attribute does not fall within the sphere of any of the other cyber phenomena such as the hacktivist who is concerned with raising awareness in relation to a topic they feel is either unjust or wrong in their opinion or to a cyberattack which is more related to organised crime. Cyberwarfare is concerned with causing damage and disruption to an enemy through directly attacking a mechanism in order to disrupt or disable essential services.

In order for an act to qualify as cyberterrorism it must still follow the same rules as conventional terrorism and does up to a certain point, whereby cyberterrorism like conventional terrorism is perpetrated by individuals, groups or governments in order to pursue a political, ideological or religious goal. The big question mark hanging over the subject of cyberterrorism is whether it can instil the same levels of fear from threat of violence or damage to life and property that has been proven time and time again through acts of convention terrorism.

The Problem with Cyberterrorism

Although conventional terrorism lacks a globally recognised definition there is no doubt as to the end result of a terrorist act: injury, loss of life, damage or destruction of property and one that places a populace in a state of fear or terror. Cyberterrorism as with conventional terrorism is just as controversial a subject but for very different reasons. There is no doubt as to the meaning of cyber, but, by adding terrorism to it creates a great deal of controversy between politicians, industry experts and the academic community. When considering the various opinions with regard to cyberterrorism there are three distinct but very different viewpoints, these are:

- **The Virtual Myth:** 'There is no such thing as cyberterrorism… nor is their compelling evidence that al-Qaeda or any other terrorist organisation has resorted to using computers for any serious destructive activity' (Green, 2002). Green also states that many security industry experts believe it is virtually impossible to use the Internet to inflict death on a large scale. (Jones, 2005) carries this argument and states that 'the only way that you will scare most people with cyber capabilities is to threaten to throw the equipment at them'. (Pollitt, 1998) takes this argument further and suggests that 'short of electrocuting one's self with the power supply or being so unfortunate as to walk under a falling machine the potential victims of cyberterrorism cannot be directly harmed or killed by a computer'. The virtual myth theory takes the view that whatever attack is perpetuated by a cyberterrorist using a digital weapon deploying ones and zeros could never have the same effect on a target as the conventional terrorist using kinetic weapons such as bullets and bombs. Cyberterrorism at best ultimately becomes a 'weapon of mass disruption' (Dwan, 2001) rather than one of mass destruction.

- **An Impending Apocalypse:** In stark contrast to the 'virtual myth' of cyberterrorism is the blunt definition of death and destruction on not just a small localised scale but potentially an attack that could have global implications and consequences. In 2002 the then Homeland Security Director Tom Ridge stated that 'Terrorists can sit at one computer connected to one network and can create worldwide havoc without the need of a bomb or explosives in order to cripple a sector of the economy, or shut down a power grid' (Weimann, 2004). Unlike a conventional terrorist, the cyberterrorist can attack a target anywhere in the world from anywhere in the world without the need for a physical local presence. As our day to day lives almost totally rely on computers we are thus vulnerable if those systems were attacked and destroyed, this threat of a potential attack is seen by some to be placing us in a perpetual state of fear. In 2010 the Federal Bureau of Investigations Chief Robert Mueller addressed a conference of computer security professionals and stated that the cyberterrorist is either politically or religiously motivated in order to create fear and alarm within a given population through disruption or destruction of critical national infrastructure (Hsu & Marinucci, 2013). Former US Secretary of Defence Leon E. Panetta in 2012 gave a cyber security speech to business executives responsible for the US National Security where he stated the threat from cyberterrorism 'would paralyze and shock the nation and create a new, profound sense of vulnerability'. Through acts of cyberterrorism extremist groups could 'derail passenger trains, contaminate the water supply in major cities or shutdown the power grid across large parts of the country' (Panetta, 2012).
- **The Dilution Theory:** This third viewpoint takes a much more pragmatic approach to the subject of cyberterrorism whereby any use of a digital device to further the goal of conventional terrorism should be considered as cyberterrorism. Evan Kohlman, an expert in terrorists' use of the Internet,

points out that the established definition of cyberterrorism is flawed and that 'any application of terrorism on the Internet' should be considered cyberterrorism (Kaplan, 2009). This viewpoint sees cyberterrorism as an extension of conventional terrorism whereby an act of cyberterrorism does not need to have the same outcome and consequences of conventional terrorism as the act is there as a support mechanism to facilitate an attack. Information gathering using a computer or device which is connected to the Internet or the distribution of propaganda through electronic means such as webcasts or communication between individuals or terrorist cells are all considered forms of cyberterrorism. This viewpoint can be further broken up into two parts: Pure cyberterrorism and regular cyberterrorism (Gordon & Ford, 2002). This notion proposed by Gordon and Ford proposes that regular cyberterrorism falls within the Dilution Theory whereby an act of cyberterrorism is an act whereby a terrorist utilises a computer to further a terrorist's political message, raise funds or research potential targets where the computer is not being used as a metaphorical weapon (Kenney, 2015). Pure cyberterrorism is a terrorist act that is generally considered as the convergence between cyberspace and terrorism.

Each of these opposing viewpoints highlighted above show how the term cyberterrorism is indeed a controversial subject like conventional terrorism but for very different reasons. This chapter will discuss why there are very different viewpoints depending on who is defining the subject and what their potential motivations may be in doing so. These three viewpoints cover the entire threat spectrum of cyberterrorism from a term that is in essence a myth to one that will threaten our security, way of life and potentially our existence.

CNI and the Cyberterrorist

The majority of previous and current research into cyberterrorism predominately focuses on attacks against a countries CI / CNI which are assets essential for the functioning of the countries society and economy. Disruption or deprivation of these assets could be catastrophic to the economy and at the least disruption to the day to day lives of society. A small list of the most commonly associated services comprised in the sphere of CI or CNI are:

- Electricity, gas and oil generation and distribution,
- Telecommunications,
- Water, including drinking supply and sewerage,
- Agriculture, food production and distribution,

- Public health services,
- Financial services,
- Security services including police and armed forces.

Some experts believe that if the target of a cyberterrorist attack is not against CI / CNI then the attack is not considered cyberterrorism. Politicians have used the threat of attack against CI / CNI to great effect in portraying the ultimate potential consequences of cyberterrorism. The fear and threat from the cyberterrorist and cyberterrorism has also been used to great effect by the media on an unsuspecting public with constant threats of death and destruction at the hands of terrorists using their computers to hijack aircraft, explode nuclear power plants and take control of military computers from anywhere in the world (Green, 2002). About 80% of the UK CNI is owned and run by the private sector making many to believe there is an inherent risk that not enough is being invested or done to ensure that these critical systems have enough adequate protection from potential cyberattack. This is perhaps why there are many out there that believe the most attractive target to the cyberterrorist is CNI as an attack could have long lasting effects on a population that is utterly dependant on the services that CNI provides.

In 2003 across eight states of the northeast coast of the USA there were power outages that left 45 million people without power for over 7 hours. The power outages were not down to a cyberterrorist attack or another form of cyberattack but rather a general computer failure. During those 7 hours without electricity the affected population did not experience high levels of fear of harm or death but saw the outage as 'one of those things that happen from time to time' (Jones, 2005, p. 4). Panic can ensue during the initially stages of an electrical power outage but this is often due to family members scrabbling around in the dark in a vain attempt in remembering where the torches and candles are kept. Later as the power outage continues this initial panic turns to annoyance and inconvenience as we are so dependent on electricity for a vast variety of uses in and around our homes. A wide scale coordinated cyberterrorist attack on a number of major power producers and suppliers could have a massive disruptive effect on a population but does not have the capacity to place that population in a state of fear or terror of potential harm that may result from a widespread power outage.

CNI is not just the subject of power and utility production and supply. On September 11, 2001 terrorists physically hijacked a number of aircraft and used them as weapons to strike the World Trade Center's Twin Towers in New York and the Pentagon building in Washington, DC. This attack was orchestrated by a large support network with months and years of preparation, planning and training to accomplish. Politicians have transposed this type of attack from a kinetic attack to a digital attack whereby the cyberterrorist could take control of an air traffic control system computer system resulting in aircraft being commanded to fly into one another or targets on the ground causing wide scale death and destruction from above. This attack does not require such a large support network nor does it relay on being in a specific location or the need to physically breach security to carry out such an attack. The cyberattack could be deployed and executed from anywhere in the world by a single accomplished cyberterrorist. This type of scenario has been banded around but how close to a possible reality is it? Although computers play a vital part in the safe operation of modern day air traffic control the system is still essentially a human operator relaying a set of commands to a human pilot. In order for the above cyberterrorist attack to work would have to rely on some form of automated air traffic control system relaying false or fabricated commands to an automated aircraft.

The list of possible cyberterrorist attack scenarios on CNI is long and varied and one that has been used to great effect by politicians, academics and industry experts for a wide range of motivations, but to date there is no actual evidence that a terrorist organisation has either the capability or the desire to attack these types of targets. Fuelling the idea of cyberterrorism maintains a heightened level of anxiety and fear in regard to terrorism in general as elevated levels of fear in politicians and the public can have an ulterior motive and one that could possibly enable a government to pass or approve a particular agenda whatever that may be. The threat of terrorism and US foreign and domestic policy along with the continuing War on Terror has maintained these elevated levels of fear in the public and now with more and more news reports coming to the fore of cyber related crime, cyberterrorism has become the new threat and fear in the public consciousness. To many of the general public when they image what a cyberattack may be often think along Hollywood plotlines of a teenage Matthew Broderick in the 1983 film Wargames where he hacked into a computer system through an unsecured backdoor which ultimately turned out to be the master computer monitoring and controlling the US nuclear weapons arsenal and almost hurling the world into a digitally initiated Armageddon (Green, 2002).

The British government have recognised this gap in security and in 2010 CNI became a Tier One risk; the highest priority for UK national security and in doing so allocated £650 million to cyber security and resilience. To date there has been a distinct lack of cyberterrorist attacks against CNI this could be because the systems are currently 'secure enough' or their potential to cause fear and terror in their indirect targets is limited even with the popularist theories of wide scale death and destruction. There is a considerable amount of literature regarding CNI being a high priority target for cyberterrorism attacks but with so many people online on a national or even global scale, the general public could prove to be an easier, more tempting direct target for a cyberterrorist attack rather than the CNI. The terrorist will always choose the easier option, the softer target to attack, as it may well increase the chance of success.

The Mimetic Virus

With most if not all of our CI / CNI being controlled and managed by computers to some degree the common belief within the academic and professional world is that the cyberterrorists' main target of attack would be against CI / CNI. This section looks at another perspective of attack and one which is possibly more vulnerable with the potential of being more influential than CI / CNI. By targeting a population for a cyberterrorist attack and implanting fear in their minds of potential harm that may come to them or their friends and family from their own personal computer equipment could indirectly influence a much wider audience which may then cause a reaction from regional or national governments.

The tool of choice of the cybercriminal is malware which includes malicious code such as the computer virus and this has been evident throughout the introduction of computers from the earliest models to today's complex high powered computing devices. The term computer virus is well known even amongst the un-technically minded computer users with their effects ranging from nuisance consequences of unwanted adware to deletion of files to more sinister effects such as personal identity theft and theft of assets. In the year 2014 more than 480 million new pieces of malware, computer viruses or other malicious software were released (Harrison & Pagliery, 2015). Many of these viruses and malware were used for digital exploitation such as cyber thieves coercing victims through theft of personal data or blackmail, or hackers using malware in order to breach target networks, databases and websites. Since the Creeper System; a self-replicating program spread through ARPANET was introduced in 1971 the number of computer viruses and malware has slowly grown along with the use of personal computers, only during the last decade has this trend changed whereby the numbers of new malware and viruses released in to the wild has grown at a colossal rate. The graphs below show the rate at which new malware has been released in to the wild from 2007 to 2016. The statistics for 2016 is not complete at the time of writing. Comparing the new malware data to the total malware released the figures for 2016 have already exceeded the total number for 2015 with six months of the year still to go (see Figure 1).

In the early 1990's there was not the widespread personal computing and internet use we see in today's modern, fast paced high technology world. Social media was non-existent with the primary means of connecting to the Internet was through a modem and telephone line and the predominant means of electronic communication was via email. The Internet was seen by most early adopters and the mainstream media as a new technology full of mystery and potential. In 1994 there were just 25 million Internet users worldwide equating to 0.4% of the global population. The true worth of the Internet was yet to be defined or realised in the eyes of the public and cyberspace was seen as a new and mysterious realm.

Figure 1. Last 10 years of Malware
Source: AV-TEST, 2015.

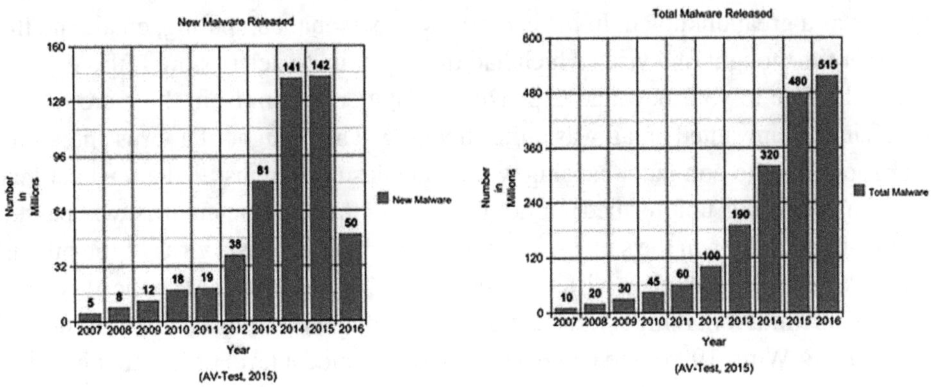

On the 15th November 1994 an email message began to circulate from user to user warning the recipient that a virus was circulating as an attachment to an email entitles 'Good Times". Initially the emails contained just a few lines of text warning the recipient that they should delete any such entitled email or it could corrupt and damage all the files stored on their hard drive, the message also came with the helpful suggestion of forwarding the email to the recipients friends and family to forewarn them of the threat, which they duly and unwittingly did. As time went on the warning messages became more and more elaborate possibly from a Chinese whisper effect or from malicious recipients who wanted to add fuel to the rapidly growing fire and as a result the effects of the Good Times virus became more and more exaggerated. The Good Times virus spread through fear and naivety because the general knowledge of computing was still considered by the masses as a great unknown and was not spread from mailbox to mailbox through self-replication or by computer code hijacking a recipients address book and propagating through the stored email addresses. The virus was replicated by the recipients themselves not just through a lack of knowledge and fear of this new digital threat of a computer virus but in an unwittingly false attempt at being helpful towards friends and family. The effects of the Good Times virus not just claimed deletion or corruption of files but actual physical damage to computer components as well. The claims suggested that if executed the virus could damage the processor by put the computer into an 'nth-complexity infinite binary loop' which in itself is a meaningless term. What was not known at the time was that the Good Times virus was in fact a hoax, the whole threat being devised and deployed by The Cult of the Dead Cow, a Texas based hacker group who claimed responsibility for the creation and release of the Good Times virus in order to expose the gullibility of self-proclaimed experts on the

Internet (Katie, 2009). The Good Times virus was spread through fear and naivety due to the general understanding of computing which was still considered by the masses as a great unknown. In the early days of personal computing, email and the Internet the thought of a virus which had the potential to delete stored files or even cause damage to your new and expensive computer system from the mere fact of receiving an unwanted email was quite shocking. The thought of a virus spreading from computer to computer causing wide scale destruction instilled a sense of fear in an unsuspecting, uninformed public. The Good Times computer virus 'was passed from human to human the same way a virus does but this was a virus of the human mind, known as a mimetic virus' (Skoudis & Zeltser, 2004). Mimetic theory is defined as imitation, mimicry, the act of expression and the presentation of the self (Gebauer & Wulf, 1995), the Good Times virus created a level of fear that infected the thoughts of the recipient making them believe that the threat and subsequent consequence was real and one which they must warn others of its destructive nature. The success of any hoax is that it 'contains pseudo-technical babble' as discussed in Don't fall for a virus hoax (Sophos Press Release, 1999) that sounds convincing and is capable of tapping into the fears and uncertainty people have about computers and the Internet.

Attacking a target computer through a virus or other means with the purpose of causing damage is not a new concept; an article by (Greenburg, 2011) proved that batteries inside Apple MacBooks could be attacked through corrupting the batteries on-board microcontroller which monitors the internal battery temperature and current power levels. When a batteries internal temperature reaches a prescribed limit or has received the required full charge the microcontroller shuts down the recharge process. By directly attacking the microcontroller these safeguards were removed which in turn rendered the battery inoperable to a point where it required replacing. The article concluded stating that the experiment could be taken a stage further whereby removing all of the batteries safety mechanism could in fact cause a catastrophic failure of the battery. In 2000 a rather dubious piece of sensational journalism from the Weekly World News entitled 'Hackers can turn your home computer in to a BOMB' was published, it had no evidence that this could actually happen but it did tap into the already existing fears that people had with computers. In 2010 with the release and subsequent discovery of the Stuxnet showed the world that a computer virus could indeed have a physical, damaging effect on computer attached equipment. In this case Stuxnet attacked the Siemens Step7 software which controlled the critical spinning speeds of the centrifuges in operation within Iran's nuclear enrichment program. Stuxnet also introduced the mainstream public consciousness to the phrase cyberwarfare and other terms such as weaponised virus and cyberattacks.

The problem with actual viruses and malware such as Stuxnet is that they require huge amounts of development, time and resources to produce. Stuxnet contained no less than four zero-day vulnerabilities. A zero-day vulnerability is a security hole in software which is unknown to the software vendor and one that can be exploited by computer hackers. Zero-day vulnerabilities although not as rare as software vendors would like to make out are still difficult to find and require time and effort to discover and exploit. There are numerous antivirus software programs available to the end user that should if maintained correctly safeguard the connected device from viruses and malicious software. For any newly created digital weapon to bypass these safeguarding programs requires a great deal of technical knowhow and resources to accomplish. A mimetic virus does not require any level of skilled programming sophistication or line after line of malicious code which needs to be obfuscated within a host programme in order to mask its true identity and malevolent nature nor does it require the use of a previously unknown zero-day vulnerability. When designing a mimetic virus there are no tangible limits as to its ultimate nature and effects when triggered as long as its effects are grounded in a pseudo-reality using terms that an unsuspecting public would have previously seen, heard or read in the media (Ayres & Maglaras, 2016). Computers are no longer used solely within a business environment we all have access to some form of computing device not just at work but at home. The cost of computing has dropped to such low costs that there are now multiple devices used by most family members from the young to the old. The mass use of computing devices by a wide and diverse cross section of the public could be a much more easily accessible and tempting target and by tapping into the fear that a hijacked battery full of highly charged chemicals in their or their children's laps could explode at any given moment from being infected by a computer virus could be an excellent opportunity for the cyberterrorist. With such a mimetic virus having been created within the mind of the cyberterrorist the next step in this particular attack vector is to find an effective method of weapon delivery in order to infect the minds of an unwary target audience.

Previously we were confined to our desks in offices or our homes in order to communicate with each other through the electronic ether, but in today's society we are no longer restrained by these limitations as technology has progressed and access to information and services can be obtained in a truly mobile environment using a whole plethora of mobile devices including laptops, tablets and mobile phones. The email has now been surpassed as the main source of digital communication and we live in a world where instant messaging and social media plays a crucial role in social communication in our day to day lives where every minute that passes over 300 hours of video footage is uploaded to YouTube (YouTube, 2015) and 15 million text messages (McVeigh, 2012) and 204 million emails are sent (Knoblauch, 2014). Facebook has 1.49 billion monthly active members and on the 28th August 2015 'for the first time ever, one billion people used Facebook in a single day' (Zuckerberg, 2015) along with 500 million tweets being sent every day (Twitter, 2015). We are constantly bombarded with information from our peers where a piece of information, picture or video which is considered humorous, outrageous, offensive or of a stimulating interest has the capability to traverse the globe within seconds with millions if not tens of millions sharing the information (Ayres et al, 2016). Conventional terrorist groups use video to great effect in order to communicate their particular message or ideology to the masses in order to either create revulsion or fear or both. A few carefully crafted videos and posts on selected social media sites highlighting the effects of a mimetic virus could swiftly spread the virus throughout a target audience and beyond; the mimetic virus then transfers from a realm of fantasy and on into reality. This mimetic transference of fantasy to reality could be achieved through the massive global use of social media. A meme is an idea or behaviour that spreads from person to person within a culture and acts as a unit or vessel for carrying cultural ideas, symbols or practices that can be transmitted from one mind to another through writing, speech, gestures, rituals or other imitable phenomena (Graham, 2002). Memes are cultural analogues to genes in that they self-replicate, mutate and respond to selective pressures. Memetics was popularised by Richard Dawkins and according to his theory, 'culture is transmitted through units known as memes' (Gebauer & Wulf, 1995). The meme has taken on a much more popularised form since the advent and widespread use of the Internet; the Internet meme which can now spread much more swiftly from person to person through email, blogs, forums, instant messaging and social media. The number of people who access the Internet on a regular basis has grown exponentially. Today there are 3 billion Internet users representing over 40% of the global population, the UK alone has over 57 million Internet users and globally almost one in four people are using social networks; by 2017 this will rise to one in three (Social Media, 2013). The term 'viral' has entered our everyday vocabulary with online videos and Internet memes having the ability to reach millions of people worldwide (see Figure 2.), not in a matter of days but in seconds.

The cyberterrorist can craft their digital weapon to execute any plausible attack on their target as long as the weapons effects have some grounding within reality, for example there are hundreds of videos on YouTube of people demonstrating the explosive nature of lithium polymer and lithium ion batteries either as a warning or just for fun, both of these types of batteries can be found in the portable computing devices we use on a daily basis. Taking the earlier detailed example of attacking a laptop battery microcontroller it is not that far a stretch of the imagination whereby a weaponised computer virus could attack these types of batteries and cause a much more catastrophic failure with the potential of explosion resulting in not just damage to the physical device itself but potentially the persons operating the device

Figure 2. The Mimetic Virus; the spread through social media

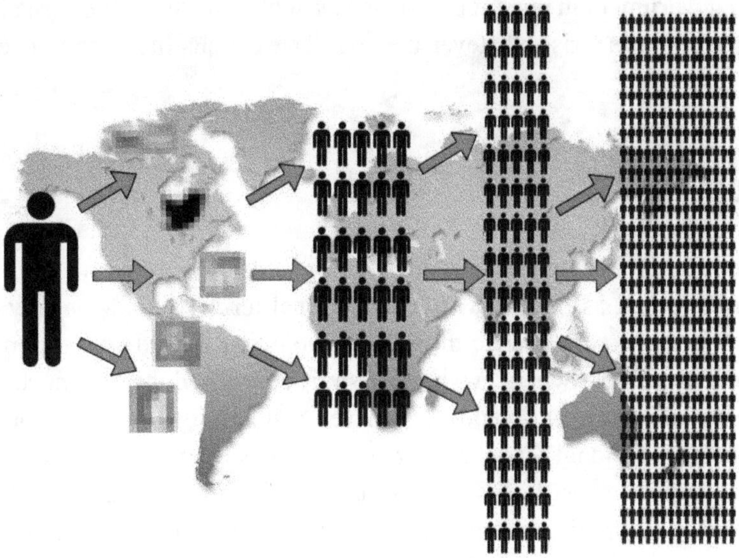

at the time. The stage has now been set by the cyberterrorist who has fabricated the mimetic virus and attributed a number of destructive effects once the virus has infected its target. Perhaps carefully crafted and staged videos and other posts on specific, targeted social media would give a visual aid to the audience of the virus effects. The plethora of choice of social media platforms could be utilised to spread the virus with similar consequences akin to the Good Times virus in the mid 1990's. After the weapons creation and ultimate propagation through the target audience and possibly to an even wider than anticipated audience the fear and threat of attack has now been instilled, but to what ends? This type of attack is still in essence fake using a mimetic virus as the digital weapon but one that has no actual physical real world effects, but if a population feels it is in threat of impending attack or in fear of its safety a government is compelled to act in order to reassure its citizens that they are safe and everything is being done to keep them from harm. Influencing a wider audience from an initial targeted attack is the true purpose of terrorism (Yonah, 2007). The U.S. invasion of Afghanistan was the principle reaction to the 9/11 terrorist attacks on the World Trade Centre and Pentagon (Thruelsen, 2006) by al-Qaeda. Many have argued that the 9/11 attacks were carried out because of the support of Israel by the United States, sanctions imposed against Iraq and the presence of US military in Saudi Arabia. Another motive for the attacks may have been to coerce the U.S. into a war that would incite a pan-Islamist revolution (Duran, 2002, pp. 22-42). Whatever the ultimate motive was for the 9/11 attacks it was an unlawful use of violence, motivated by ideological beliefs, which instilled fear and coerced a government in pursuit of al-Qaeda's goal. This therefore must also be the goal of the cyberterrorist whatever methods or weapons they use to coerce their target.

CONCLUSION

Highlighted in this chapter is the surrounding controversy over the definition of cyberterrorism just as there is with conventional terrorism but for very different reasons. There is no doubt as to the destructive intent behind conventional terrorist acts of violence but where cyberterrorism is concerned there is much debate as to whether a similar result could be achieved using a digital weapon on a digital target. The conventional terrorist operates in the real world with tangible effects whereas the cyberterrorist does not and has to rely on attacking its target through the medium of computing. There is also the view that the only viable target for the cyberterrorist must be against CNI because of the almost dependant integration with technology and thus attacking this technology could in turn have a subsequent effect on a wider population. This chapter moved away from the popular theories of predicated targets and looked at a population as the direct target for the cyberterrorist. With the mass growth of personal computing and readily access to the Internet we all enjoy we may have potentially made ourselves vulnerable to an electronic attack by would be assailants. New malware released into the wild has grown at alarming rates whereby in the 8 years from 2007 the total of new malware has increased by almost 3000% and total malware in the first 6 months of 2016 has already exceeded the overall figure for the entire previous year. Stuxnet showed the world that the humble computer virus could have a dramatic and damaging effect on computer attached hardware and it is not so far a stretch of the imagination that internal computer components could also be attacked with equally, if not more damaging effects on the hardware or potentially the actual operators. The mimetic (fake) virus has in the past caused widespread fear and panic on numerous occasions including the 1994 Good Times virus which was spread solely through the medium of email. Malware and the computer virus will no doubt be the digital weapons of choice for the cyberterrorist; it is much easier for the cyberterrorist to craft a fake weapon than a potentially complex real one which may require specific skills and expertise in order to create and requiring an equally elaborate method to deploy to its chosen target. There is no reason why the cyberterrorist may not take advantage of a mimetic virus to plant the seed of doubt and fear in the minds of individuals who are so interconnected through the mass use of social media could inadvertently spread such a virus to a much wider audience.

REFERENCES

Anon. (2015). *What is terrorism?* Retrieved July 2, 2015, from http://http://www.terrorism-research.com/

AV-TEST. (2015). *Malware*. Retrieved from AV Test The Independant IT Security Institute: https://www.av-test.org/en/statistics/malware/

Ayres, N., & Maglaras, L. A. (2016). *Cyberterrorism targeting general public through social media*. Security and Communication Networks.

Ayres, N., Maglaras, L. A., Janicke, H., Smith, R., & He, Y. (2016). The mimetic virus: A vector for cyber security. *International Journal of Business Continuity and Risk Management.*

C, K. (2009). *Goodtimes Virus*. Retrieved from Know Your Meme: http://knowyourmeme.com/memes/goodtimes-virus

Collin, B. (1997). *The Future of Cyberterrorism. In Criminal & Justice International* (pp. 15–18). Retrieved from The Future of Cyberterrorism.

Denning, D. E. (2000). *Cyberterrorism Testimony before the Special Oversight Panel on Terrorism*. Georgetown University.

Duran, M. S. (2002). Somebody else's civil war. *Foreign Affairs, 81*(1), 22-42.

Dwan, B. (2001). Cyber-terrorism: Virtual for Who? *Computer Fraud & Security, 11*, 12–14.

Gebauer, G., & Wulf, C. (1995). *Mimesis: culture, art, society*. University of California Press.

Gordon, S., & Ford, R. (2002). Cyberterrorism? *Computers & Security, 21*(7), 636–647. doi:10.1016/S0167-4048(02)01116-1

Graham, G. (2002). *Genes: a philosophical inquiry*. Routledge.

Green, J. (2002). *The Myth of Cyberterrorism. Washington Monthly.*

Greenburg, A. (2011). *Tech*. Retrieved 8 25, 2015, from Forbes: http://www.forbes.com/sites/andygreenberg/2011/07/22/apple-laptops-vulnerable-to-hack-that-kills-or-corrupts-batteries/

Harrison, V., & Pagliery, J. (2015, April 14). *Nearly 1 million new malware threats released every day*. Retrieved from CNN Money: http://money.cnn.com/2015/04/14/technology/security/cyber-attack-hacks-security/

Hoffman, B. (2013). *Inside terrorism*. Columbia University Press.

Hsu, F. D., & Marinucci, D. (2013). *Advances in Cyber Security: Technology, Operation, and Experiences*. Fordham University Press. doi:10.5422/fordham/9780823244560.001.0001

Jones, A. (2005). Cyberterrorism: Fact or Fiction. *Computer Fraud & Security, 6*, 4–7.

Jones, D. A. (2005). Cyber Terrorism: Fact or Fiction. *Computer Fraud & Security*, 4.

Kaplan, E. (2009). *Terrorists and the Internet*. Retrieved July 3, 2015, from http:// http://www.cfr.org

Kenney, M. (2015). *Cyber-Terrorism in a Post-Stuxnet World*. Foreign Policy Research Institute.

Knoblauch, M. (2014). *Internet Users Send 204 Million Emails Per Minute*. Retrieved 8 25, 2015, from MashableUK: http://mashable.com/2014/04/23/data-online-every-minute/

McVeigh, T. (2012). *Text messaging turns 20*. Retrieved 8 25Text, 2015, from The Observer: http://www.theguardian.com/technology/2012/dec/01/text-messaging-20-years

National Research Council. (1990). *Computers at Risk Safe Computing in the Information Age*. National Academies Press.

Panetta, L. E. (2012, October 11). *Remarks by Secretary Panetta on Cybersecurity to the Business Executives for National Security*. Retrieved July 8, 2015, from US Department of Defense: http://www.defense.go

Pollitt, M. M. (1998). Cyberterrorism: Fact or Fancy? *Computer Fraud & Security, 1998*(2), 8–10. doi:10.1016/S1361-3723(00)87009-8

Release, Sophos Press. (1999, November 23). *Don't fall for a virus hoax*. Retrieved July 10, 2015, from Sophos: https://www.sophos.com/en-us/press-office/press-releases/1999/11/va_hoaxes.aspx#cost

Schmid, A. P. (2011). *The Routlaedge handbook of terrorism research*. Taylor & Francis.

Seymour, G. (1975). *Harry's Game*. London: Collins.

Skoudis, E., & Zeltser, L. (2004). *Malware: Fighting malicious code*. Prentice Hall Professional.

SocialMedia. (2013). *Social Networking Reaches Nearly One in Four Around the World - SSocial Networking Reaches Nearly One Four Around World.* Retrieved from eMarketer: http://www.emarketer.com/Article/Social-Networking-Reaches-Nearly-One-Four-Around-World/1009976

Stats, I. L. (n.d.). *Internet Users*. Retrieved July 4, 2015, from Internet Live Stats: http://www.internetlivestats.com/internet-users/

Thruelsen, P. D. (2006). *From Soldier To Civilian: Disarmament Demobilisation Reintegration in Afghanistan.* DIIS Report.

Twitter. (2015). *Twitter usage / company facts*. Retrieved 8 25, 2015, from Twitter: https://about.twitter.com/company

Weimann, G. (2004). *Cyberterrorism, How Real Is the Threat?* Washington, DC: United States Institute of Peace.

Yonah, A. (2007). *Evolution of US counterterrorism policy*. Greenwood Publishing Group.

YouTube. (2015). *Statistics*. Retrieved 8 25, 2015, from YouTube: https://www.youtube.com/yt/press/en-GB/statistics.html

Zelikow, P. (2011). *The 9/11 commission report: Final report of the national commission on terrorist attacks upon the United States*. Government Printing Office.

Zuckerberg, M. (2015, August 28). *Mark Zuckerberg*. Retrieved from Facebook: https://www.facebook.com/zuck

Chapter 9
The Fight for Cyber Thoreau:
Distinguishing Virtual Disobedience from Digital Destruction

Matthew D. Crosston
Bellevue University, USA

ABSTRACT

This analysis sheds light on where cyber disobedience fits in to larger society positively and does not deserve to be summarily thrown in with all types of illegal cyber acts. Making these legal delineations does not solve all of the problems a state will face in the new cyber age. But they will go a long way in helping a democratic state honor its foundation of civil liberties and freedom and avoid becoming a perpetuator of virtual hyperbole and digital panic. In so doing, it guarantees the people greater freedom for themselves while also having greater confidence in the fairness of their own government. This, in place of the chaotic, ambiguous, and uninformed legal and political environment governing cyber action today, would be a dramatic improvement.

DOI: 10.4018/978-1-5225-1938-6.ch009

INTRODUCTION

The intent of this chapter is to re-imagine and apply the term 'civil disobedience' to the cyber domain. The full extent of how it can apply will be analyzed in a framework of the evolutionary rights of protest. Traditional civil disobedience is a 'performance' via visual and physical mediums, while cyber disobedience is essentially a digital performance with difficulty in attributing true ownership of the act. This is an interesting antithesis to the physical protestor, who uses strategic visibility, threat of arrest, and popular backlash against state behavior as vital aspects of the disobedience. These variables are noticeably absent in cyber disobedience. But this new space also has a low cost of entry and regimes will find it difficult to physically control. Another distinction is how actors intending to cause wide-scale state damage are a very real cyber threat, while cyber disobedience can appear far more menacing than it is in reality. Thus delineating *cyber disobedience* that does not intend to fundamentally damage the state from *cyber-attacks* that aspire to is incredibly important but has proven elusive within formal cyber policy. Perhaps the most obvious delineation will be in determining the self-declared intent and motivation of cyber civil disobedience groups, along with their ultimate overall goals.

In some ways this is a completely new application of studies that consider innovation within activism for the purpose of social change. Some initial research has considered the impact of the internet on social activism in recent years. (Earl & Kimport, 2011) But this is the first application of the legal debate across global norms within such research. Newer studies are becoming more and more intrigued by the transnational potentiality of social activism and the impact it can have across state structures. (Ilcan & Lacey, 2013) As will be seen in this chapter, the essence of cyber activism is by default transnational, boundary-absent no less, and thus the arguments being made here should be seen as an important addition to this growing literature. Finally, a third sub-category within the literature is directly benefited by this chapter, namely how social media is used for social change. To date this has been largely used for the analysis of student groups at a youth level. (Velasquez & LaRose, 2015) This chapter's findings enhance the scope and reach of such literature by expanding the applicability beyond students and youth. Indeed, social media, cyber activism, and societal change will likely only become more prevalent in future studies. Thus, this chapter represents something of the front edge of the coming digital social activism wave.

Some might also find this chapter extremely pertinent to Rawls' Law of Peoples, which conceptualized a particular liberal understanding of social justice and defined legitimacy in a political sense. The Rawlsian 'emergency exemption' got rid of what been a long-standing traditional aspect of *jus in bello,* that which prohibited the targeting of civilian populations. It has been argued that this view of order posited a non-inclusive conception of justice in a wildly diverse world and thus would actually *encourage* rather than discourage conflict, resistance, and violence, possibly even enhancing the power of autocratic government beyond what is considered the liberal zone of peace. (Martin, 2015) The consequences discussed within this chapter of *not distinguishing* cyber civil disobedience for social change from cyber-attacks threatening actual state integrity and/or survival is a critical addition to this philosophical debate between Rawls and his challengers.

The State: The Ambiguity of Law and Politics

Political coherence and consensus, with regard to cyber policy, has been slow and disjointed across the globe. This has allowed draconian measures to be haphazardly applied to cyber violators with little distinction as to act and/or intent. While traditional civil disobedience carries the threat of physical violence, jail sentences are typically very light. Conversely, a hacktivist may find him/herself facing multiple years in prison and fines reaching six figures for participating in DDoS attacks. This interesting dichotomy is further evidence that the language and formal punishment guidelines of traditional civil disobedience need to be considered when cyber policy is drafted covering the digital expression of the same acts. The threat of cyber civil disobedience is relatively low when juxtaposed against state-threatening actors, which is why policy needs to clarify explicitly the scope of cyber civil disobedience, while still protecting personal civilian information, utilities, and other important internet privacy structures. There is no denying that 150 years after the term 'civil disobedience' was coined by Thoreau, the idea, reality, and impact of civil disobedience has been drastically changed with the emergence and potential of the cyber vigilante. Acknowledging and adapting wisely to this brave new world, without impinging on fundamental freedoms, is essential for the global community.

A cyber vigilante works in obfuscated digital shadows, attacks without warning, gathers mass DDoS attacks of willing and unwilling participants, and organizes real world protests. Where traditional activism was limited by opportunity and awareness, cyber civil disobedience has few boundaries, does not always demand a crowd following, and is only limited by technical ability. Where crowds can be oppressed by force, cyber vigilantism can only be reacted to in defensive manners, giving an additional edge to the cyber civil disobedient. It's a disturbing reality to lawmakers, corporations, and governments. "The impossibility of forming any comprehensive, consistent picture… is precisely what makes the [groups] so unsettling to governments." (Serracino-Inglott, 2013) While many civil disobedient groups, such as Anonymous, directly claim responsibility during DDoS or other types of virtual disobedience, some groups do not, creating large-scale policing problems with attribution. Thus, attribution of attack is the single most unnerving characteristic of the cyber vigilante and is certainly the largest obvious divergence with traditional civil disobedience.

With the burgeoning industry of cyber security, cyber civil disobedience is being met with a dedicated force that is both corporate and state-sponsored. Technological solutions to cyberspace security risks are currently being passionately debated and innovations will no doubt continue. Conversely, there exist debates in scholarship and the political sphere of what is considered prosecutable behavior on the internet, no matter what ambiguous virtual walls were constructed. The difficult legal considerations about cyber vigilantism extend deep into philosophical discussions of the virtual versus the real:

In the conventional world it is perfectly acceptable for half a million people to organize a march through the streets of a major capital city to protest about war if the authorities are notified in advance, despite the fact that all the retail outlets would suffer loss of business during the protest. Similarly, if half a million people were to organize simultaneous attempts to log-on to a number of merchants' websites then this would almost certainly be considered unacceptable. (Gresty, Taylor, & Lunn, 2005)

The US Supreme Court has consistently upheld the people's right to protest and the right to engage in peaceful and orderly political demonstrations is, under appropriate conditions, a fundamental characteristic of liberty. (Harlan, 1969) How this may apply, however, in the cyber realm is far from understood. The *Boston College International & Comparative Law Review* concluded that forms of cyber vigilantism that are primarily expressive, ie, that do not involve obtaining or exploiting illegal access to computers or networks for commercial advantage or financial gain and cause little or no permanent damage, should receive at least some protection as a legitimate form of protest. (Hampson, 2012) This shall theoretically only be described as activism. The Computer Fraud and Abuse Act of 2006 (CFAA), however, does not apply this to the federal government. (Anderson, 2013) Currently, the distinction between cyber disobedience vs. state cyber-attacks is inherently vague.

There also exists little consensus within politics on how far the state may penetrate into the cyber sphere in order to apprehend rogue cyber vigilantes. Interestingly, libertarian-leaning Republicans joined Democrats to fight for more restrictions on the capability to collect data. As noted by Benjamin Wittes, 'to put matters simply, the politics in this space are strange.' (Leithauser, 2014) Thus, with no united front and little societal standard or boundaries constraining cyber protests, precise legal definitions and distinctions between cyber-attacks on the state and cyber civil disobedience against state policy will remain vague and opaque.

The Perpetrators: A Permanent Wild Wild Virtual West?

With the advent of the internet, three different categories of cyber civil disobedience groups erupted onto the scene in the cause of so-called political activism. First, when it came to organizing early cyber protests, original groups incurred heavy costs. Costs to publicize the event, send out messages, and even pinpoint the location could add up quickly. Cyber protests today can be organized more efficiently and at much less cost. Newer groups, however, thought that such traditional political activism over the internet was inadequate and wanted more attention-getting methods. This second group managed to create "virtual sit-ins" done in the form a distributed denial of service (DDoS). This disruption incurred high costs to reinstate service on the organization being attacked. They have been described in the past as cyber vandals, but are more accurately described as cyber squatters. The third group (hacktivists) took virtual civil disobedience to an even higher level, collecting and releasing names and passwords of innocent civilians who had nothing to do with the issue they were protesting. This extreme third form teeters on the verge of becoming civil cyber terrorism.

As the internet became more and more dominant in the everyday lives of government, organizations, and private citizens, it became the perfect way to make information readily available for public consumption. As budding hacktivist organizations grew around the internet, taking issues with various claims and stories being provided directly to the public by governments and corporations, they would invariably contact news media organizations to push their counter-agendas. More often than not, the news companies would choose to not emphasize the priorities of the hacktivist groups. It took several attempts of this failed back-and-forth between media and hacktivists before perhaps the most famous group of all, Anonymous, stated, 'the people don't care.' (AnonymousPrime99, 2012). Ultimately, it wasn't until the PayPal DDoS by Anonymous that both press and people began to pay attention to this new phenomenon called hacktivism. Because of that DDoS, Anonymous finally got truly global attention and was able to more formally publicize its objectives. Apparently, Anonymous came to the conclusion that the only way to get the attention of both people and press was to *threaten* them. This had far-reaching consequences across all hacktivist groups, as they became more aggressive in spreading their message. This is what began the transition where basically all cyber squatters would also dabble in being full-on hacktivists and subsequently get broadly labeled by corporate and government culture as virtual terrorists.

Like any group operating in the void of substantive and explicit law, cyber vigilantes often work within their own self-defined morality. Anonymous members are supposedly obligated to abide by three principles: don't attack media; don't attack infrastructure; and work non-violently for internet freedom. (Paganini, 2012) Interestingly, the group also claims for itself the mantle of protector over those that are attempting to use the free internet to further noble causes. When a hacker defaced a charity website, for example, the group tracked the person to Spain and passed his name to authorities and the CEO of the charity. (Paganini, 2012) The group also played a role in the Arab Spring, defacing dictator websites and facilitating avenues of communication for the protestors. In the eyes of a cyber-vigilante, the moral system of the internet should be higher than that of the state:

The meager efforts at Open Government, or the means deployed to counter oppressive regimes, are as ineffective as the courts and the jails on the frontier. No matter how crude deviant methods for justice might seem, their superior effectiveness easily trumps the nominal availability of formal alternatives. (Serracino-Inglott, 2013)

This indicates that cyber vigilantes will continue to operate within the internet not only because the current political atmosphere cannot deter them, but because the global political system is seen as having utterly failed regular civilians. Thus, cyber vigilantes are currently operating self-righteously within a relative morality that operates outside the law. Given the fact that there has been a remarkable lack of global consensus on establishing unified norms of cyber behavior, and this status quo doesn't seem likely to change anytime soon, the moral relativism of various cyber perpetrators is going to continue to carry the day.

Discourse and Discipline: Instituting Control over Cyber Disobedience

The evolution and control of cyber civil disobedience may not rest solely with future policy decisions but might ultimately end up being an arms race between hackers and security providers. 25 years ago Andrew Ross foresaw the future where there would be a shift in the relation of countercultural activity to technology, a shift in which a software-based technoculture would replace typical vigilantes and be organized around the demonizing of abject hardware structures. (Ross, 1990) This in a way can be seen as a cyber-decentralized army, one built online to defend and organize hacktivism and cyber civil disobedience somewhat randomly. These activists will be difficult to ever unite into one cohesive cyber identity with military-like discipline and loyalty. Factions almost always seem to split off, but the idea that they could be potentially united by the general principle of the necessity for chaotic cyber civil disobedience is at least plausible. Unlike any other 'outlaw' group in history, cyber vigilantes are 'radically democratic in their structure. There are no leaders, anyone can join, and members are located almost everywhere in the world.' (Wong & Brown, 2013)

Today's advanced technological world has witnessed the overwhelming speed in which public discourse has moved from traditional physical protests to the anonymous digital shadows that can strike without warning and then immediately lurk back into the darkness. This transformation is little understood by the psychology of mass uprisings: while emotions rise high in mass physical protests and often lead to unintended violence, cyber activism usually has a very clear limited goal and disappears when the goal is achieved, cold and calculated. It may only be time, however, before a cyber faction of some larger movement breaks off and operates divergently from the original moral code of the parent group with destructive consequences: 'while fortunately cyber conflicts have not yet escalated into greater military engagements, this may change as nefarious activity continues without diplomatic, economic, military repercussion or consequence.' (Iasiello, 2014)

Thus, the political discourse and debates over what should be allowed within cyber civil disobedience and how far the state should be able to react to apprehend unlawful actors has remained ambiguous and unclear, but also rather heavy-handed and punitive. The modern world needs to determine how far the right of cyber civil disobedience extends and whether the type that is currently displayed by these various hacktivist groups will fall positively or negatively under clearer cyber legal boundaries. Whether it is the United States or Russia or China, the dangers inherent in an experienced cyber group acting in anonymity without regard for justice are potentially very high, but these would be actions not of cyber civil disobedience, but of cyber state-terrorism. Making a distinction between virtual terrorist and digital hacktivist in the legal, moral, and political sense is thus quite important, for it is the necessary extension of the principles of freedom into the digital world. This is not just for the sanctity of state sovereignty and security, but also for the fundamental right of civilian protest everywhere. For the most part, that distinction does not yet exist and thus hacktivists are automatically grouped into the same danger category as cyber terrorists. There even exists a popular school of thought that believes this is the right thing to do.

The Emergence and Evolution of Hacktivism as State Danger

What makes hacktivism such a big threat to states? Quite simply, the fact that cyberspace has no real boundaries. During the Arab Spring, Egypt literally turned the internet off to its civilian population and it was hacktivists that were able to help civilians regain access to cyberspace. On 16 September 2012, a group of Mexican hacktivists that called themselves 'Ciber Protesta Mexicana' took down dozens of government-owned websites and replaced those sites with a black screen and message that stated the July presidential election was fraudulent and the election results should be rejected by the people. They also spoke against corruption, drug-related violence, and the country's stagnant economy. (Rial, 2012) Thus, hacktivists for some are now the patriots of the modern age. They see the expansion of the government and its ever increasing encroachment on the private lives of its citizens as a threat. Which begs the question: *who* is the real threat? Those who conduct virtual civil disobedience to institute political/economic/social change, the government agencies trying to stop it, or yet other government agencies tacitly approving such acts in order to push their own political counter-agendas?

If one thing has been made clear about the internet in the 21st century it is that it is incredibly difficult to determine who are wearing the white cowboy hats and who are donning the black ones. Hacktivism is not completely pure as some vigilante groups would like people to believe, as some hacktivists in countries like Russia and China feel the need to defend their government from what they perceive as unjust assaults from foreign enemies. For Russian hacktivists many seem to do it for profit as well, where the government hires them for interference missions, while Chinese 'honkers' use hacktivism as a form of virtual patriotism to protect their country from its enemies in a purely geostrategic consideration. (Crosston, 2011) Indeed, looking over the various options available, the techniques will likely only become more sophisticated and refined, as shown in Table 1.

In a nod to this coming dangerous reality, the United States has attempted to equip its national security infrastructure with new legal tools to fight the anticipated onslaught. The National Defense Authorization Act of 2012 and 2013 gave the United States Intelligence Community the right to declare war on an attacker who executes a severe cyber-attack on the US. It is clear that this can be citizen or foreigner. Anonymous has protested the NDAA, which allows for the indefinite detention of US citizens, but so far the law stands as is. The reaction from the hacktivist community is not wholly unsurprising:

Table 1.

Traditional Civil Disobedience	Hacktivism
Sit-ins	Website Defacements
Barricades	Website Redirects
Political Graffiti	Denial of Service Attacks
Strikes	Information Theft
Underground Presses	Site Parodies
Political Theater	Virtual Sit-ins
Sabotage	Virtual Sabotage

Free speech in the land of the free? Arms in the home of the brave? Privacy in a state of wiretaps and government intrusion? Unreasonable searches? We are a little behind our 1984 deadline, but working slowly one amendment at a time. It is hard to trick hundreds of millions of people out of their freedoms, but we should be complete within a decade. (Warnick, 2012)

Despite the obvious fact that hacktivism is not just a local American issue, the United States has decided to take the lead in attempting to develop policing policies regarding cyberspace. According to the White House document, 'International Strategy for Cyberspace: Prosperity, Security, and Openness in a Networked World,' the goal of America is:

The United States will work internationally to promote open, interoperable, secure, and reliable information and communications infrastructure that supports international trade and commerce, strengthens international security, and fosters free expression and innovation. To achieve this goal, we will build and sustain an environment in which norms of responsible behavior guide states' actions, sustain partnerships, and support the rule of law in cyberspace. (White House, 2011)

The problem with a statement this broad is that the policies would not only be inadequate for determining what is and is not responsible behavior, but also would begin to infringe on the rights of users who use the internet responsibly. To be clear, such policies would not be effective against any nation that felt it did not wish to abide by it. This can be seen over and over again as China and Russia continue to use hacktivism to protest United States military interventions as well as attempt to collect classified information to further their own national security agendas. A more subtle but no less important point in all of this is how damning it looks to the cyber civil disobedience community when the United States seems quite enthusiastic about creating 'punishment norms' for limiting cyber capability in broad general terms but then is arguably the last in line when it comes to establishing global governance and international norms over state cyber behavior.

The country and target go a long way in determining whether or not hacktivism has the support or condemnation of the government in which it is conducted. Foreign countries will approve a hacktivist group hacking another government's website. This is evidenced by the indifferent response of the Chinese government when the hacking group, Honker Union, conducted a series of attacks on American assets after the accidental bombing of the Chinese embassy in Yugoslavia in the 1990s. While the Chinese government has strict laws against any kind of cybercrime, no action was taken against this group, which defaced several United States governmental and business websites. Thus, it would appear such activity has the tacit approval of the Chinese government, as it not only sent a statement about perceived American wrongdoing, it also gave the government in China virtual plausible deniability because of the aforementioned attribution problem.

The United States is just as guilty of this. During then Senator Obama's presidential campaign in 2008, he stated that government whistleblowers are part of a healthy democracy and such acts of courage and patriotism should be encouraged rather than stifled. This was apparently forgotten, however, when Private First Class Bradley Manning released a video that showed US military impropriety and other compromising documentation to Wikileaks. He was tried on charges of treason and aiding the enemy. In July 2011, 14 people were arrested after being suspected in participating in a DDoS attack on PayPal, after it decided it would no longer send funds to Wikileaks. Ironically, earlier that year, members of Anonymous around the world helped the citizens of Egypt when the Egyptian government cut off access to the internet and also conducted a DDoS attack on the Egyptian government's websites. The American government did not arrest any member of Anonymous for that DDoS attack or even issue a declaration as to its illegality. Apparently, relative cyber morality is not just the exclusive domain of those wishing to perform virtual civil disobedience, but is commonly employed by states as well.

This is not a new concept, however. Since 2010, the Obama administration has been trying to institute an internet 'kill switch,' which would cut off the nation's access under reasons of national security. Interestingly enough, the bill was reintroduced in 2011 right after the Egyptian government cut off its citizens. While not a global trend just yet, countries like Egypt, Libya, and China have switches already to disconnect their populations from the internet. This is seemingly odd company for the United States to wish to join. This is not an effective policy to have, though, as it sets an unhealthy doublestandard precedent, basically telling citizens that governmental tools to prevent cyber civil disobedience will be pursued *preemptively,* thus constraining a fundamental freedom of any legitimate civil society. The US should be careful what it wishes for: Egypt's actions did nothing but wake up the ire and anger of the international hacking community. Anon2World, a member of Anonymous, stated, 'I put myself in their place and I found myself in a desert of nothingness because he [Mubarak] just wiped out everything...That just showed me and everybody else that the same thing can happen at any time, at anywhere, by any government.' (Anonynops, 2012) In short, not finding legal space for legitimate cyber civil disobedience to operate could only enflame more damaging or violent cyber actions against the state down the road. This is why the need for a legal distinction between types of cyber civil action is so crucial.

While many countries struggle to find ways to legislate laws regarding hacktivsm/cyber civil disobedience, there is one country (Germany) that seems to have embraced it. Ultimately, to write effective legislation against legitimate potential cyber threats requires an understanding of the culture and technology behind it and a willingness to *allow* some forms of disobedient cyber action. In the new millennium, the German airline Lufthansa used its airplanes to summarily deport immigrants and a group of German hacktivists decided to protest this action. Lufthansa's website was DDoS'd and the hacktivists suspected in the attack were subsequently arrested. The German courts ruled that the DDoS was a legitimate form of political protest. Germany thus became both a poster child for and an anathema to both schools, pro and con, of the cyber civil disobedience debate. By allowing DDoS as a legal form of protest in certain contexts, Germany is allowing subtlety and nuance to inform its cyber laws. Down the road this should give it far greater flexibility and citizen approval whenever a cyber event does occur that runs afoul of German state interests.

It seems that technology, as a thing of modernity, has more often than not among the governing authorities managed to create panic, resulting in knee-jerk policy reactions instead of logical, balanced analysis. It is understood that state secrets should be kept secret, particularly when human life is at stake. It is also understood that state secrets in terms of weapons development should be protected. The worry, therefore, that the cyber civil disobedience community is completely absent any national security ethical grounding or strategic morality is off-base. LulzSec, for example, went on a 50-day hacking spree which caused great division within the hacking community. Joshua Corman stated, 'some of the things LulzSec did in the Court of Public Opinion were less noble. Private citizens were being were being dox'd or the credit card numbers could have been redacted. You could still make your point without actually giving out people's personal details. So the Court of Public Opinion swayed that LulzSec had crossed the line that maybe the previous manifestation had not.' (Anonynops, 2012) There is no denying that such threats are a global issue. However, policies and strategies must focus on making distinctions across explicit threats and should never address the fundamental freedom of political activism that often acts as a spur of positive change in many countries.

When Congress proposed to pass the Stop Online Piracy Act (SOPA) and Protect Intellectual Property Act (PIPA), a site was created called sopastrike.com. This site advertised well and was able to garner the support from big internet names such as Google, Wikipedia, Mozilla, Flickr, and Craigslist. Because these big names signed on, the 18 January 2012 internet boycott was heavily promoted and garnered the support of over 75,000 other websites that participated in a 24-hour internet blackout. Each site that participated put up a banner provided by sopastrike.com stating that the site was down to protest SOPA and PIPA, thus educating approximately one billion people about the flaws of these two pieces of legislation. On 24 January 2012, when it came to vote, both Acts failed and the protest was successful. This form of cyber activism was organized inexpensively, effectively, and without causing a damaging or dangerous disruption to the state. Despite this though, the government is still attempting to find ways to pass these two acts, albeit calling them something different. This type of political subterfuge only drives a deeper societal wedge between the two camps and should be discouraged. If America wants to lead on this issue, it should lead from a more positive and freedom-embracing position, not a freedom-constraining one.

The Scholarly, Policy, and Revolutionary Debates on Cyber Civil Disobedience

While cyber dissidence overall around the globe has been slowed by draconian and haphazardly applied punitive policy in regard to hackers and DDoS attacks, the future may look very different. As public opinion turns against the cyber hand of the government intrusively collecting information, a smarter and more balanced global cyber policy will need to be made clear. Cyber civil disobedience, arguably, has played a positive role, creating much awareness and progression on the internet. To silence that progress now would be a political and technological mistake. It is to the benefit of everyone if the 'Wild West' cyber myth is discontinued and a cyber order of mutual respect and subtlety replaces it. This goal is complex, without doubt, but is still better than the current system of being incoherent and indiscriminately punitive.

Cyber vigilantes see the governance of the digital shadows in a very different way. They envision a digital space of policy democracy without corporate influence or governmental pressure. As one of the most infamous hackers in the world, Adrian Lamo, has stated:

I hoped and believed that I could [hack systems] in a way that would set a precedent that would allow people to come forward in good faith to try and do the right thing, to let them believe that maybe motives did matter, that it wasn't all black-and-white. I think this is symptomatic of something we're seeing in the government today. In many ways they're eliminating shades of grey. They want to polarize people. It's important to our national agenda today to see good guys and bad guys. Because as soon as we start to believe that maybe it's not all black-and-white, that someone can do wrong for a good reason, then not every action of law is inherently infallible. (Greenberg, 2012)

Misunderstanding the motivations of cyber vigilantes, and their abilities, only creates further motivation for their causes. As Lamo described, their aspirations come from a desire for cyber freedom and a legitimacy that justifies cyber protests. Thus, attempting to silence them through indiscriminate law and societal labeling would only provide more ammunition for worldwide recruitment to the various causes of reckless cyber vigilantism. This would also discourage the development of a system of progressive and positive cyber civil disobedience.

A considerable amount of the dialogue openly antagonistic toward cyber vigilantes has focused on comparative analogies with the nuclear age and devolutionary processes. Joseph Nye has spoken against cyber vigilantes, saying, 'the bad news is that cyber technology gives much more power to non-state actors than does nuclear technology, and the threats such actors pose are likely to increase.' (Nye, 2013) Thus, Nye is perpetuating that the existence of cyber vigilantes is a devolutionary process of the nuclear age, in which the power of war and its threat is transferred back to those with the technological ability to wage it. He is also deconstructing the debate to a more primitive level, where subtlety and nuance in the identification of cyber disobedience is removed. Similarly, the director of the Homeland Security Policy Institute has stated:

Non-state actors, terrorists, and criminals are also leveraging cyberspace to their own ends, benefiting from a domain that levels the playing field and allows smaller and even individual actors to have a disproportionate impact. This asymmetry gives rise to an ecosystem that is fraught with a range of perils that did not previously occupy the focus and energies of major powers. (Cilluffo, 2012)

'Disproportionate impact' here is referring to the devolution of power that has taken place in the 21st century due to the proliferation of cyber capability. Notably, there is also acknowledgement that this transfer of power has given rise to the difficult situation of cyber vigilantes occupying the 'focus and energies of major powers.' Finally, non-state actors, terrorists, and criminals are all promoted to the same category in the analysis, leading to policy that directs punitive action equally and broadly. Such policy bluntness could possibly stifle peaceful cyber activists and legitimate cyber civil disobedience. Thus, while the internet has been praised as giving rise to the 'politics of no one,' this devolution of power has also been seen as threatening to governments. (Wong & Brown, 2013)

Contrary to the nuclear analogy debate, there exist arguments that cyber vigilantism is more like a coming age of revolutionary internet activism. While these arguments are equally-based on the post-nuclear power devolutionary processes, their outcomes are focused on positive social influence. Perhaps the best scholarly example of such argumentation is Molly Sauter's *The Coming Swarm*, where she states, 'by ignoring the potential legitimacy of these out-of-the-mainstream disruptive tactics, critics are contributing to this systematic disenfranchisement by artificially and harmfully restricting what political speech and conduct is acceptable and, by extension, whose.' (Sauter, 2014) Essentially, this argument rejects the notion that all cyber disobedience should be considered automatically a criminal action and therefore should not be prosecuted under the law in an arbitrary and indiscriminate way.

Aligned with this sentiment is the rejection of blanket illegality attributed to online DDoS attacks that act as a form of political activism, as this would be the limiting of free speech. It is true that where the rules of the real world become disconnected from the virtual is a space where laws become difficult to discern, where the nature and intent of political activism becomes convoluted. But letting this drive an assumption that participating in regional, international, corporate, or other cyber activism is de facto *automatically* unjustified is an 'empirical misrepresentation' perverts the innovative, positive legacy cyber civil disobedience can have on society. (Paganini, 2012) As of now, no US law considers DDoS or other directed cyber-attacks as being political free speech, legislatively painting with a broad punitive brush that might create more problems than it solves. Thus, the debate is not so much about allowing the Wild Cyber West to reign, but rather to allow some core fundamental civil liberties to inform and infuse the global cyber realm. There should be a legal space for accepted cyber civil disobedience protests. So far, the law has not caught up with the more progressive scholarship.

These two contrasting viewpoints are wildly divergent and no compromise consensus seems to be near in sight. As Molly Sauter acknowledged with respect to cyber-attacks, "DDoS is deployed as a tactic of criminality much more than it is a tactic of activism. This means that each use of DDoS as an activist tactic must first prove that it is not criminal before it can be accepted as activism.' (Sauter 2014) Since the cyber realm contains no borders, these legal distinctions will be clearly beneficial in describing international cyber activism against cyber terrorism. To strategically delineate and properly designate laws and norms to criminal (and thus, conversely, legal) activities will be of great importance to scholars and policymakers alike. Without an attempt by current policymakers to create political discourse to at least begin to establish legality and illegality, a societal assumption space emerges that adds to the perception that nearly all cyber civil disobedience will be illegal and categorized equivalent to virtual terrorist activity that threatens the state.

None of this, by the way, gives carte blanche approval to cyber-attacks against the state. Leon Panetta, the former US Defense Secretary, stated that a cyber-attack 'could be as destructive as the terrorists attack on 9/11.' (Farnsworth, 2012) While this has never been proven in a physical sense to be accurate, shutting down important infrastructure is a much more realistic threat and does carry grave consequences to societal functionality. While groups like Anonymous have shown to wield quite significant online force, they are largely unable to cause the sort of cataclysmic event described by Panetta and have never shown the proclivity to do so. Instead, as argued by Nathan Sales, real cyber dangers nearly exclusively lie with powerful foreign governments. (Sales, 2013) Importantly, with this division there exists an opportunity to apply policy in a much more beneficial and direct way.

If lawmakers around the world, but most importantly beginning with America, can concede that cyber vigilantism can and likely will include allowable digital forms of civil disobedience, then policy must reflect that change, as it would represent a major shift in how governments currently view the issue. This would also be no small change in attitude as 'the dual forces of commercialization and the defense interests of states have combined to foster an online environment that is increasingly hostile to innovative or disruptive modes of political engagement and dissent.' (Sauter, 2014) Policy must define a legal space for allowable civil disobedience or at least define strictly illegal practices. Ambiguity and indecision has allowed corporations, states, and citizens to fight for themselves in terms of protection and hackers and activists to continue to act in undefined spaces at great and unnecessary risk.

CONCLUSION

The acceleration of cyber vigilantism in a haphazard patchwork of digital innovation has caused politicians and lawmakers to define online parameters inaccurately as 'black-and-white.' Policy has largely treated cyber vigilantes, terrorists, criminal hackers, civil activists, and online theft as the same. While personal information, classified materials, and all utilities should be strictly off-limits, there exists a need for civil disobedience on the internet, even when it is directly protesting the state, perhaps *especially* when it is. Without a minority cyber protest voice in a majority cyber state world, dissent will be pushed further from the realm of normalcy and the decline of civil society protest will pervade. It will no doubt be a difficult line to define, between acceptable and unacceptable, legal and illegal, but the broad outlawing of cyber civil disobedience would be a much more dangerous precedent to set in a democratic system.

There are many issues today that put the creation of international policies over cyberspace at a standstill. Each nation is its own sovereign state with its own culture, government, and laws. Some countries have filters which censor internet access as a matter of course while other countries encourage the flow of information whether it is truthful or slanderous. Perhaps the largest problem in creating global policy is the fact that most consider it nearly impossible to enforce. Dorothy E. Denning stated that, 'international law does not carry the same weight as domestic law. Under international law states…assume international legal obligations only by affirmatively agreeing to them…They are free to decline participation and they are free to back out later…Under domestic laws, the citizens of a country are vulnerable to prosecution for violating any laws, regardless of whether they agree with them, and regardless of whether the laws are even just.' (Himma, et all, 2008) This is a crucial and interesting point that deserves further consideration.

Some countries view DDoS as an acceptable form of protest while others view it as a prosecutable crime. Where is the middle ground in this? If policy defines a certain cyber activity as criminal, it must apply to all activities, not specialized groups or targets. Michael Schmitt from the International Law Department of the United States Naval War College proposed criteria to help with creating such a standard: 'the best states can do is to engage in educated conjecture as to how the international community is likely to view the cyber operations in question as a matter of law... The criteria were meant to assist in that effort...they are merely factors that can be expected to influence states when making use of force appraisals.' (Schmitt, 2012) Schmitt's Criteria are as follows:

1. **Severity:** Was anyone killed or wounded? Any property damaged?
2. **Immediacy:** How quickly did the consequences of the attack take effect? How long will it take for the consequences to be felt?
3. **Directness:** What is the relationship between the attack and its effect?
4. **Invasiveness:** Did the attack have to cross borders to the target country?
5. **Measurability:** What are the measureable effects of the attack?
6. **Presumptive Legitimacy:** Does the international community consider the attack legitimate?
7. **Responsibility:** Who is responsible for the attack? A country? A political group? An individual?

Add a score for each criterion and, if the score is high, it is plausible to consider that the cyber action elevates above civil disobedience and instead is a use of force and should thus be considered illegal. This would give any nation the ability to implement Article 39 of the United Nations Charter (self-defense, in essence). It is a novel idea that at least begins the process of standardizing the definition of 'use of force' and the ability for a nation to defend itself without devolving to indiscriminate legal cyber oppression.

The speed at which civil disobedience has moved from physical protests and clashes with police to cyber action, sometimes with the two forms fusing together, has been an astonishing feat of the 21st century. The advent of the cyber vigilante has confused the line between justified actions of the cyber protester with that of the more malicious conduct of virtual terrorists and criminals motivated only by nefarious intent. This analysis has attempted to shed light on where cyber disobedience fits in to larger society positively and does not deserve to be summarily thrown in with all types of illegal cyber acts. Making these legal delineations does not solve all of the problems a state will face in the new cyber age. But they will go a long way in helping a democratic state honor its foundation of civil liberties and freedom and avoid becoming a perpetuator of virtual hyperbole and digital panic. In so doing, it guarantees the people greater freedom for themselves while also having greater confidence in the fairness of their own government. This, in place of the chaotic, ambiguous, and uninformed legal and political environment governing cyber action today, would be a dramatic improvement.

REFERENCES

Anderson, A. (2013). The Computer Fraud and Abuse Act: Hacking into the Authorization Debate. *Jurimetrics*, *53*(4), 447–466.

Anonvnops. (2012, September 24). *We are Legion – The Story of the Hacktivists* [Video file]. Retrieved from http://www.youtube.com/watch?v=8TMtbNJ81X0

AnonymousPrime99. (2012, May 12). *We Are Legion: Digital (R)Evolution* [Video file]. Retrieved from http://www.youtube.com/watch?v=bLdhRP8iB6E

Armerding, T. (2012, September 13). *Anonymous Had Bad Month, But No Less "Reliable"*. Retrieved from http://www.csoonline.com/article/716092/anonymous-had-bad-month-but-no-less-reliable-

Arquilla, J., & Ronfeldt, D. (2001). *Networks and Netwars: The Future of Terror*. Crime, and Militancy.

Brafman, O., & Beckstrom, R. A. (2006). *The Starfish and the Spider: The Unstoppable Power of Leaderless Organizations*. New York: Penguin Publishing.

Bunt, G. R. (2003). Islam in the Digital Age: E-Jihad, Online Fatwas and Cyber Islamic Environments. London: Pluto Press. Retrieved from http://www.bandung2.co.uk/books/Files/Religion/Islam%20in%20the%20Digital%20Age%20(E-Jihad,%20Online%20Fatwas%20and%20Cyber%20Islamic%20Environments)%20-%20Gary%20R.%20Bunt.pdf

Cilluffo, F., Cardash, S., & Salmoiraghi, G. (2012). A Blueprint for Cyber Deterrence: Building Stability through Strength. *Military and Strategic Affairs*, *4*(3), 11.

Crosston, M. (2012). Virtual Patriots and a New American Cyber Strategy: Changing the Zero-Sum Game. Strategic Studies Quarterly, 6(4).

Crosston, M. (2013). Duqu's Dilemma: The Ambiguity Assertion and the Futility of Sanitized Cyber War. *Journal of Military and Strategic Affairs, 5*(1).

Crosston, M. D. (2011, Spring). World Gone Cyber MAD: How "Mutually Assured Debilitation" Is the Best Hope for Cyber Deterrence. *Strategic Studies Quarterly*.

Earl, J., & Kimport, K. (2011). *Digitally Enabled Social Change: Activism in the Internet Age*. Cambridge, MA: The MIT Press. doi:10.7551/mitpress/9780262015103.001.0001

Farnsworth, T. (2012). US Officials Detail Cyber Policy. *Arms Control Today*, *42*(9), 32.

FedSolutions. (2012, May 24). *Are You Paying Too Much For DDoS Attacks?* Retrieved from http://www.fedsolutions.com/2012/05/cost-of-ddos-attacks/

Fitri, N. (2011, April). *Democracy Discourses Through the Internet Communication: Understanding the Hacktivism for the Global Changing*. Retrieved from http://www.academia.edu/1438665/Hacktivism_for_Global_Changing_Nofia_Fitri_ojcmt.net_

Gjelten, T. (2011, July 20). *FBI Tries To Send Message With Hacker Arrests*. Retrieved from http://m.npr.org/news/Technology/138555799

Greenberg, A. (2012). *This Machine Kills Secrets: How WikiLeakers, Cypherpunks, and Hacktivists Aim to Free the World's Information*. New York: Penguin Publishing.

Gresty, D., Taylor, M., & Lunn, J. (2005). Legal Considerations for Cyber-Activism. *Systems Research and Behavioral Science*, *22*(6), 566–567. doi:10.1002/sres.625

Hakar, V. (2003). The Right to Civil Disobedience. *Osgoode Hall Law Journal*, *41*(2/3), 411.

Hampson, N. (2012). Hacktivism: A New Breed of Protest in a Networked World. *Boston College International and Comparative Law Review*, *35*(2), 531–532.

Himma, K. E., & Tavani, H. T. (2008). *The Handbook of Information and Computer Ethics*. Hoboken, NJ: John Wiley & Sons, Inc. Retrieved from http://www.e-reading.org.ua/bookreader.php/141465/The_Handbook_of_Information_and_Computer_Ethics.pdf

House homeland security subcommittee on oversight, investigations, and management hearing. (2012). Lanham, MD: Federal Information & News Dispatch, Inc.

Iasiello, E. (2014). Hacking Back: Not the Right Solution. *Parameters*, *44*(3), 113.

Ilcan, S., & Lacey, A. (2013). Networks of Social Justice: Transnational Activism and Social Change. *Studies in Social Justice*, *7*(1).

Leithauser, T. (2014, November 10). Elections Muddle Prospects for Cyber, Surveillance Bills. *Cybersecurity Policy Report.*

Libicki, M. C. (2007). Conquest. In *Cyberspace: National Security and Information Warfare*. New York, NY: Cambridge University Press.

Martin, R. (2015). Rawls on International Economic Justice in the Law of Peoples. *Journal of Business Ethics*, *127*(4), 743–759. doi:10.1007/s10551-014-2184-x

McGhee, W. E. (2008). *Information technology governance: An exploratory study of the impact of organizational information technology security planning* (Doctoral Thesis, Capella University). ProQuest Dissertations and Theses.

Nye, J. Jr. (2013). From Bombs to Bytes. *The Bulletin of the Atomic Scientists*, *69*(5), 12. doi:10.1177/0096340213501338

Olson, P. (2012). *We Are Anonymous: Inside the Hacker World of LulzSec, Anonymous, and the Global Cyber Insurgency*. New York: Little, Brown and Company.

Paganini, P. (2012, August 8). Anonymous Ethical Code. Back to the Origins. *Security Affairs.* Available at http://securityaffairs.co/wordpress/7986/cyber-crime/anonymous-ethical-code-back-to-the-origins.htm

Picket, K. (2011). *Read Demands of "Occupy Wall Street"...And Try Not to Laugh.* Retrieved from http://nation.foxnews.com/occupy-wall-street/2011/10/04/read-demands-occupy-wall-street-and-try-not-laugh. October 3

Ratliff, T., & Hall, L. (2014). Practicing the Art of Dissent: Toward a Typology of Protest Activity in the United States. *Humanity & Society*, *38*(3), 271–272. doi:10.1177/0160597614537796

Rial, N. (2012). *Mexican Hackers Attack Official Sites*. Retrieved from http://www.neurope.eu/article/mexican-hackers-attack-official-sites

Romeo, J. (2012). Hacktivism endures. *SC Magazine*, *23*(3), 24–25.

Ross, A. (1990). Hacking Away at the Counterculture. *Postmodern Culture*, *1*(1), 2. doi:10.1353/pmc.1990.0011

Sales, N. (2013). Regulating Cyber-Security. *Northwestern University Law Review*, *107*(4), 1515.

Sauter, M. (2014). *The Coming Swarm*. New York: Bloomsbury.

Schmitt, M. N. (2012). *The 'Use of Force' in Cyberspace: A Reply to Dr. Ziolkowski*. NATO CCD COEP Publications. Retrieved from http://www.google.com/url?sa=t&rct=j&q=michael%20schmitt's%20criteria&source=web&cd=3&cad=rja&ved=0CDwQFjAC&url=http%3A%2F%2Fwww.usnwc.edu%2Fgetattachment%2Fe80ed2a4-675b-4278-8321-840a2c58a27c%2FCyber-ziolkowski-response&ei=RkWgUMS6LeejiAKX0oHgBw&usg=AFQjCNEEUQ2A8ewpyeybf3vyYeyft3-7VA

Schurman, K. (2012). *LulzSec: 50 Days of Lulz*. San Francisco, CA: Hyperink.

Serracino-Inglott, P. (2013). Is It OK to an Anonymous? *Ethics & Global Politics*, *6*(4), 220. doi:10.3402/egp.v6i4.22527

Solyom, C. (2012). *Insider Tells Why Anonymous "Might Well Be The Most Powerful Organization On Earth"*. Retrieved from http://news.nationalpost.com/2012/05/12/insider-tells-why-anonymous-might-well-be-the-most-powerful-organization-on-earth/

Sterner, E. (2012, April 23). Hacktivists' Evolution Changes Cyber Security Threat Environment. *World Politics Review*.

TheAnonMessage. (2012). *Anonymous: Operation Defense Phase II* [Video file]. Retrieved from http://www.youtube.com/watch?v=Hdv5xR5YqFM&feature=plcp

TheAnonMessage. (2012). *Anonymous: Operation V (Nov. 5th 2012)* [Video file]. Retrieved from http://www.youtube.com/watch?v=jAaCZWQndqA&feature=plcp

Thoreau, H. D. (1849). *Civil Disobedience*. Available at http://www.gutenberg.org/files/71/71-h/71-h.html

Velasquez & LaRose. (2015). Social Media for Social Change: Social Media Political Efficacy and Activism in Student Activist Groups. *Journal of Broadcasting and Electronic Media, 59*(3).

Warnick, B., & Heineman, D. S. (2012). *Rhetoric Online: The Politics of New Media* (2nd ed.). New York: Peter Lang Publishing.

White House. (2011). *International Strategy for Cyberspace: Prosperity, Security, and Openness in a Networked World*. Retrieved from https://www.whitehouse.gov/sites/default/files/rss_viewer/international_strategy_for_cyberspace.pdf

Wong, W., & Brown, P. (2013). E-Bandits in Global Activism: WikiLeaks, Anonymous, and the Politics of No One. *American Political Science Association*, *11*(4), 1024.

ADDITIONAL READING

McCormick, T. (2013). Anthropology of an Idea: Hacktivism. *Foreign Policy*, *5*(200).

Ogunye, T. (2015). *Global Justice and Transnational Civil Disobedience* (Vol. 8). Ethics & Global Politics.

Smith, S. A. (2003). The Price of Dissent: Testimonies to Political Repression in America. *Labor Studies Journal*, *28*(3), 10. doi:10.1353/lab.2003.0060

Smith, W. (2012). Policing Civil Disobedience. *Political Studies*, *60*(4), 12. doi:10.1111/j.1467-9248.2011.00937.x

KEY TERMS AND DEFINITIONS

Civil Disobedience: The active, professed refusal to obey certain laws, demands, and commands of a government, or of an occupying international power. Civil disobedience is a symbolic or ritualistic violation of the law, rather than a rejection of the system as a whole. Civil disobedience is sometimes, though not always, defined as being nonviolent resistance.

Cyber Law: A term that encapsulates the legal issues related to use of the Internet. It is less a distinct field of law than intellectual property or contract law, as it is a domain covering many areas of law and regulation. Some leading topics include internet access and usage, privacy, freedom of expression, and jurisdiction. There are laws on censorship versus freedom of expression, rules on public access to government information, and individual access to information held on them by private bodies. There are laws on what data must be retained for law enforcement, and what may not be gathered or retained, for privacy reasons. A major problem of cyberlaw lies in whether to treat the Internet as if it were physical space (and thus subject to a given jurisdiction's laws) or to act as if the Internet is a world unto itself (and therefore free of such restraints).

Cyber-Terrorism: The act of Internet terrorism in terrorist activities, including acts of deliberate, large-scale disruption of computer networks, especially of personal computers attached to the Internet, by the means of tools such as computer viruses. Some authors choose a very narrow definition, relating to deployments, by known terrorist organizations, of disruption attacks against information systems for the primary purpose of creating alarm and panic. By this narrow definition, it is difficult to identify any instances of cyberterrorism. Cyberterrorism can be also defined as the intentional use of computer, networks, and public internet to cause destruction and harm for personal objectives.

Hacktivism: The subversive use of computers and computer networks to promote a political agenda. With roots in hacker culture and hacker ethics, its ends are often related to the free speech, human rights, or freedom of information movements. Due to the variety of meanings of its root words, hacktivism is sometimes ambiguous and there exists significant disagreement over the kinds of activities and purposes it encompasses. Some definitions include acts of cyberterrorism while others simply reaffirm the use of technological hacking to affect social change.

International Regimes: International processes and collection of rules, and sometimes, when formally organized, many of them can transform into intergovernmental organizations. They are, however, not actors or non-governmental organizations. International regimes often form in response to a need to coordinate behavior among countries around an issue. Powerful regimes are considered by some scholars as independent actors in international politics. Although ultimately states create and sustain regimes, once institutionalized, regimes can exert influence in world politics that is practically independent of state sovereignty.

Political Dissent: Refers to any expression designed to convey dissatisfaction with or opposition to the policies of a governing body. Such expression may take forms from vocal disagreement to civil disobedience to the use of violence. Historically, repressive governments have sought to punish political dissent. The protection of freedoms that facilitate peaceful dissent has become a hallmark of free and open societies.

Rule of Law: The legal principle that law should govern a nation, as opposed to being governed by arbitrary decisions of individualgovernment officials. It primarily refers to the influence and authority of law within society, particularly as a constraint upon behaviour, including behaviour of government officials. Rule of law implies that every citizen is subject to the law, including law makers themselves. In this sense, it stands in contrast to anautocracy, dictatorship, or oligarchy where the rulers are held above the law.

State Sovereignty: The principle of international law that each nation state has sovereignty over its territory and domestic affairs, to the exclusion of all external powers, on the principle of non-interference in another country's domestic affairs, and that each state (no matter how large or small) is equal in international law. Scholars of international relations have identified the modern, Western-originated, international system of states, multinational corporations, and organizations, as having begun at the Peace of Westphalia.[2] Both the basis and the conclusion of this view have been attacked by some revisionist academics and politicians, with revisionists questioning the significance of the Peace, and some commentators and politicians attacking the Westphalian system of sovereign nation-states.

Chapter 10
Twitter Use in Student Protests:
The Case of South Africa's #FeesMustFall Campaign

Trishana Ramluckan
University of KwaZulu-Natal, South Africa

Sayed Enayat Sayed Ally
University of KwaZulu-Natal, South Africa

Brett van Niekerk
University of KwaZulu-Natal, South Africa

ABSTRACT

In October 2015 a series of protests at South African universities against proposed fee increases escalated into a nation-wide protest with marches globally in support. Twitter proved to be a major communication and information sharing tool during these protests. The protests provided a platform to investigate the use of social media during a political crisis in South Africa. The chapter analyses the role and use of Twitter in the student protests and discusses this phenomenon from a crisis, advocacy and social information warfare perspective.

DOI: 10.4018/978-1-5225-1938-6.ch010

INTRODUCTION

South Africa is a country, not unfamiliar with a "protesting society". From the pre-apartheid era into a new democracy, protests have engulfed the South African landscape on a global scale. With the advent of democracy in South Africa since 1994, public protests focused on areas such as service delivery, against undemocratic laws or the issue of land redistribution- these forms of protests have become embedded into South Africa's democratic society. According to Reynolds (2015) in the past decade of democracy, there has been a sturdy escalation of popular protests, increased militancy that is reminiscent of the anti-apartheid regime. South Africa's political landscape has been a tumultuous one - The student protests that had brought South Africa's top universities to an academic standstill, erupted in October 2015, with what became known globally as the #FeesMustFall Campaign.

However, in recent years with the evolution of collaborative enabling technologies the mobilization of the masses for the purpose of protest has become easier than ever. There has been, as what is described, an explosion of social media and related user generated content for ordinary citizens. Journalists now utilize social media platforms for the sharing of news, events and other incidents of interest. Social media is used to the extent of effecting political change to aiding in emergency response during crises and natural disasters (Heinzelman & Waters, 2010). There have been numerous incidents supporting the role of social media in crisis and disasters. After the Japanese earthquake and tsunami, social media websites, which were accessed through smart phones, enabled individuals to reassure citizens of their safety when landline communication was impossible as a result of earthquake and tsunami damage (Gaudin, 2011).

In the aftermath of the Haiti earthquake, emergency first responders were believed to have benefited from the geo-location information provided as well as crowdsourcing information regarding outbreaks of violence. Citizens used it to locate food aid.

What had been termed the social media revolution which ultimately proved to be the app that helped overthrow the incumbent president Hosni Mubarak in Egypt in 2011, and thereafter lead to the mass mobilization of protestors across the Middle East and North Africa became known as the Arab Spring. The Arab Spring had not occurred because of social media but it remained the defining characteristic, escalating the peoples cause for a new regime. Social media afforded the masses the opportunity of using a wide-reaching collaborative platform to communicate, something that the traditional media could not previously do (van Niekerk, Pillay & Maharaj, 2011).

Kotsiopoulos, Hagen, Watson, and Kalemaki (2014) provides insight into the various platforms that facilitate the use of social media and other Web 2.0 collaborative tools in recent crisis as illustrated in Table 1.

Table 1. Stakeholders' use of social media applications

Name	Type of Application	Stakeholders Using Application	Example of Crisis Where Application Was Used
Twitter	Social Networking	• Government/Media/ • CSOs/Industry/ • Public	• Hurricane Sandy (Oct 2012) • Chile Student Protests • Hong Kong Student Protests • EuroMaidan (Dec 2013) • Arab Spring (2010/2011)
Facebook	Social Networking	• Government/Media/ • CSOs/Industry/ • Public	• Hurricane Sandy (Oct 2012) • EuroMaidan (Dec 2013) • Arab Spring (2010/2011)
Google Public Alerts	Web Tool	• Government/CSOs/ • Industry	Severe Thunderstorm Warnings (Missouri, Aug 2013)
Google Person Finder	Crowd-sourcing	• Government/Media/ • CSOs/Industry/ • Public	Japan Earthquake & Tsunami (March 2011)
Custom Google Maps	Crowd-sourcing		New Zealand Earthquake (September 2010)
Google Fusion Tables	Web Tool	Media/Industry	The Guardian –London Riots (Aug 2011)
Airbnb	Web Tool	Public	Hurricane Sandy (Oct 2012)
Reddit	Social Networking	Public	Boston Marathon Attacks (Apr 2013)
You Tube	Social Networking	• Government/Media/ • CSOs/Industry/ • Public	• Eyjafjallajokull Volcanic Eruption (Mar-Apr 2010) • EuroMaidan Protests (Dec 2013)
Ushahidi	Crowd-sourcing	Media/CSOs/Public	Crisis Mappers UK London 2012 Olympics
Ubalert	Web Tool	• Government/Media/ • CSOs/Industry/ • Public	New Zealand Earthquake (Sept 2010)
AMBER alert Europe	Crowd-sourcing	Government/CSOs	On-going

Source: Adapted from Kotsiopoulos Hagen, Watson, and Kalemaki (2014).

As in the case of Haiti and the Arab Spring, social media has demonstrated its prevalence and resilience as an effective communications tool in crisis and disasters. Table 1 provides examples of the growing use of social media platforms in numerous other crisis. Its efficiency is so much so that some of the stakeholders include governments, CSOs and the media. However, even with its proven success in a number of crisis situations, social media usage in crisis still seems to be a contentious debate – governments appear to take the cautious route when considering its potential as a crisis communications tool, the exploitation of social media communication remains a contentious issue as the areas such as informational ethics, public safety, individual liberties, and consequences of misuse must be taken into consideration (Gaudin, 2011).

Social media and mobile phones have allowed political or social concerns to be transmitted rapidly through networked communities and inter-personal online networks. This magnifies the mobilization of protests and riots, as can be seen in Arab Spring and the 2011 London riots (van Niekerk, Pillay & Maharaj, 2011; Potter, 2011). Social media has also formed a platform for advocacy, with civil society organizations running campaigns on social media. Examples of advocacy include climate change awareness by 'tcktck', reaching an estimated eleven million people in 2009, and various campaigns by Greenpeace (Pillay, 2012; van Niekerk, Pillay & Maharaj, 2011).

International Social Media Protests

In 1994, the first online protest, or "social netwar" (Ronfeldt & Arquilla, 1998) was conducted in Mexico by the Zapatista movement. The online campaign showed significant success compared with their initial physical protests which were effectively dealt with by the Mexican security forces (Mann, 2008). With the advent of social media, the online component of protests became more prevalent due to the convenient, real-time communications available on mobile platforms. Consequently, since 2008 there have been a number of protests and uprisings where social media played a part.

In 2008 protests in Greece utilized Twitter and Facebook following a police shooting, and in 2009 both Moldova and Iran experienced 'Twitter Revolutions' due to concerns of election fraud. These were followed by protests in the Chinese province of Xinjiang (van Niekerk and Maharaj, 2013; World Movement for Democracy (WMD), 2009). The protest in Moldova could be considered a 'social netwar' as they were considered leaderless and more of a popular movement (WMD), 2009). Both Iran and Xinjiang protests resulted in the governments shutting down mobile and Internet access in an attempt to curb the protests (van Niekerk and Maharaj, 2013; WMD, 2009).

At the end of 2010 and early 2011 a series of anti-government protests spread across North Africa and the Middle East, known as Arab Spring. The protests started in Tunisia, led by young unemployed graduates frustrated with perceived government corruption (Bay, 2011). Shortly after the fall of the Tunisian government similar protests erupts in Egypt who were encouraged by the apparent successful uprising in Tunisia (Hendawi, 2011). This was followed by Libya and Syria, both countries falling into civil war. Other Middle Eastern countries also experienced limited protests (Black & Chulov, 2011; Sky News, 2011). Both Facebook and Twitter provided convenient tools for protestors to co-ordinate, share information, pressure the government against which the protests were directed, and raise global awareness. In some cases the government took steps to block social media, however the protests had already gained momentum (van Niekerk, Pillay & Maharaj, 2011).

Inspired by the Arab Spring events, the hacktivist (using computer hacking for activism) group Anonymous organized a number of peaceful protests globally through social media, which became known as the Occupy Movement (Kamzi, 2011). These protests styled themselves as being leaderless

In 2011 the Chilean student movement, made a powerful impact of democratic governance in Chile, while laying out its vision for a new social construct in the South American country. The protest was considered as the most widespread protest in Chile's history and all it required was the mobilization of 320,000 followers on twitter. According to García, Chauveau, Ledezma, & Pinto (2013) the 2011-2012 Chilean Student Movement offered an excellent foundation to study the dynamics of network generation and diffusion through social media – especially that of Twitter.

The students focused on targeting their message and inviting the international media to sit at their online lunch table. They then created an all-English Twitter account @Mobilized2011 to state their side of the story without any editorial filter (Manning, 2011). The utilization of these collaborative social networks, not only formed a basis in the educational space but also formed a background and sustainable platform for the broader political space. The Twitter data obtained for the study was indicative of the relevance, interconnectivity and entanglement in both the political and media spheres - that is that the collective student based movement has infiltrated and infected both spheres by mobilizing resources, information and people, combining both the digital and physical worlds (García, Chauveau, Ledezma, & Pinto, 2013). The once infamous student protest leader Camila Vallejo, was incidentally voted into Chile's Lower House Parliament. Constantine (2013) regards this as a victory for Chilean education, and accredits this victory to effective information sharing lead mobilization.

In December 2013 protests against the pro-Russian government in Ukraine began, gaining leading to one of the leaders in the student protests being offered a position into the Chilean government) momentum in early 2014. Being able to compare against previous protests in Ukraine, the Orange Revolution, clearly illustrated the impact of social media on such protests movements (Satell, 2014). These protests, known as EuroMaidan, made extensive use of social media. Multiple Facebook pages existed in Ukrainian and Russian as well as English for an international audience. Along with Facebook, Twitter and YouTube were used to distribute information and videos of the protests and police action. Twitter usage peaked at 2500 tweets per hour (Tucker, Metzger, & Barbera, 2014). In Ukraine, the social media was used for more than co-ordination and "command and control" of the protests; support services, such as legal support for arrested protestors, food and medical supplies could be organized through a variety of Facebook pages (van Niekerk, 2015). There was also strong similarities between phrases in placards and being used online in social media (Onuch, 2014). Whilst outside the scope of this article, it is also worthwhile to note that social media was also prevalent in the military engagements, such as the Russian annexure of Crimea (van Niekerk, 2015).

In September 2014, students in Hong Kong invaded government headquarters (Civic Square) urged on by the Occupy Central Movement. Students boycotted classes to protest the selection process of the new leader of Hong Kong. The Occupy Central Movement is a politically motivated pro-democracy Hong Kong based group – and it was propelled into the spotlight through social media. Within a month of the protests approximately 1.3 million posts were placed on a number of social media websites, which brought the world's attention on Hong Kong. The protests were said to be the most topical world event (Cheung, Kao, Chan & Lau, 2014). According to Lee (2014) tweets increased to approximately 12 tweets every second after a student had shared her experience on being sprayed with tear gas and pepper spray on Twitter. With the high number on posts on Twitter and other social media platforms this movement which was known as "Occupy Hong Kong" was seen on a global platform, illustrating the viral nature of social media communications especially in a crisis such as this. Furthermore Scott Liken (as quoted in Lee, 2014) acknowledged the similarities to the organization of the Arab Spring but stated that the mobilization of Hong Kong pro-democracy demonstrators occurred rapidly through social media which was never before experienced. The Occupy Central with Love and Peace became the fastest growing leaderless social media collaborative movement in the East with approximately100, 000 supporters on Facebook and approximately 25,000 followers on Twitter.

In a key search for the popular hashtags such as "Occupy Hong Kong", more than 110,000 images appeared and due to the sheer volume of photos posted on Instagram lead to its formal banning in China on September 28 - the platform which was once considered to have been the most popular social media platform is China (Lee, 2014).

A Political and Social Backdrop to the Protests

South Africa has a long history of student protests dating back to the anti-Apartheid marches prior to 1994. Policies implemented by the Apartheid government, such as the Bantu Education Act of 1953 eventually led to events that began the student uprising on 16 June 1976, known as the Soweto Uprising. During the uprising, police opened fire on the protesting students with live ammunition, and the images from this resulted in international condemnation and started liberation movements that were in exile (South African History Online, 2016). However, since the inception of a new democratic regime lead by an ANC government students still maintain their struggles. In recent years the number of student protests has increased both in veracity and frequency (Feni & Mama, 2015).

The reasons for the protests in October 2015 began due to the proposed fee increases, but these protests were considered to be a result of the culmination of other areas of discontent which included better student accommodation, efficient transportation services and the academic non-exclusion of students. At tertiary institutions such as University of Cape Town (UCT), Stellenbosch University, and Cape Peninsula University of Technology (CPUT) students demonstrated against the slow transformation process, the Afrikaans language policy as well as other academic policies.

Students at the University of Cape Town were disgruntled with the Cecil John Rhodes' statue in 2015 and protested for its removal- they viewed the non - removal of the statue as the slow transformation phase at the university. This lead to the creation of the "Rhodes must fall "campaign which resulted in social media propelling this campaign on a global stage and the ultimate removal of the much debated statue on 9 April 2015.

Protests at the University of the Witwatersrand began due to the perceived race and class discrimination on October 14. The students organized a mass demonstration against what they deemed to be exorbitant fee increases. The Twitter posts went viral across South African university students and began a nationwide student campaign which featured hashtags like #FeesMustFall and #Nationalshutdown. The Twitter platform helped student protestors collectively organize and share information throughout South Africa, also bring global attention to the crisis (Beauchamp, 2015). The protests did take a violent turn with students damaging property and inciting acts of violence. A recent analysis by the South African Institute of Race Relations states that the underlying reason for youth protests begins with the high level of youth unemployment and leads to slow economic growth and the blame rests with the government.

An example of some of the economic tensions is illustrated in Figure 1, which began with the Nkandla debate in 2012. The debate revolved around the public's money that was spent on the private residence of President Jacob Zuma's. This was compared to the status of student fees in South Africa as shown in Figure 1. This also included the comparison in costs to the South African e-Toll initiatives, the South African Airlines Bailout, and the proposed nuclear deal.

Methodology

The study aims to assess the usage of Twitter and social media during the protest. Secondary data in the form of news reports and actual tweets and images from the

Figure 1. Post regarding perceived wasteful expenditure

Nkandla (R246m) = 2'000 Free 2 Year Degrees
SAPostOffice 2014-2015 Loss (R1.37bn) = 11'117 Free 3 Year BSc Degrees
3xVIP Jets (R2bn) = 16'227 Free 3 Year BSc Degrees
SAA Jan '15 Bailout (R6.5bn) = 57'7241 Free 3 Year BSc Degrees
eTollProject (R10bn) = 81'301 Free 3 Year BSc Degrees
PetroSALosses (R14.5bn) = 781'706 Free 3 Year BSc Degrees
NuclearDeal (R1.2trn) = 64.7 Million Free 3 Year BSc Degrees

Just putting it out there....

protest, including the wording of placards that appeared in the images. This provides contextual meaning to the analysis, as well as the ability to assess the type of role that Twitter and social media platforms played in the protest.

Hashtags and data relating to the hashtags are key to the analysis. Two independent efforts were made to retrieve this data. The first was collecting data from Followthehash.com. The portal provided the hashtag frequency or usage, the rate of the tweets, and the estimated audience and reach. The number of tweets were constrained to 1500 per query of the platform due to API limitations. Due to the number of tweets, the 1500 can be considered a sample of the tweets received by the query. Multiple queries were made to maximize the sample. According to Sekaran (1992), a sample of 384 is required for a population of 1 million. Given that the sample of tweets received exceeds 24 000 for the main period of protest, the sample can be considered sufficient for the purpose of this study. The geographic location of hashtags related to #feesmustfall were accessed using Trendsmap, and trends of search terms relating to protests were accessed using Google Trends.

The second set of mining the tweets took place in 2015 between the 25th October and 2nd December by means of a custom R script, which was used throughout the process to download tweets using the Twitter API. The script was executed daily collecting approximately 1500 tweets per day using the #feesmustfall hashtag only, and was stored in CSV format. However there were instances such as on Sundays towards the end of November, were the tweet count was not greater than 1500.

Once the mining of tweets was complete, the collection of CSV files was compiled into a single Excel file. During this process, cleaning of the data was deemed necessary. Some of the issues that were addressed were aspects such as special characters not appearing correctly, removal of rows with duplicate IDs and the separation of Retweeted tweets from non-Retweeted tweets. In instances where the text field was not legible due to special characters, the tweet was removed. This process took the longest, as each correction involved writing out scripts and then verifying that incorrect data was indeed removed.

Using the list of non-Retweeted tweets, the analysis was performed primarily in R. When using twitter data, the typical types of analysis that can be carried out include sentiment analysis or even cartographic data visualizations using latitude/ longitude data from a tweet. However the analysis chosen in this project focused on the hashtags that users adopted in their tweets, with the exception "feesmustfall". The reasons for not adopting sentiment analysis was that firstly the researcher found that the vocabulary used in twitter by users included slang words and non-English words/phrases, for example: "LOL khaw'yeke abanye abantwana… "@Mila_Landani: Emva kwalo #FeesMustFall naba serious kakhulu:/" (Anon, 2015), which is not typically found in freely available lexicons used in sentiment analysis. Secondly, the entire #feesmustfall movement was formed from mainly disgruntled students and outsourced workers and it was assumed that the sentiment would seemingly be negative. A cartographic data visualization was not possible also as the majority of tweets did not contain latitude/longitude data.

The chosen analysis method focused in on hashtags, and predominantly the top 10 hashtags per day. Using various packages in RStudio, the entire list of tweets was filtered for hashtags and sorted by date. The final list comprised of the top 10 hashtags per day accompanied by their frequency. This was then visualised using the online tool Tableau.

Limitations

Whilst this form analysis provides an interesting overview into what events were receiving the most attention via twitter, it does not provide the analysis as one would find in a quantitative study. This form of qualitative research was deemed appropriate as much of the data generated was primarily opinions of many individuals. As such, the visualization created from this data provides a good perspective into events that took place during the movement and received higher amounts of attention on Twitter.

Twitter Use during the Protests

During the South African student protests, Twitter seems to have been the preferred method of communication with the number of tweets reaching up to 300 per minute (Bath, 2015).

Analysis Using R and Tableu

The chosen analysis for the data collected was in the form of a data visualization. This data visualization can be found on Tableau Public (Sayed Ally, 2015). The amount of tweets after cleaning the data ended up at approximately 14,400. The use of a Bubble map was adopted, wherein each hashtag was represented by a bubble, the size and colour of the bubble represented the hashtag frequency dimension. Furthermore, certain tweets were included into bubbles where possible. This is illustrates in Figure 2

The data visualization created, provides visual summary of the top trending hashtags within the #feesmustfall trend by means of a bubble chart (see Figure 2). Each Tableau card located at the top of the visualization provides a means to navigate through the visualization. Apart from the frequencies of each hashtag, the use of actual tweets from the data are included in the visualization (see Figure 3). These provide the reader with an example of the popular statements been made on the Twitter platform and some insight for the associated hashtag.

Figure 2. #feesmustfall Data Visualisation
Source: Sayed Ally, 2015.

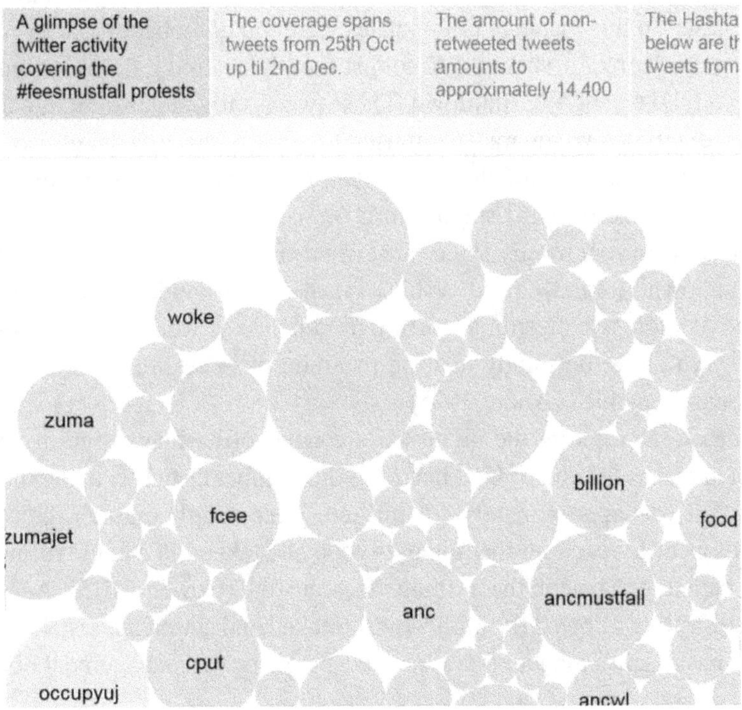

Figure 3. Tweets illustrating hashtag use
Source: Sayed Ally, 2015.

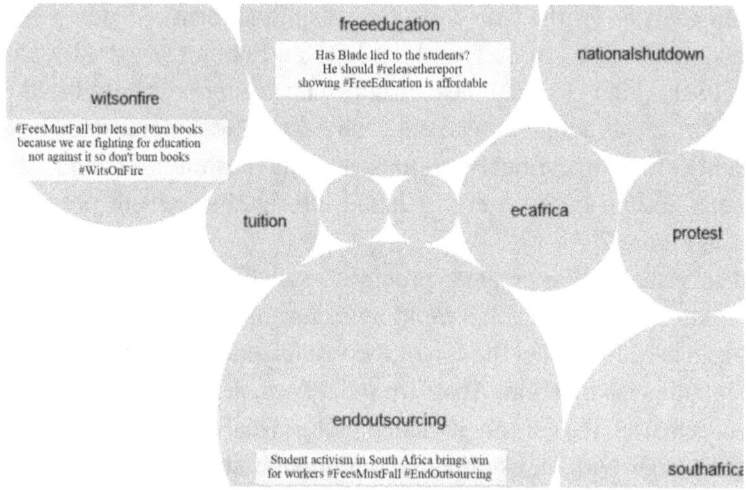

The #feesmustfall hashtag came in at the highest frequency, whilst the "endoutsourcing" hashtag came in at second highest. This hashtag stemmed from the outsourced workers that were also protesting as a result of their working conditions. It is unsurprising that the #endoutsourcing hashtag was used throughout the duration of the protests. In this instance both parties were mutually beneficial to each other with a certain degree of success. Amongst the high ranking hashtag frequencies, were #occupyUJ and #UWCshutdown. These tweets provided insight into the events taking place at these two separate campuses. One use of these hashtags was to spread information immediately to fellow students. For example the following tweet provides information to students staying in UWC residence: "Please retweet for your own safety. Stay in your rooms. Police are armed #Feesmustfall #UWCFeesMustFall #UWCshutdown" (Anon, 2015). Whilst on the other hand, these same hashtags were used to address protestors resorting to violence as well: "To those protesting at UWC… you have definitely lost the meaning of #FeesMustFall. The violence shown is unacceptable" (Anon, 2015).

Staying on the topic of the highest frequencies of hashtags used, it was previously stated that at the end of October 2015 the number of tweets using the hashtag #feesmustfall was approximately 1.5 million. Interestingly enough the university authorities prior to reaching that many tweets, had taken out a court ruling that effectively called for a ban on the hashtag #feesmustfall (Cowen, 2015). According to reports, this ruling allowed the authorities to take legal action against Twitter users using this hashtag in a manner that is not acceptable by the management bodies. The fact that the amount of tweets using the #feesmustfall ran into the millions, proves that the court ruling was a failure. This seemingly defiant use of the #feesmustfall hashtag leads one to believe that the government can do little to effectively stop the use of online social platforms in such uprisings. This also shows a lack of understanding or clarity of how laws apply to social media. The hashtag is an idea or construct which is used to relate or the content of a tweet with that idea; therefore the attempt to ban the hashtag is effectively attempting to ban an idea. This idea can also spread to areas where the South African legal system or the universities involved have no jurisdiction. The sheer volumes of tweets would make it virtually impossible to enforce the court ruling.

Analysis Using Open Access Twitter Tools

Figure 4 shows online statistics revealing the total number of tweets for the third day which was the 23rd of October 2015 in particular were of the highest number peaking at close to 16 000, with the student march to the Union Buildings in Pretoria (Tshwane)- the administrative capital of South Africa. Figures 1 and 2 included other contentious Tweets including the #Witsonfire on the 28th of October 2015 (The Witwatersrand University being set on fire by protesting students) which followed

Figure 4. Tweets per day for selected search terms
Data source: http://www.followthehashtag.com.

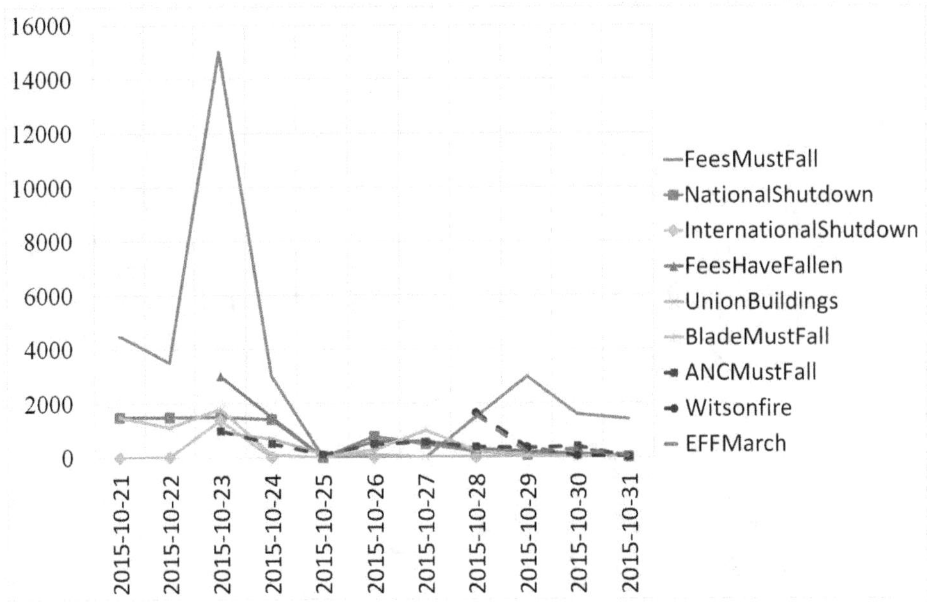

on from the #Feesmustfall to illustrate the comparison between the various issues trending on twitter.

Figure 5 depicts the tweets referring to the national shut down, to the EFF march (the Economic Freedom Fighters are a political party in South Africa, who rallied their supporters in a political freedom march in Johannesburg, South Africa on the 27th October 2015), excluding the #FeesMustFall. What is evident from Figure 5 is that even without the #Feesmustfall, the #Feeshavefallen, #Blademustfall, #Unionbuildings, #Ancmustfall and #Nationalshutdown began declining in the number of Tweets on the 23rd of October 2015 after the no increase statement was released outside the Union Buildings and declined further to approximately 100 Tweets per day on the 25th of October 2015. However, from the 26th of October 2015, there appeared to be a resurgence of the number of Tweets per day with reference to the #Nationalshutdown, #Feeshavefallen, #Ancmustfall and #Blademustfall campaigns, whilst the #Unionbuildings continued to decrease in the number of Tweets per day.

Table 2 is reflective of the various hashtag campaigns for the period 21-23 October 2015. These included the #Feesmustfall, #Nationalshutdown, #Internationalshutdown, #Feeshavefallen, #Unionbuildings, #Blademustfall and the #Ancmustfall. The highest number of Tweets for this period revolved around the #Feesmustfall with the total number being 24498 in the three day period. The campaign with the lowest number of Tweets was the #Internationalshutdown with a recorded 1398 Tweets, but over a one day period.

Figure 5. Tweets per day for selected search terms, without FeesMustFall
Data source: http://www.followthehashtag.com.

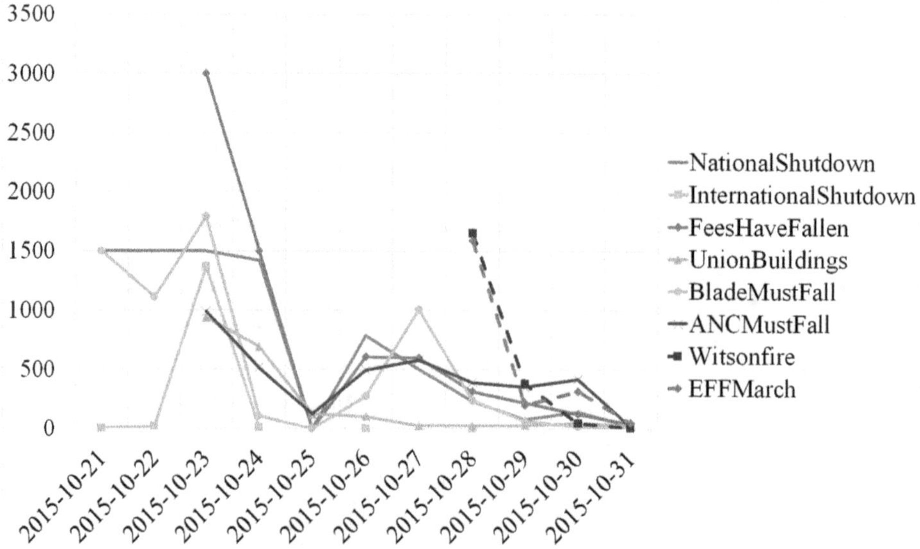

Table 2. Twitter usage samples for the period 21-23 October 2015

Keyword	#FeesMustFall	#National Shutdown	#International Shutdown	#FeesHave Fallen	#Union Buildings	#BladeMustFall	#AncMustFall
Total tweets	24498	5919	1398	4500	1482	4510	1500
Tweets/hour	343.8	85.1	20.7	186.0	75.1	60.8	71.3
Total audience	38 756 918	8 626 097	3 548 252	8 309 060	2 689 259	5 680 524	2 950 913
Contributors	13919	3992	1212	3348	1063	3314	900
% original	29.68	16.73	9.87	30.16	18.89	42.77	26.33
% replies	1.87	0.95	0.50	1.22	0.20	3.95	5.6
% RTs	68.44	82.33	89.63	68.62	80.90	53.28	68.07
% contain images/ links	32.93	44.57	84.33	30.24	28.48	26.87	29.93
Total potential impressions	105399710	13053787	4347189	14769021	3657178	8315689	4048040
Measured time in days	3	3	3	1	1	3	1

Source: http://www.followthehashtag.com.

Table 2 illustrates that the original Tweets decreased, whilst the Retweets increased overall compared to the first three days. The #AncMustFall Retweets decreased and the original Tweets & replies increased– demonstrating a staying power or momentum of the campaign. This also led to increased dialogue over the ANC in government. The localized themes including #Feeshavefallen and #Unionbuildings did not gain momentum as much as the main themes did including the #FeesMustFall and #Nationalshutdown. As compared to Table 3 and Table 4, Table 2 is reflective of the various campaigns over a three day period whilst Table 3 and Table 4 are reflective of the number of Tweets per campaign over a ten day period. The #Feesmustfall increased even further over the ten day period (21-31 October 2015), as shown in Table 3, two other linked campaigns that is the #Blademustfall and the #Ancmustfall gained momentum even after the fees had fallen.

Table 3 is reflective of the same campaigns but over a ten day period. The campaigns remained on a steady increase as per the number of total Tweets. As for the percentage of original Tweets the #Blademustfall and the #Ancmustfall remained at the highest total percentages over a ten and eight day period respectively.

Table 4 illustrates the percentage of sampled Tweets in the first three days of the student protests (21st -23rd October 2015). The highest percentage of Tweets for this period was the calling for #Internationalshutdown, #Nationalshutdown and the #Unionbuildings. The total percentage of the #Blademustfall remained higher than

Table 3. Twitter usage samples for the period 21-31 October 2015

Keyword	#Feesmustfall	#National Shutdown	#International Shutdown	#FeesHave Fallen	#Union Buildings	#BladeMustFall	#AncMustFall
Total tweets	33522	7609	1402	6406	1994	6122	3923
Tweets per hour	138.2	31.8	8.6	33.2	10.5	25.2	20.6
Total audience	54135425	13258122	3549570	14743451	3544261	7362142	4581682
Contributors	17573	4702	1216	4683	1411	3925	1898
% original	30.52	22.01	9.99	34.41	22.47	38.63	34.49
% replies	2.17	1.08	0.50	1.47	0.45	3.90	8.16
% RTs	67.31	76.91	89.51	64.13	77.08	57.46	57.35
% images/ links	31.87	43.71	84.17	30.21	32.35	22.77	19.53
Total potential impressions	181795479	21471474	4348467	29698420	4820061	22902608	8818825
Measured time in days	10	10	7	8	8	10	8

Source: http://www.followthehashtag.com.

Table 4. Percentage of sampled tweets in the first three days

Keyword	#FeesMustFall	#National Shutdown	#International Shutdown	#FeesHave Fallen	#Union Buildings	#BladeMustFall	#AncMustFall
% total tweets	73.08	77.79	99.71	70.25	74.32	73.67	38.24
% audience	71.59	65.06	99.96	56.36	75.88	77.16	64.41
% original tweets	71.09	59.109	98.579	61.57	62.5	81.56	29.19
% replies	63.14	68.29	100	58.51	33.33	74.48	26.25
% RTs	74.31	83.27	99.84	75.17	78.01	68.31	45.38
% images and links	75.51	79.31	99.92	70.34	65.43	86.94	58.62

Source: http://www.followthehashtag.com.

that of the #Feesmustfall and the #Feeshavefallen campaigns, possibly reflecting a politically motivated undertone to the student protests.

What began as student protests with students exercising their right to "freedom of expression" as provisioned for by the South African Constitution to the proposed fee increases in South African Public Universities evolved into a political stance rallying for the ousting of key political leaders and the current South African government. As explained by a study conducted by Dynamix in 2011 it was discovered that from the 1800 South African youth interviewed, the general consensus was that the majority of young South Africans (89%) between the ages of 16-24, had lost faith in the ANC government- believing that the government had failed in its promises of free education (Naidoo, 2011). Furthermore it may be seen that this may have been an influencing consideration with the #Feesmustfall campaign which progressed into an anti-government drive with a group of citizens calling for the removal of the ruling party. The hashtag #ANCMustFall has gained momentum on social media feeds after tensions heightened between students and government.

Geographic location of tweets is useful in illustrating the viral nature and reach that protests with an online presence can have. Figure 6 is representative of the spread of tweets in a South African National context to an African context including countries across borders like Zimbabwe, Malawi and Madagascar. The extent of the international support gained by the protests is illustrated in Figure 7, with the hashtag spreading to Australia, North and South America, Europe and the East. Again the most prevalent Twitter campaign remained to be specifically aligned to the #Feesmustfall campaign. The screenshots of Trendsmap shown in Figures 6 and 7 illustrate the viral nature of the protest hashtags through the use of a social media platform like Twitter. From a national context, it spread into an African context as illustrated in Figure 6. The hashtags of #Feesmustfall, #Unionbuildings and #Nationalshutdown were contained within a national context, while the related Tweets that emerged globally were more aligned to the #Feesmustfall as illustrated

Figure 6. TrendsMap of related tweets in southern Africa
Source: Trendsmap.com, used with permission.

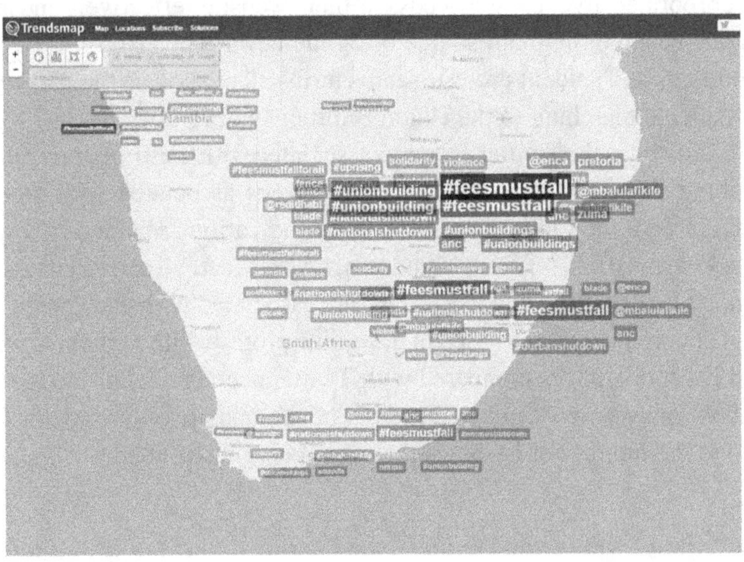

Figure 7. TrendsMap of related tweets showing international support
Source: Trendsmap.com, used with permission.

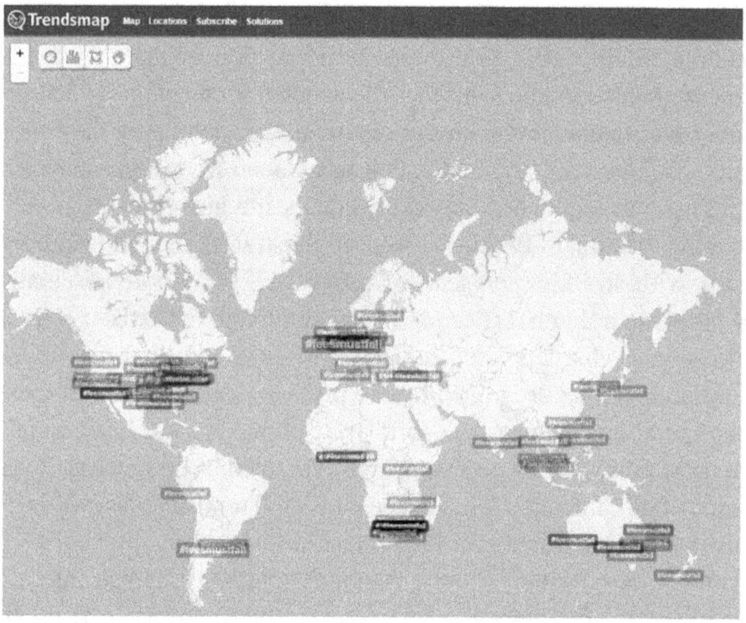

in Figure 7. The #Feesmustfall was a common theme for the solidarity of students from North America to the Australian continent.

Figure 8 illustrates the popular search terms over the period of the protests. The eNCA news channel covering the protests was the most popular search item. The eNCA incorporated live Twitter feeds on their website, effectively meaning that social media became the main source of online news. The eNCA, #Feesmustfall, Union buildings and student protests search terms all peaked on the 23rd of October with the students marching to the Union Buildings in Pretoria.

Table 5 illustrates the Twitter accounts associated with the #Feesmustfall protests whilst Table 6 is reflective of the Facebook pages associated with the protests. Table 7 represents the top hashtags related to the various search terms which included the #feesmustfall, #nationalshutdown, #ANCMustFall and #Endoutsourcing. What is notable is that the major hashtags are associated with each other, showing a small set of common phrases used in the online protest. Although there were more dedicated Facebook pages compared with Twitter accounts relates to the protests, the Facebook pages were limited in terms of the audience views with few likes compared to the number of hashtags that were being distributed.

Figure 8. Google search trends for the protest
Source: http://trends.google.com.

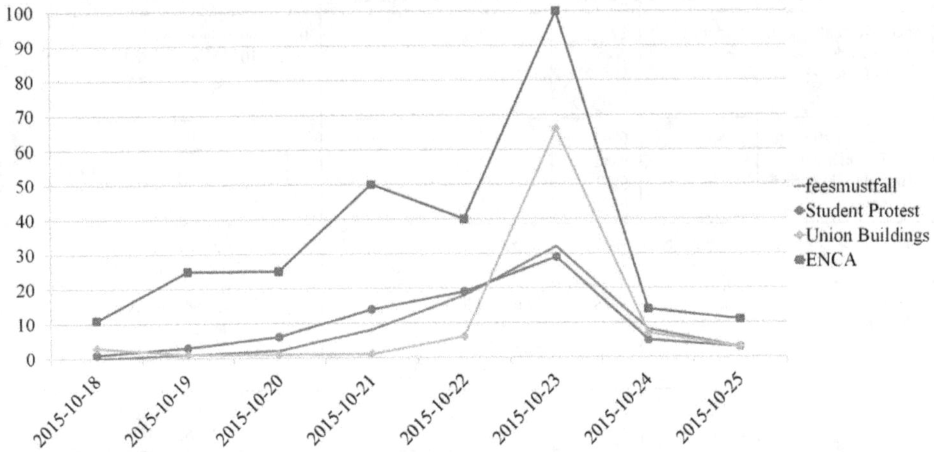

Table 5. Twitter accounts associated with the #FeesMustFall protests

	FeesMustFall	#FeesMustFall	FeesMustFall News	National Shutdown
Handle	@Fees_Must_Fall	@FeesWillFall	@FeesMustFallZa	@FeesMustFall15
Tweets	369	49	4	23
Following	154	14	1	65
Followers	318	10	22	20
Favourites	14			14
URL	https://twitter.com/ Fees_Must_Fall	https://twitter.com/ FeesWillFall	https://twitter.com/ FeesMustFallZa	https://twitter.com/ FeesMustFall15

With the situation in South Africa becoming more contentious, fellow British students in London gathered in the capital to voice their frustrations at plans to scrap student grants and what they say is the Government's concerted attack on free education. Approximately 10 000 students marched through the city streets, with the rally having been centred on the department for Business and Skills. The Metropolitan Police battled to control the carnage left behind rom three separate student protests in the capital alone. The protestors demanded free education. The Labour Party leader Jeremy Corbyn supported the protesters by demanding the abolition of tuition fees in a statement read out at the gathering (Morgan, 2015), whilst the Duell and Glanfield (2015) reported that John McDonnell (a member of the British Labour Party) was greeted with cheers as he climbed up a small platform to address the students through a megaphone. McDonnell was quoted as saying: "Your gen-

Table 6. Facebook pages associated with the #FeesMustFall protest

Page Name	Type	Stats	Associated Webpage	URL
#FeesMustFall #NationalShutdown Solidarity Night	Event (24 Oct)	17 going, (157 invited)		https://www.facebook.com/events/1015439878478842/
#NationalShutDown #UCTShutDown #HiddinghShutDown	Event	634 went (629 invited)		
Nation Feesmustfall	Community Organisation	3940	http://www.nationfeesmustfall.co.za/	https://www.facebook.com/Nation-Feesmustfall-577615595727127/
FeesMustFall News – FMFN	Community	3928	https://twitter.com/FeesMustFallZa https://candystick.co.za/c/donate-for-damaged-goods/	https://www.facebook.com/FeesMustFallNews
#FeesMustFall	Public Group	54 members		https://www.facebook.com/groups/847399868690916/
#FeesMustFall	Event (23 Oct)	934 went (3.9k invited)		https://www.facebook.com/events/1149594975070392/
Feesmustfall	Cause	62 Likes	http://www.feesmustfall.co.za	https://www.facebook.com/Feesmustfall-737325626401062/
Feesmustfall	Community Organisation	466 Likes	https://twitter.com/FeesWillFall	https://www.facebook.com/Feesmustfall-1517805708531923/

eration has been betrayed by this Government in increases to tuition fees, in scrapping the education maintenance allowance and cuts in education. This Government is betraying you and future generations. You need to oppose it and I'm here in solidarity with that opposition." With this message rampaging rioters took over the city streets bringing transport to a complete standstill.

On November 4, 2015 students across London mobilized to march for "free education and cuts in university tuition fees" (Taylor, 2015). The London protests had some similarities to that of the South African student protests in October 2015, in that the protests became political with students obtaining the endorsement of the Labour Party Leader Jeremy Corbyn (Taylor, 2015). It may be argued that the international sympathy protests for the #Feesmustfall campaign encouraged the London protests, and the #FreeEducation hashtag is present in both sets of protests. Table 8 represents the number of tweets for the London protests. The tweets during the London protests contained a higher percentage of images as compared to those during the South African protests. This could imply a greater degree of comfort of the London users, as there have been previous social media led protests in the country. Data costs may also play a part.

Table 7. Top hashtags for various search terms

FeesMustFall	National Shutdown	International Shutdown	FeesHave Fallen	Union Buildings	BladeMustFall	AncMustFall
FeesMustFall (33772)	NationalShutDown (7612)	InternationalShutdown (1402)	FeesHaveFallen (6508)	UnionBuildings (1993)	BladeMustFall (6437)	ANCMustFall (6003)
NationalShutDown (3313)	FeesMustFall (4241)	FeesMustFall (1297)	FeesMustFall (2003)	FeesMustFall (821)	FeesMustFall (2512)	FeesMustFall (891)
UnionBuildings (1859)	UnionBuildings (304)	NationalShutDown (67)	UnionBuildings (362)	FeesHaveFallen (346)	ANCMustFall (687)	ZumaMustFall (390)
EndOutsourcing (1333)	uctshutdown (199)	UnionBuildings (47)	UPrising (113)	UniversityFees (126)	NationalShutDown (488)	BladeMustFall (175)
FeesHaveFallen (693)	FeesMustFallForAll (167)	WitsFeesMustFall (20)	FillUpTheDome (95)	TUTstudents (88)	ZumaMustFall (369)	ANCWLmarch (145)
FeesMustFallForAll (285)	EndOutsourcing (175)	FeesMustFallForAll (19)	FreeEducationForAll (86)	NationalShutDown (80)	StudentsMustFall (298)	SouthAfrica (113)
FreeEducation (257)	FeesHaveFallen (131)	Asijiki (13)	Leadership (80)	habibmustfall (42)	FeesMustFallForAll (276)	ANC (103)
ANCMustFall (249)	FreeEducation (128)	FillUpUnionBuilding (12)	SouthAfrica (64)	UPrising (39)	FeesHaveFallen (123)	FeesHaveFallen (86)
Sabcnews (225)	Uwc (108)	FeesWillFall (10)	TUTstudents (61)	ZumaMustFall (39)	Blade (86)	EFFMarch (75)
Wits (220)	UPrising (82)	AddressUs (10)	FreeEducation (50)	Pretoria (24)	sabcnews (82)	zuma (72)

Table 8. Twitter usage samples for the London student protests, 4 November

Keyword	#FreeEducation	#StudentMarch	#GrantsNotDebts
Total tweets	1500	934	218
Tweets per hour	75.50336	4.313673	1.017006
Total audience	4046916	5089966	517630
Contributors	1182	655	174
% original	13.4	33.40471092	46.78899083
% replies	1.4	1.713062099	2.293577982
% RTs	85.2	64.88222698	50.91743119
% images/links	66.06666667	50.64239829	43.57798165
Total potential impressions	5060698	5692468	735664
Measured time in days	1	9	9

Source: http://www.followthehashtag.com.

Analysis of Posts and Online News Articles

To gain insights into a protest with a strong online presence, the content of the tweets themselves, and the images that are tweeted or noticed by online news media and writers often have a profound impact on the spirit of the protests. One such placard, shown in Figure 9 (top right) illustrated the sense of betrayal felt by students, stating that were there to collect a refund from the promises the ANC made when being voted into government during the first democratic elections in South Africa in 1994. The poster implies what is perceived as empty promises by the ANC government, which aligns to other concerns of wastage and corruption, as indicated in Figure 1. Images of students carrying placards with the #FeesMustFall, #EndOutsourcing, and #FreeEducation hashtags (an example is shown bottom right of Figure 20), amongst others, implies very strong alignment on physical and online protest action. The #Endoutsourcing occurred due to public universities outsourcing services to external companies, which may have led to the perception that funds were being spent outside the universities. Many of those impacted by the outsourcing joined the students, and both groups mutually supported each other. Not only do these hashtags and placards deliver the demand, but they also provide an associative term, giving protestors a sense of belonging. Examples of these are provided by Cassim (2015), Furlong (2015), Geach (2015), Makhati (2015), and Mkwanzi (2015).

Despite the focus on the key hashtags and themes of the protests, social media was used to provide support services to the protestors, in the form of medical advice (see top left of Figure 9), contacts for free legal advice, supplies and tutoring for protesting students (Cassim, 2015). These examples illustrate that social media was

Figure 9. Examples of tweeted images
Sources: Brady (2015), Cassim (2015), Geach (2015), Mkwanzi (2015).

used in a supporting capacity in addition to the information distribution and coordination roles. Instances of good will were also publicised through social media, such as the numerous clean-up efforts, and various religions protecting each other during prayers (Cassim, 2015).

With the onset of the protests and statements that there was a student funding crisis, South African Minister of Higher Education, Blade Nzimande responded by denying that there was a crisis in the South African education system. According to the best practice principles of effective crisis management, one of the main principles is "never deny a crisis" (Reynolds, 2015). An image of a student placard reading "is this still not a crisis blade?" was tweeted by (@RefilweI), shown in Figure 20, illustrates the negative reaction of the protestors to the denial of the perceived crisis. As the protests were increasing in intensity and students marched on the Parliament complex in Cape Town, Minister Blade Nzimande angered South Africans after stating in jest that his department would start a "students must fall" campaign if students don't accept his proposed 6% university fee increase cap (Manete, 2015). Social media was key to this as it was recorded and quickly went viral online. There was strong reactions online and in the protests to this comment, as illustrated in Figure 20. According to Reynolds (2015) a key principle for effective crisis management is the demonstration of mutual respect and open dialogue between the parties. From the comments and reaction by both the minister and the protesting students, this principle was not considered. Combining the crisis denial and the "students must fall" comment, both online and physical protests began using the hashtag #Blademustfall in response. These incidents may have also contributed to the increasing political nature of the protests, which saw the introduction of the #ANCmustfall hashtag and mass marches on the Union Buildings in Pretoria, when President Zuma was meeting with student leadership and university leaders. In comparison to the student protests in Chile and Hong Kong which were purely political, the #Feesmustfall campaign became political in reaction to the government responses.

According to radio host Redi Tlhabi at Radio 702, as quoted by Wesi (2015), there seemed to have been two splits in the protests. Redi's tweet implied that the student protests may have been "hijacked" by a smaller political group wearing political t-shirts and who were actually responsible for the damage to university property.

There are allegations of a deliberate sabotage of the Fees Must Fall march Mlambo (2015) provides background to the various 'leaders' of the protest from the University of Cape Town, two from the University of the Witwatersrand, and one from the University of KwaZulu-Natal who was in prison during the FeesMustFall campaign due to his actions in a previous set of protests. All four of these leaders are members of the students' representative councils on their respective campuses. An article in the Business Week (Staff Writer, 2015) states that the protests were led by a group of individuals but that the students' #FeesMustFall protest at the Union Buildings in Pretoria was "hijacked" by political leaders – notably the ANCYL and SASCO and the EFF wearing political t-shirts – who co-opted the movement and caused chaos and violence to a peaceful protest.

Inevitably the widespread attention that the entire movement received attracted various members of public to the twitter platform, this included users that were using the #feesmustfall hashtag for other reasons. Some Twitter users used the #feesmustfall hashtag in an effort to bring attention to their own causes. One example encountered whilst creating the data visualization on Tableau was that of a media agency using the #feesmustfall hashtag to highlight headlines from their website which had no relation to the entire movement, these instances were removed from the data visualization. This was not an isolated case as many other individuals and organisations followed similar actions. There were even cases when political parties used this hashtag in a similar manner to bring attention to their own movements. The publicity of using this hashtag was effective, as some of the political party names appear in the top 10 hashtags on certain days which can be viewed in the Data Visualisation (see Figure 2).

Ultimately the use of social media and Twitter as a form of protest provides a highly accessible platform to the public with ability to rapidly disseminate information from most places. It is highly effective and has been proven at a nationwide and even at a regional level in the past (O'Donnell, 2011). However, these platforms can be monitored by anyone with an internet connection and in that sense creates a platform for all parties that adopt it without exclusivity. From a monitoring point of view, this type of analysis involves a great amount of data cleansing and this is a challenge, to reach a point where data could be transformed into a form that allowed analysis. The amount of effort it takes to reach this point may seem prohibitive and even tedious, however the knowledge derived from this type of analysis is certainly valuable. It is amongst some of the reasons that authorities around the world are adopting this data source as a source for intelligence despite the effort of the entire process (Kopstein, 2013; Williams, 2011).

The Protests as a Crisis

According to Fearn-Banks (2007) a crisis is defined as a disruption in the "economic, cultural, social and political lifecycle." Crises often increase the role of communication creating more complex information environments. Furthermore it results in events which can threat the strategic goals of an organisation (Fearn-Banks 2007).

From the definition above, protests that cause an interruption in the the reproduction of economic, cultural, social and political life can be deemed a crisis. On an international scale, the 2016 protests in Macedonia provided an example of protest as a form of crisis. The streets of Skopje were filled with both protestors and riot police, all concerning the election dates. The protestors were calling for the resignation of Macedonian President Gjorge Ivanov. The actual build up to the protests began in 2015, when the then Prime Minister Nicolas Gruevski was accused of wiretapping approximately 20 000 people – including journalists and politicians trying to inhibit or control the freedom of the media (BBC, 2015). In a similar context protests were staged in Venezuela calling for the Ousting of President Nikola Maduro. The opposition party is said to promote these mass protests as a strategy to oust Maduro by recalling him through a constitutional referendum to shorten his term in office (BBC.Com, 2015).

In an African context, Nyamnjoh, Nkwi and Konings (2007) contend that African universities face a far deepening crisis: disgruntled African students have shown growing activism and militancy. They have become accustomed and engage in numerous, often violent strikes for the improvement of their deteriorating living and education conditions. Students often form antagonistic views to the university management and the government- as they are perceive that they merely exerting their democratic liberties.

In South Africa, the protests gained momentum after the South African Minister of Higher Education Blade Nzimande announced that the state of higher education in South Africa was not experiencing a crisis but rather a challenge (Presence, 2015). The placard shown in Figure 10 with the question "Is this still not a crisis Blade?" illustrates that the students did not agree with the statement by the Minister. One of the key points of effective crisis management is never to deny a crisis. Blade Nzimande firstly denied the education crisis, then continued by remarking that the government should begin the #Studentsmustfall campaign. This statement again goes against another key element for effective crisis management which is to empathize with the aggrieved party. This two comments potentially contributed to the escalating protests.

The South African student protests began an unprecedented moment in the history of the country. Thousands of students brought universities to a halt, and took their concerns to the political capital of South Africa on October 23, 2016 - students from across Pretoria and Johannesburg mobilised outside the Union Buildings,

Figure 10. Students reaction to Blade Nzimande's #Studentsmustfall comment
Sources: Hirayama (2015) and @ Refilwel (2015).

requesting a meeting with President Jacob Zuma. Even though the terms were later accepted and it was announced that there would be no fee increase for 2016, there were still skirmishes with the police in Pretoria, and many across the country continued to protest well into the night.

The similar political undertones used in the London protests has another implication: that political crises can spread amongst countries through the use of social media. The dissatisfaction surrounding student fees in South Africa and London both ultimately led to political crises as shown in the case of the Arab Spring, where political protests spread from country to country. What normally began as student protests formed part of a larger political agenda. The #AncMustFall dialogue appeared to have increased in the second week, indicating that the hashtag had more 'staying power' or 'momentum', even though it was not the primary hashtag. This "staying power" persisted despite the fact that the government had stated that fees would remain the same, with no increase. The reason for the staying power for the #feesmustfall campaign could well be the mismanagement of the initial crisis- by the Minister's denial of the higher education crisis followed by his #Studentsmustfall comment. For the duration of the protests, social media was used to organise and disseminate messages. Twitter was the main social media platform utilized for various purposes which first and foremost was used to mobilise the protestors but also to provide "aftermath" assistance e.g. medical supplies and legal counsel for students who were arrested. According to Thomas (2015) an important characteristic worth noting was that without the instant connectivity provided by social media, the #FeesMustFall campaign would not have spread as quickly as it did. Initially when protests began at Rhodes University social media, and in particular Twitter, provided the protesters and student publications a national and international platform in reporting and disseminating information on the protests whilst local journalists were still attempting to get to the scene.

Throughout the student protests social media proved to be a powerful tool, not only used by the students to mobilise and coordinate the protests but by the general public as a fast accurate source of information into the situation. Early signs of logistical support to sustain the protests began emerging during the South African protests. The Ukrainian protests in 2014 provide a similar example of the power of social media (Gruzd, 2015), however the logistical support appeared more advanced, possibly due to the protests being more protracted. Had the South African government not prevented fee increases, the social media usage was indicating the students preparing for protracted protests, which may have been disastrous for the government as the political crisis escalated.

The Protests as Advocacy and Social Netwar

The social netwar aspect of the FeesMustFall campaign can be seen in that the protests were intended to be leaderless. The #Occupy hashtag was used in some cases, indicating an affinity to the Anonymous-style of open and equal membership. There was a mixed reaction when a student leader from the University of Witwatersrand appeared on the cover of a magazine; some believed that it was unfair that a single person be recognised for a collective movement (Ncokazi, 2015). Other students recognised the importance of the magazine cover (Ncokazi, 2015), and the article by (Mlambo, 2015) does indicate some leaders. It is likely that there was some leadership during the protests at some level. The social netwar aspect of the FeesMustFall campaign seems to fall between those of the leaderless #Occupy related protests in Hong Kong and the led protests from Chile.

Whilst there were attempts at dedicated Facebook or Twitter groups, these were poorly supported. This is the opposite of the protests in Ukraine, Chile, and Arab Spring, all of which had dedicated webpages. In the case of Ukraine, dedicated Twitter or Facebook groups were used to provide specific services. In the FeesMustFall campaign this was rarely seen although some attempt was made, and most groups were focused on the distribution of information through personal networks and to main stream news media. Therefore the command and control aspect was immature compared to Ukraine, however the information sharing and the spread of ideas illustrates a high level of personal networking to mobilise the protests.

The #FeesMustFall protests had a primary focus of preventing fee increases, and promoting free education, whereas the Chile and Hong Kong student protests were primarily for political reasons. The FeesMustFall campaign did bring in political aspects, and there was an underlying anti-government dialogue, as is evident from the #ANCMustFall hashtags. These Tweets, whilst not as numerous, appeared to have more momentum. As the protests targeted the very seats of government, namely the parliament buildings and Union Buildings, there was the potential for the protests to rapidly escalate to if there were any overly-harsh responses by security forces. There is the possibility of these protests resurrecting and possibly becoming a protracted social netwar via social media. A concern for the South African government is that whilst these protests did not show the full sophistication of those in Ukraine, the showed an attempt at introducing the support for protracted protest actions, even though this was the first major nation-wide protest of its kind in South Africa.

The protests can also be seen as advocacy, as it sought to create awareness of an issue surrounding higher education in the country. In this it was successful; both parents and academics joined the students in peaceful marches, particularly in Cape Town, and global support was garnered with marches held in numerous countries. Through social media, media group Independent Media was challenged to allow the students to tell their story; this was accepted, and students were allowed to co-edit an edition of the *Cape Argus* newspaper (Independent Online, 2015). As advocacy, social media allowed pressure to be placed on the leadership of the institutions, the government, and the media. Ultimately, the primary objective of preventing fee increases was successful. What started as a couple isolated issues at universities quickly became a national crisis, then even gained global importance, and may have inspired similar protests in London.

From the analysis section, the ability to monitor and analyse online movements and civil disobedience through social media is evident. This indicates two sides to the social netwar: the offensive aspect of coordinating protests, and the defensive side of intelligence gathering. Through analysis of the interactions on social media, security services can be more proactive in determining the intent of the protesters.

CONCLUSION

A network society, is defined by García, Chauveau, Ledezma & Pinto (2013) as a "social structure which is based on networks operated by information and communication technologies".

The student protests of 2015 that took the country by force and forced significant concessions by university authorities and the government was a lesson for the use of technology in such a globally networked society. Most commentators to the protests have focused on whether the students' demands were practical with regard to the serious monetary constraints. Whilst others have analysed as to whether the movements experienced last year are going to continue in the future.

These protests and the corresponding campaigns and responses on the Twitter social media platform is an illustration the power of social media- it is evident that social media has become an integral element of any major protest or upheaval. Social media can be defined as websites and or applications which enable end users the ability to create and share content on social networks. In addition, social media allows users to join communities or form networks with other people. According to Tucker, Metzger, Penfold-Brown, Bonneau, Jost, & Nagler (2015) social media has the ability to affect the development of protest movements in various ways which includes the building of a protest movement at a rapid pace whilst encouraging mass participation as illustrated in the #Feesmustfall campaign. As in the case of the #Feesmustfall protests as soon as the protests were in full action, social media was able to disseminate information about the movements quickly. With the advent of the social media age, it has become easier to acquire necessary information about protests i.e. the potential costs as well as benefits of participation in protests.

As illustrated by the data obtained for the #Feesmustfall social media can be used to provide updates about protests and other political issues as they develop, in one of the most efficient sharing information tool that encompasses different purposes customized to the audiences. Furthermore it can and as illustrated in this chapter, share the logistical information about a protest as well as provide medical and legal assistance to those affected. With reference to Tucker et al. (2015) a key audience comprises of virtual communities, who can be located outside the country, of potential protest supporters- this is evident in the #Feesmustfall protests as the protests was resurrected in London providing support for the South African protestors. As in many protests, where there's fear that the national media will fail to report on the protests accurately, the international community or media becomes the important target audience which explains the chosen language of English (Tucker, Metzger, Penfold-Brown, Bonneau, Jost & Nagler, 2015)

With reference to Suttner (2016) the #FeesMustFall protests highlighted the broader reaching questions that went beyond just the educational sphere and defined the role of a social media platform i.e. Twitter which provided the larger communities with a view to the nature of a technological society.

REFERENCES

Bath, C. (2015). *Twitter Bombarded By #Feesmustfall Tweets*. Retrieved 19 August 2016 from http://ewn.co.za/2015/10/23/More-than-140000-FeesMustFall-tweets-overnight

Bay, A. (2011, January 18). *Tunisia's remarkable revolt*. Retrieved from http://www.strategypage.com/on_point/20110118224752.aspx

BBC. (2015). *Macedonia corruption: Fourth night of protests as snap election called*. Retrieved from http://www.bbc.com/news/world-europe-36058173

Beauchamp, Z. (2015). *South Africa's huge student protests*. Retrieved from http://www.vox.com/2015/10/24/9607236/south-africas-huge-student-protests-explained

Black, I., & Chulov, M. (2011, February 18). *Bahrain, Libya and Yemen try to crush protests with violence*. Retrieved from http://www.guardian.co.uk/world/2011/feb/18/bahrain-libya-yemen-protests-violence

Brady, L. (2015). Tweet. Retrieved from https://twitter.com/BradyLeanne/status/657421743892615168/photo/1?ref_src=twsrc%5Etfw

Cassim, B. (2015, October 24). *10 unforgettable moments of solidarity from #FeesMustFall*. The Daily Vox. Retrieved from http://www.thedailyvox.co.za/10-unforgettable-moments-of-solidarity-from-feesmustfall/

Cowen, N. (2015). *High Court Issues Interdiction against A Hashtag In #Feesmustfall Doc, 20 October*. Retrieved 11 April from http://www.htxt.co.za/2015/10/20/high-court-issues-interdiction-against-a-hashtag-in-feesmustfall-doc/

Duell, M., & Glanfield, E. (2015, November 4). Scuffles break march against university tuition fees shadow chancellor John McDonnell accuses Government betraying students. *The Daily Mail*. Retrieved from http://www.dailymail.co.uk/news/article-3303774/Scuffles-break-march-against-university-tuition-fees-shadow-chancellor-John-McDonnell-accuses-Government-betraying-students.html

Fearn-Banks, K. (2007). *Crisis Communications: A Casebook Approach*. Mahwah, NJ: Lawrence Erlbaum Associates.

Feni, M., & Mama, S. (2012). *Student protests turn violent*. Available at: http://www.iol.co.za/news/south-africa/western-cape/student-protests-turn-violent-1.1214650#.VXAgO8-qqko

Furlong, A. (2015, October 19). *Students shut down UCT, Ground Up*. Retrieved from http://groundup.org.za/article/students-shut-down-uct_3412

García, C., Chauveau, P., Ledezma, J., & Pinto, M. (2013). What can social media teach us about protests? Analyzing the Chilean 2011-12 student movement's network evolution through twitter data. In *Proceedings of the 4th International Conference on Collaborative Innovation Networks COINs13*.

Gaudin, S. (2011, March 11). *Twitter, Facebook become lifelines after Japan Quake, Computerworld*. Retrieved from http://www.computerworld.com/s/article/9214213/Twitter_Facebook_become_lifelines_after_Japan_quake

Geach, C. (2015). *FeesMustFall: 10 powerful placards, Independent Online*. http://www.iol.co.za/news/south-africa/feesmustfall-10-powerful-placards-1933343

Gregory, P. R. (2014b). You Tube shatters Russian lies about troops in Ukraine: Putin denies truth to Obama. *Forbes*, *14*(April). Retrieved from http://www.forbes.com/sites/paulroderickgregory/2014/04/14/you-tube-shatters-russian-lies-about-troops-in-ukraine/

Gruzd, A. (2015). *The Use of Social Media During the 2014 Crisis in Ukraine*. University of Pennsylvania. Retrieved from https://www.asc.upenn.edu/news-events/events/use-social-media-during-2014-crisis-ukraine-anatoliy-gruzd

Heinzelman, J., & Waters, C. (2010). *Crowdsourcing Crisis Information in Disaster-Affected Haiti, Special Report 252.* United States Institute of Peace.

Hendawi, H. (2011, January 27). *Egypt protests pose threat for regime.* Retrieved from http://news.yahoo.com/s/ap/20110127/ap_on_re_mi_ea/ml_egypt_protest_66

Hirayama, L. (2015, 22 October). Tweet. Retrieved from https://twitter.com/Lalla_Hirayama/status/657216095347200000?ref_src=twsrc^tfw

Independent Online. (2015, October 22). *Giving #FeesMustFall a voice.* Retrieved from: http://www.iol.co.za/capeargus/giving-feesmustfall-a-voice-1.1934338#.VioK634rLRY

Kamzi, A. (2011, September 27). How Anonymous Emerged to Occupy Wall Street. *The Guardian.* Retrieved from http://www.guardian.co.uk/commentisfree/cifamerica/2011/sep/27/occupy-wall-street-anonymous

Kopstein, J. (2013, June 26). UK police secretly monitoring 9,000 political campaigners using social media surveillance. *The Verge.* Retrieved from http://www.theverge.com/2013/6/26/4467056/uk-police-monitoring-9000-political-activists-using-social-media

Kotsiopoulos, I., Hagen, K., Watson, H., & Kalemaki, E. (2014). *Towards Modelling Crisis Concepts and Dynamics*, Deliverable D1.4.1, COSMIC. Retrieved from www.cosmic-project.eu

Lee, D. (2014, October 30). The role of social media in Occupy protests, on the ground and around the world. *The South China Morning Post.* Retrieved from http://www.scmp.com/news/hong-kong/article/1628305/role-social-media-occupy-protests-ground-and-around-world

Makhati, T. (2015, October 30). *19 of our favourite posters from the #FeesMustFall protests, Live Mag.* Retrieved from http://livemag.co.za/feesmustfall/19-of-our-favourite-posters-from-the-feesmustfall-protests/

Mann, A. (2008). Spaces for talk: Communication technologies (ICTs) and genuine dialogue in an International Advocacy Movement. *Asian Social Science*, *4*(10), 3–13.

Mkwanazi, N. (2015, October 26). *A One-Sided Story: My Thoughts on the #FeesMustFall Movement, exPress impress.* Retrieved from http://www.expressimpress.org/2015/10/26/a-one-sided-story-my-thoughts-on-the-feesmustfall-movement/

Mlambo, S. (2015, November 6). The Young & the Restless. *The Daily News*, p. 13.

Morgan, J. (2015, November 4). Student fees and grants protest: Jeremy Corbyn gives support. *The Telegraph*. Retrieved 22 September 2016 from https://www.timeshigh-ereducation.com/blog/student-fees-and-grants-protest-jeremy-corbyn-gives-support

Nyamnjoh, F., Nkwi, W. G., & Konings, P. (2007). *University Crisis and Student Protests in Africa*. Retrieved from http://www.africanbookscollective.com/books/university-crisis-and-student-protests-in-africa

O'Donnell, C. (2011). *New study quantifies use of social media in Arab Spring*. Retrieved 11 April 2016 from http://www.washington.edu/news/2011/09/12/new-study-quantifies-use-of-social-media-in-arab-spring/

Onuch, O. (2014, January 2). Social networks and social media in Ukrainian "Euromaidan" protests. *Washington Post Monkey Cage Blog*. Retrieved from http://www.washingtonpost.com/blogs/monkey-cage/wp/2014/01/02/social-networks-and-social-media-in-ukrainian-euromaidan-protests-2/

Pillay, K. J. (2012). *Do Web 2.0 Social Media Impact Transnational Social Advocacy? A Study of South African Civil Society and Greenpeace* (PhD Thesis). University of KwaZulu-Natal.

Potter, N. (2011, August 9). *London riots 2011: Protestors use Blackberry Messenger; hackers support them*. ABC News. Retrieved from http://abcnews.go.com/Technology/london-riots-2011-protesters-blackberry-messenger-hackers-back/story?id=14264839

Presence, C. (2015, October 20). *Blade: Student protests not a national crisis*. Retrieved from http://mg.co.za/article/2015-10-20-blade-student-protests-not-a-national-crisis

@ RefilweI. (2015, 21 October). Tweet. Retrieved from https://twitter.com/RefilweI/status/656694365604204544?lang=en

Reynolds, A. (2015, June 22). Denial is the enemy of Crisis Management. *Public Relations Weekly*. Retrieved 21 September 2016 from http://www.prweek.com/article/1352151/denial-enemy-crisis-management

Ronfeldt, D., & Arquilla, J. (1998). *The Zapatista Social Netwar in Mexico*. Santa Monica, CA: RAND Corporation.

Satell, G. (2014, January 18). If you doubt that social media has changed the World, take a look at Ukraine. *Forbes*. Retrieved http://www.forbes.com/sites/gregsatell/2014/01/18/if-you-doubt-that-social-media-has-changed-the-world-take-a-look-at-ukraine/

Sayed Ally, S. E. (2015). *Tableau Public Data Visualization. #feesmustfall.* Retrieved from https://public.tableau.com/profile/ally5527#!/vizhome/feesmustfall/Story1

Sekaran, U. (1992). *Research Methods for Business: A Skill-Building Approach* (2nd ed.). New York: John Wiley.

Sky News. (2011). *Middle East: See facts about all the countries in the region.* Retrieved from http://news.sky.com/skynews/Interactive-Graphics/middleeast

South African History Online. (2016, July 21) *The June 16 Soweto Youth Uprising.* Retrieved from http://www.sahistory.org.za/topic/june-16-soweto-youth-uprising

Staff Writer. (2015, October 23). Political parties hijack #FeesMustFall protest: reports. *Business Tech.* Retrieved from http://businesstech.co.za/news/government/102050/political-parties-hijack-feesmustfall-protest-reports/

Suttner, R. (2015). Op-Ed: Why is the state not a target of student protests? *The Daily Maverick.* Retrieved from http://www.dailymaverick.co.za/article/2015-10-20-op-ed-why-is-the-state-not-a-target-of-student-protests/

Taylor, M. (2015, November 1). Students Protest Our Free Education expected to draw thousands. *The Guardian.* Retrieved from https://www.theguardian.com/world/2015/nov/01/student-protest-attacks-free-education-to-draw-thousands

Thomas, S. (2015, October 22). *#FeesMustFall: how SA students are using social to subvert traditional media.* Memeburn. Retrieved 21 September 2016 from http://memeburn.com/2015/10/feesmustfall-how-sas-students-are-using-social-to-subvert-traditional-media-narratives/

Tucker, J. A., Metzger, M., & Barbera, P. (2014). *SMaPP Lab Data Report: Ukraine Protests 2013-2014.* Social Media and Political Participation Lab, New York University. Retrieved 31 March 2014 from http://smapp.nyu.edu/reports/Ukraine_Data_Report.pdf

Tucker, J. A., Metzger, M., Penfold-Brown, D., Bonneau, R., Jost, J., & Nagler, J. (2015). *Protest in the Age of Social Media.* Carnegie Corporation of New York. Retrieved from https://medium.com/carnegie-corporation-international-peace-and/protest-in-the-age-of-social-media-7ae9fd940b06#.n857xqa1l

van Niekerk, B. (2015). Information Warfare in the 2013-2014 Ukraine Crisis. In J. Richet (Ed.), *Cybersecurity Policies and Strategies for Cyberwarfare Prevention* (pp. 307–339). Hershey, PA: IGI Global. doi:10.4018/978-1-4666-8456-0.ch012

van Niekerk, B., & Maharaj, M. (2013). Social Media and Information Conflict. *International Journal of Communication*, *7*, 1162–1184.

van Niekerk, B., Pillay, K., & Maharaj, M. (2011). Analysing the Role of ICTs in the Tunisian and Egyptian Unrest from an Information Warfare Perspective. *International Journal of Communication*, *5*, 1406–1416.

Wesi, T. (2015). *Fees Must Fall protest in Pretoria 'hijacked'*. Retrieved 22 September 2016 from http://citizen.co.za/833478/fees-must-fall-protest-in-pretoria-hijacked/

Williams, C. (2011). *Facebook and Twitter to help police track riots*. Retrieved 11 April 2016 from http://www.telegraph.co.uk/technology/social-media/8723038/Facebook-and-Twitter-to-help-police-track-riots.html

World Movement for Democracy (WMD). (2009). *Twitter Case Study*. Retrieved 17 June 2010 from: http://www.wmd.org/resources/whats-being-done/information-and-communication-technologies/case-study-twitter

KEY TERMS AND DEFINITIONS

Advocacy: Public support for or recommendation of a particular cause or policy.

Crisis: An interruption in economic, cultural, social and political activities that threatens the existence of the current social, organisational, or national.

Hashtag: A hashtag is word or phrase which is used on social media sites such as Twitter as a means of identification on a specific topic which is preceded by a hash (#).

Protest: Action expressing disapproval of or objection to something.

Social Media: Includes different types of electronic communication e.g. Web sites for social networking through which users create online groups for the purposes of information sharing and collaboration.

Social Netwar: A netwar is considered as a type of low-intensity conflict that is waged by internet users which may include different organisations, terrorists, social movement groups and activist groups.

Chapter 11
The Roots of Terror:
The Lesser Evil Doctrine under Criticism

Maximiliano Emanuel Korstanje
University of Palermo, Argentina

ABSTRACT

Risk would serve as a fertile ground to move resources, otherwise would be stagnated, so that elite may centralize and solidify "extractive institutions" to enhance the economic performance. The war of all against all, predicated by Hobbes sets the pace to the war of few blocs to yield a supreme authority over the rest. The theory of globalization is reluctant to explain how the world tends to a centralization of resources and violence. Here we come across with a paradox, if the XXth century posed a lot of states making the war to forge their own identity (as it was the case in Europe and US who participated in two total wars), within the state a sentiment of nationhood persisted over other counter-reactions. Citizens not only were twinned to embrace a same history and heritage, but also suspended the internal violence against their brothers.

INTRODUCTION

Contemporarily, the flourishing of fear corresponds with the production of media information and the adoption of new technologies. As never before, we live a culture of fear that marked a turning point respecting to other times (Timmermann, 2015). Whilst Latina America poses its concerns in the problem of crime and delinquency

DOI: 10.4018/978-1-5225-1938-6.ch011

(Kessler 2009; Damnert & Arias, 2007), in Anglo-Saxon countries terrorism becomes the main threat (Howie, 2007; 2012; Skoll 2007; Altheide 2004; 2006; Soyinka, 2005; Achcar, 2006; Ignatieff 2013). There is a clear cultural matrix in Latin and Anglo-culture to understand the risk which was widely examined by Korstanje (2014b; 2015). The reform introduced the concept of predestination, unknown by Catholicism, reconstructing in this way a bridge between present and future. The risks in English speaking countries represent a platform to conquest the future by embracing the "precautionary principle". The needs of preventing risk, in the local crime or the war against terror has been situated as the main agenda of US government (Gray, 2007; Sunstein, 2002). As Gregory Flaxman & Ben Rogerson observed, "the culture of fear" may be defined as a symptomatology of our contemporary society, even paradoxically living safer than earlier generations. Capitalism has reproduced over last years new tactics to colonize our feelings. The psychological fear, likely, and its logic remain obscure for our understanding. What 9/11 inaugurated, was a deep dissociation between mediated disasters and the probabilities of risk. We are subject to countless threats as car accidents, heart disease, or even cancer, most of them ignored, but what we frighten is terrorism. In terms of authors, "our fears are misplaced" (p. 334). Not surprisingly from Hobbes on, the modern state passed this original fear of death to moderated levels of trust. Logically, fear inoculates changes that help regulating the production and economies within each state. Thus, we need to speak on the "economy of fear" instead of a basic emotion (Flaxman & Rogerson, 2011).

In this context, a great variety of scholars have addressed the problem of post modernity and fear from diverse angles. In these approaches, there was a strong focus on linking risk with society as Richard Sennett (the corrosion of character), Ulrich Beck (Society of Risk), Giddens (risk and attachment), Sunstein (the laws of fear), Niklas Luhmann (the sociology of risk), Naomi Klein (the doctrine of shock), Diken Bulent (the comedy of terror) or Geoffrey Skoll and Maximiliano Korstanje (the fetish of risk). The conceptual discussion in the ways the academicians understand the crises. If risk is enrooted in the economic system, how may understand the last Wall Street collapse in 2008?, is the introduction of risk conducive to the decline of nation-state?.

In this essay review, we hold the thesis that in the society of terror, risk not only alludes to the formation of much broader alliance towards the configuration of the unique state, a type of neo-universalism, based on the industry of hospitality and tourism as it has been anticipated by Kant and Leibniz, but also it gives further legitimacy to elite to change the old meaning of jurisprudence. To set a clear example, terrorism exhibits a fertile ground not only US government undermines its relationship internally with the unions, but also celebrating pacts and covenants otherwise would be rejected. Doubtless, terrorism triggers uncertainty because the state never knows where and when the next attack will hit. It opens the doors to embrace a logic of "preemption", that leads to a tyranny. In order for the war-machinery works, the state needs a juridical right to intervene in other autonomous states. As Alex Bellamy (2005) puts it, the preventive intervention in the geo-politics fields re-formulates the legal background of nation-states conferring to US the right to take direct action in case of real or potential hazard is found. Beyond the alliance to strengthen the power of inter-states in their war against terror, the United States reserve the legal concept of "self-determination" to intervene in other international scenarios when the homeland security runs some risk. As an allegory, terrorism paves the ways to centralize the international authorities of states into one main power.

The Psychological Nature of Fear

Every culture has developed ways to adapt to its environment. One method is the construction of feared object which serves as a mechanism to adjust social perceptions of danger. Elements which instill fear vary from one society to another (Korstanje, 2011). Fear can be defined as a basic emotion, which protects the survival of an organism. Not just human beings, but all animals experience fear of external threatening stimuli. Alerted by fear, the organism has three possible reactions: paralysis, attack, or withdrawal (Fraisse, 1973; Panksepp, 1982; Levenson, Ekman & Friesen, 1990; Strongman, 1996). Nonetheless, the concepts of behavioral psychology have not embraced by other social sciences. Anthropology has developed its own sense of what fear means. Although, recognizing a strong neurobiological basis that reduces or enhances the fear, ethnologists evaluate the social factors by which some fears are over-valorized while other rejected (Malinowski, 1967). Therefore, culture plays a vital role not only conferring a specific meaning to objects, but also to fears.

Mary Douglas, a pioneer scholar interested in exploring the connection of fear, evil, and risk, argued that psychological fear represents an attempt to react when faced with a hostile situation. To some extent, the preservation of culture is at stake in contexts of uncertainty or instability. When socialized, fear unites a society. Without fear, Douglas added, societies would experience substantial fragmentation (Douglas, 1992). In subsequent studies, Douglas developed a new thesis arguing that risk, danger, and sin are intertwined social constructs. Sin and risk give further legitimacy to the status quo, which would otherwise discredit privileged groups if they did not give solutions to lay people. Risk and sin both provide rationalizations for how the world works. The potentiality of threat provides legitimation for social solidarity and status hierarchies (Douglas, 2007).

Anthropological perspectives gained less notice in their treatment of the problem of fear for two reasons. First, psychological explanations advanced considerably with recourse to neuro-dynamic explanations for emotions. Second, sociologists tended to devote attention to risk. Consequently, the qualitative meaning and narrative of fear failed to expand to other social areas of study. In psychology, fear represents a basic emotion. Academic psychologists have largely eschewed qualitative approaches to study fear. They have left the study of fear as a subjective experience in the hands of philosophy.

Philosophically, the self experiences anxiety when faced with a decision. Existentialism defined anxiety as a result of freedom or uncertainty. The self opts for a way out of choosing. While fear corresponds with a specific object or stimulus, anxiety has an abstract nature produced by the presence of nothingness (Heidegger, 1997, Kierkegaard, 2003). Following the observations of K. Tierney, risks should be defined as any probability of damage resulting from an event where the integrities of victims are at stake. In recent decades, sociologists have treated risk as a social construction. Within sociology, two contrasting waves have historically discussed the nature of risk (Tierney, 1994). One group explored the probability of harm, focusing on the effects of unseen risk for social systems. Another paid attention to the perception of citizens and the paradoxes this generates (Duclos, 1987). The specialized literature in risk management took the pragmatic perspective that bad evaluations of risk may lead to bad decisions. In this sense, efforts to mitigate risks open new ones (Oliver-Smith, 2002).

Zygmunt Bauman (2011) explained that risks are social constructions to try to control the future. In the Middle Ages, happiness was thought to be restricted to few people, who can attain it only through suffering and expiation. The American Revolution introduced a radical change in the way that happiness was conceived, as suggested by Thomas Jefferson's claim in the Declaration of Independence (1776) that the pursuit of happiness is a self evident truth of the human condition. This assertion of a global right to happiness broadened its possibilities, but linked the possibility of happiness to freedom and choices. Risk, then, came to regulate the uncertainty of the future, but also conferred on the subject the liability for failure to be happy.

Ulrich Beck has argued that modernity opened new global risks, which were alien to the medieval world view. Chernobyl in the Ukraine was the symbolic of the role played by technology in fabricating new risks. In Beck's view, technology had helped enhance security, but today it generates new and dire risks that threaten human existence. In Beck's "risk society" the old modes of production, which fabricated commodities, have turned into methods that produce risks (Beck, 2006; 2011). One of the limitation in Beck`s development consists in ingenuity respecting to the process of class formation. The fact that everyone is in danger in the late-modernity does not imply that higher-classes monopolizes their resources to mitigate the produced risks. This means that, among many other things, the introduction of risks does not homogenize society, as Beck notes. It creates a more asymmetrical community between two main classes, those who are exploited and subject to immobility, and another more mobile class that exerts the power. At time elite generates new risks to enhance their quality of life, the oppressed workforce is subject to face those risks others created (Skoll & Korstanje, 2012).

Parallel to Beck, Anthony Giddens acknowledged globalization as a project based on two key factors. The first is that money has come to serve as a mechanism of connecting presence with absences, or needs with their satisfaction throughout the world. The second element is a network of experts, who not only evaluate potential risks but also devise ways for mitigating risks. Starting from the premise that experts monopolize the trust of lay people, for Giddens, risk is what society creates to sustain its efficient functioning (Giddens, 1991; 1999).

In opposition to Giddens's argument, Niklas Luhmann has criticized the thesis of risk society because of the increasing alarmism it spreads in public consciousness. Certainly, Luhmann adds, risks always are rooted into a previous profits or benefit, whereby the subject should decide. It corresponds with the principle of contingency. Unfortunately Beck did not contemplate the distinction between risk and threat. While risk signifies a previous decision by the self, threat refers to something external to the self. A terrorist attack, an airplane accident, or a natural disaster are threats, since the victims have no way to reverse the situation. The passengers in an airplane crash have no way of avoiding the harm. In contrast, for the air travel company owner, who opted to reduce costs, the accidents are a risk. Generally, those who make the decision are generators of risks. They are not the same as those who face the risks (Luhmann, 2006). J. Richardson (2010) says that threats which jeopardize society are introduced in the social system by means of knowledge. Risk, in these terms, would be the efforts to intellectualize the future by offsetting costs and benefits. The final decision made on the possibility to face or avoid the damage is given by the degree of contingency, with respect to the problem to be solved (Richardson, 2010). This seems to be the reason why technology designed to mitigate risks under some conditions of uncertainty, generates new risks. A discussion of this nature, coined in the core of social sciences, has not been duly evaluated in tourism fields. In the next section, some of the more relevant studies in tourism risk are scrutinized. Is technology conducive to create unreal fears, which only have presence in fiction?.

The Terror of Terror(ism)

Is terrorism inextricably linked to State?. The question whether insurgents lack any representation in government or election system is one of the aspects that characterize terrorism. These cells not only have been excluded from democracy, but also from the parliamentary participation (Piazza; 2007; 2008). This point of entry is very interesting since suggests that terrorism derived from democracy. Nor poverty neither psychological frustration are key factors that explain the radicalization of terrorists. Doubtless, they are political agents who have developed a radical mind respecting to outer world (McCauley & Moskalenko, 2008). Additionally, the problem of mobility, tourism and the current system of transport, which needs from certain level of freedom to work, has been played a fertile ground for bombing or terrorist attacks over the last decades (Diken & Lausten, 2002; Urry, 2007; Korstanje & Clayton, 2012).

Michael Brown alerts that the agenda of security has changed for governments in this new century. We are witnessing specific problems which are not necessarily linked to the classic war-fares or clash among states. We now face new "threats" which are embedded within nation-states. The governments struggle not only to prevent the non-military attacks, as the case of terrorism, but regulating the media without affecting democratic rights. Luke Howie in his seminal text, Witnesses to Terror, addresses the problem of terrorism and media. The fact is that the world and economies have changed forever after 9/11, trying to predict what in nature unpredictable is. Even if the obsession for gaining further security remains in United States, Howie's research shows how years change the interviewees' viewpoints depending on its degree of exposition. In this token, it is important not to lose the sight that there is a strong complicity between terrorists and journalism. The knowledge, which supposedly makes from this life a safer place, becomes in a double-edge sword. To be more precise, Howie adds, terrorism me be defined as more than a political technique or strategies to dissuade the states of certain claims, terrorism is stronger in the witness's terror.

Terrorism works this way for witness. If there was one way to describe the outcomes of the research that I have conducted for this book, I would say that terrorism causes people to feel terror. Terror is the name we give to the uncertainty we feel in the feel of global violence in some of the world's most populous cities. If Terrorism does not cause terror, the it is not terrorism. (Howie, 2012: p. 12)

The definition described above is of paramount importance to understand the connection of terrorists and eye-witnesses. Basically, the targets are not necessarily selected to create mass-death, as many pseudo-specialists suggest, but also to lead an extreme panic in the rest of population. At some extent, promising findings can be obtained if we pay attention to the psychological effects of terrorism in daily life as well as how lay-people intellectualized and changed their behavior post 9/11. At time more attention is given to terrorism, more violence emerges. It begs a more than interesting question, how can we deal with terrorism?, is preventive platform a valid resource in the struggle for peace?.

Jean Baudrillard recognizes that late-modernity has destroyed not only the social ties, but the symbolic foundation of society. This apocalyptic thesis leads us to think the boundaries between consumers and their consumed goods have been blurred. The current nature of fear escapes to our biological emotions, rather, it is only enrooted in the future. The allegory of disasters, which today entertain a great portion of western audience, is carefully designed not only to serve as a disciplinary mechanism of control but also to create a "pseudo-reality", where events set the pace to "pseudo-events".

A whole strategy of deterrence that does service today for a global strategy. Steven Spielberg's recent film, minority report, provides an illustration of such a system. On the basis of brains endowed with a gift of pre-cognition (the precogs), who identify imminent crimes before they occur, squads of police (the precrimes) intercept and neutralize the criminal before he has committed his crime ... ruptural events, unforeseeable events, unclassifiable in terms of history, outside of historical reasons, events which occur against their own image, against their own simulacrum. Event that breaks the tedious sequence of current events as relayed by the media, but which are not, for all that, a reappearance of history or Real irrupting in the heart of the virtual. (Baudrillard, 2006: 2; 8)

As Douglas Kellner puts it, Baudrillard is the clear reminder the end of history, formulated by neo-liberal scholars, equals to the end of humankind. For capital owners and global capitalism, history is a serious obstacle to overcome. In the realism of hyper-reality, games are something else than a space of competition, or rivalry, or simply an act of submission, they goes beyond the hegemony of history to break the rules of any game. This was exactly what in 9/11 happened. In this vein, Kellner writes,

Shortly after September 11 terrorist attacks, Baudrillard wrote a paper L'Espirit du terrorisme in Le Monde. He argued that the assaults on the World Trade center and Pentagon constituted a strong event, that the attacks were the ultimate event, the mother of all event, the pure event within itself all the events that have never taken place. The event strike, Baudrillard declared, was over and since this time he has continued to focus intensely on the dynamics and happenings of contemporary history ... For Baudrillard, the 9/11 attacks represent a new kind of terrorism, exhibiting a form of action which play the game, and lays hold of the rules of the game, solely with the aim of disrupting it. They have taken over all the weapons of the dominant power. That is, the terrorists in Baudrillard reading uses airplanes, computer networks, and the media associated with Western societies to produce a spectacle terror. The attacks evoked a global specter of terror that the very system of globalization and Western capitalism and culture were under assault by the spirit of terrorism and potential terrorist attacks anytime and anywhere. (Kellner, 2005: 2)

Is terrorism a result of globalization?. Kellner, reading Baudrillard, has an alternative answer to this. Although social theorists are accustomed to think globalization as a type of new matrix, an economic matrix based on the encounter of the market, the technology, tourism, migration with culture, for Baudrillard, globalization represents an attempt to undermine the democracy and human rights. Once globalization cemented a much broader process of homogenization, which means "standardization", it contradicts the nature of democratic life, the respect for minorities and subjects. To put this in other terms, globalization exhibits a systematic attempt to efface the individual nature of human beings (Kellner, 2005). In this respect, the postmodern state seems to be a creator of paradoxes. By intervening in one direction to prevent the worse others unexpected risks arise. This was for example, what triggered the conflict in Afghanistan that led towards 9/11. As Mohammed Ayoob observed,

It is indeed ironic that the terrorist attacks of 9/11 were in large part the direct consequence of the externally induced collapse of the state of Afghanistan, for which the United States bore much of the responsibility. It was the absence of political order in Afghanistan that provided Al-Qaeda with the opportunity and the space to plan and execute the attacks on the United States. (Ayoob, 2009: 109)

Nobel Prize in literature, W Soyinka admitted that 9/11 did not surprise him. From that moment onwards, international public opinion (even in Africa) experienced a new climate of fear, in spite of the previous experiences of political terror. Soyinka believes the world has faced extreme situations of panic before 9/11 ranging from Nazism and the Second World War to nuclear weapon testing. One of the aspects of global power that facilitates this feeling of uncertainty seems to be the lack of a visible rivalry once the USSR collapsed. The politic terror promulgated by states diminishes the dignity of enemies. These practices are rooted inside a territory but paved the way for a new form of terrorism which ended in the World Trade Center attacks. It is incorrect to see 9/11 as the beginning of a new fear but as the latest demonstration of the power of an empire over the rest of the world. Mass communications, though, transformed our ways of perceiving terrorism even if it did not alter the conditions that facilitate the new state of war. Soyinka examines the current connection between power and freedom. Unlike classical totalitarian States which are constructed by means of material asymmetries, the quasi-States construct their legitimacy by denouncing the injustices of the World. Quasi-States are not only terrorist cells but also mega-corporations which work in complicity producing weapons for one side or the other. Making profit of human suffering is a primary aspect that characterizes these quasi-states. The uncertainty these corporations engender denies the minimum codes of war by emphasizing the inexistence of boundaries and responsibilities. Once rectitude has been substituted by the right to exercise power, pathways towards a moral superiority are frustrated. Unlike the disaster of the Napalm-bombing of non-combatants by the United States in Vietnam, this new war-on-terror is characterized by targeting innocents as a primary option. In opposition to conventional wars, war-on-terror expands fear under the following two assumptions: a) hits can take place anywhere and anytime, and, b) there is no limits on brutality non-combatants. Wars depend on the capacity to control others based on the principle of power. Governments often need the material resources of their neighbours. Where the expropriation method of capitalist trade fails, war finds success. One might speculate that war should be understood as an extension of economic production (Soyinka, 2005)

It is important to clarify, while the nature between terrorism and disasters differ, not surprisingly, the ways audience receives the news are alike. Then, the psychological impact of terror in western citizens is handled following the same dynamic.

CONCLUSION

After the financial crash of New York that changed the economical geographies of the world, analysts are discussing to what extent the old dichotomy between centre and its periphery is experiencing a new re-feudalization, similarly to the Middle Ages post Roman Empire's collapse or the main powers are being coalesced into blocs. Far from being circular dynamics, these blocs will be united in only one. Centered on the latter hypothesis, we pose the dilemma of risk, as it has been studied by social scientists, as an ideological discourse consistent to the permeation of market into nation-state. Secondly, risk would serve as a fertile ground to move resources, otherwise would be stagnated, so that elite may centralize and solidify "extractive institutions" to enhance the economic performance. Through the former centuries witnessed a stage of decentralization, where the scattered nations struggle each other to prevail, two World wars and the onset of XXIth, brought another reality. The war of all against all, predicated by Hobbes sets the pace to the war of few blocs to yield a supreme authority over the rest. The theory of globalization is reluctant to explain how the world tends to a centralization of resources and violence. Here we come across with a paradox, if the XXth century posed a lot of states making the war to forge their own identity (as it was the case in Europe and US who participated in two total wars), within the state a sentiment of nationhood persisted over other counter-reactions. Citizens not only were twinned to embrace a same history and heritage, but also suspended the internal violence against their brothers. At time the boundaries of states were liberated, to adopt globalization as a main doctrine, external states woven more pacts and alliances to protect their citizens, but paradoxically, it created an internal point of conflicts as never before. The economies of post liberal societies had big problem to regulate the conflicts internally. As Big Brother or the Hunger Game, the neo-capitalist state stimulates the extreme competition where the social Darwinism reigns. As a result of this economies precaritize the conditions of life (work) of workforces. The allegory of capitalist systems alludes to the facts that few regulate the life of the whole. Those who participate in this game do not know that only one will be winner, so they are over-confident of their own skills and possibilities. Undoubtedly, this is the illusory discourse of capitalism to keep the workforce under control. Although externally, states decline to make the war as an option to redeem disputes, internally, peoples are pitted against their neighbors. This seems to be exactly the grounds where terrorism operates. As Howie explained, one of the main problems of terrorism is not the surprise-factor, but the tendency to distrust of our neighbors. After all, terrorists are like us, live like us, are similar to us. Gradually, states will tend to coalesce into other state forming stronger structures of powers, while internally a wealth of riots, conflicts and disintegrations will come into effect in the next decades.

REFERENCES

Abeles, M. (2006). Globalization, Power and Survival: An Anthropological Perspective. *Anthropological Quarterly*, *79*(3), 483–508. doi:10.1353/anq.2006.0030

Achcar, G. (2006). *The Clash of Barbarisms. The Making of the New World Disorder.* London: Saqi.

Altheide, D. L. (2004). Consuming terrorism. *Symbolic Interaction*, *27*(3), 289–308. doi:10.1525/si.2004.27.3.289

Altheide, D. L. (2006). Terrorism and the Politics of Fear. *Cultural Studies Critical Methodologies, 6*(4), 415-439.

Ayoob, M. (2009). State making, state breaking, and state failure. In C. Crocker, F. O. Hampson, & P. Aall (Eds.), *Leashing the Dogs of War, conflict management in a divided world* (pp. 83–114). Washington, DC: United States Institute of Peace press.

Baudrillard, J. (1989). *The Transparency of Evil: essays on extreme phenomena.* Barcelona: Anagrama.

Baudrillard, J. (1995a). *The systems of the objects*. Mexico: Siglo XXI.

Baudrillard, J. (1995b). *The Gulf War Did Not Take Place*. Sydney: Power Publications.

Baudrillard, J. (2006). Virtuality and Events: the hell of power. *Baudrillard Studies, 3*(2). Available at http://www.ubishops.ca/BaudrillardStudies/

Bauman, Z. (2011). *Society under Siege*. Buenos Aires: Fondo de Cultura Económica.

Bauman, Z., & Lyon, D. (2013). *Liquid surveillance: A conversation*. Cambridge, UK: Polity Press.

Beck, U. (2006). *Risk Society. Towards a new modernity: hacia una nueva modernidad*. Buenos Aires: Paidos.

Beck, U. (2011). Convivir con el Riesgo Global. In *La Humanidad Amenazada: gobernar los riesgos globales* (pp. 21–32). Madrid: Paidos.

Bellamy, A. J. (2005). Is the war on terror just? *International Relations, 19*(3), 275–296. doi:10.1177/0047117805055407

Brown, M. (2009). New Global Dangers. In C. Crocker, F. O. Hampson, & P. Aall (Eds.), *Leashing the Dogs of War, conflict management in a divided world* (pp. 39–52). Washington, DC: United States Institute of Peace press.

Castel, R. (1983). From dangerousness to risk. *Actes de la recherche en sciences sociales, 47*(4), 119-127.

Chavez, N. (2008). *Cuando Los Mundos Convergen, terrorismo, narcotráfico y migración post 9/11* [When the worlds converge, terrorism, narco-traffic and migration post 9/11]. Quito: FLACSO Ediciones.

Dammert, L., & Arias, P. (2007). El desafío de la delincuencia en América Latina: diagnóstico y respuestas de política. *Seguridad y violencia: desafíos para la ciudadanía. FLACSO*, 21-66.

Diken, B. (2011). Militarization as comedy of t(error). *Economia Autonoma, 4*(7), 72–92.

Diken, B., & Laustsen, C. B. (2002). Zones of indistinction security, terror, and bare life. *Space and Culture, 5*(3), 290–307. doi:10.1177/1206331202005003009

Douglas, M. (1992). *Risk and Blame*. New York: Routledge. doi:10.4324/9780203430866

Douglas, M. (2007). *Pureza y Peligro: un análisis de los conceptos de contaminación y tabú*. Buenos Aires: Nueva Visión.

Duclos, D. (1987). *Le Risque: une construction sociale?. In La Societeé Vulnerable* (pp. 91–92). Paris: Ecole Normale Superieure.

Flaxman, G., & Rogerson, B. (2011). The Economy of fear. *Symploke*, *18*(1-2), 333–336.

Foucault, M. (1992). *Microfísica del Poder*. Madrid: La Piqueta.

Foucault, M. (2000). *Los Anormales. Curso en el College de France. 1974-1975*. Buenos Aires: Fondo de Cultura Económica.

Foucault, M. (2006). *Seguridad, Territorio y Población*. Buenos Aires: Fondo de Cultura Económica.

Foucault, M. (2008). *El Nacimiento de la Clínica*. Buenos Aires: Siglo XXI.

Fraisse, P. (1973). *Las Emociones. In Motivación, Emoción y Personalidad* (pp. 106–187). Buenos Aires: Editorial Paidos.

Giddens, A. (1991). *Modernity and self-identity: self and society in the late modern age*. Stanford, CA: Stanford University Press.

Giddens, A. (1999). *Consecuencias de la Modernidad*. Madrid: Alianza Editorial.

Gray, C. S. (2007). *The implications of preemptive and preventive war doctrines: A Reconsideration*. Army War Coll Strategic Studies Inst Carlisle Barracks PA.

Hobbes, T. (1994). *Leviathan: with selected variants from the Latin edition of 1668* (Vol. 2). New York: Hackett Publishing.

Howie L (2012) *Witnesses to Terror: Understanding the Meanings and Consequences of Terrorism.* Hampshire, UK: Palgrave-Macmillan.

Ignatieff, M. (2013). *The lesser evil: Political ethics in an age of terror*. Princeton University Press.

Kellner, D. (2005). Baudrillard, Globalization and Terrorism: some comments in recent adventures of the Image and Spectable on the occasion of Baudrillard's 75th birthday. Bishop's University. Retrieved from http://www.ubishops.ca/BaudrillardStudies/

Klein, N. (2007). *The shock doctrine: The rise of disaster capitalism*. London: Macmillan.

Korstanje, M. E. (2011). Why risk why now? Conceptual problems around the risk perception in tourism industry. *Revista Brasileira de Pesquisa em turismo*, *5*(1), 4-22.

Korstanje, M. E. (2014). Why Risk-Research is More Prominent in English Speaking Countries in the Digital Society. *International Journal of Cyber Warfare & Terrorism*, *4*(1), 8–18. doi:10.4018/ijcwt.2014010102

Korstanje, M. E. (2014a). Chile helps Chile: Exploring the effects of earthquake Chile 2010. *International Journal of Disaster Resilience in the Built Environment*, *5*(4), 380–390. doi:10.1108/IJDRBE-01-2012-0002

Korstanje, M. E. (2015). El papel del riesgo en la configuración de las sociedades anglosajonas. *Estudios Sociales: Revista de investigación científica*, *23*(45), 283-307.

Korstanje, M. E., & Clayton, A. (2012). Tourism and terrorism: Conflicts and commonalities. *Worldwide Hospitality and Tourism Themes*, *4*(1), 8–25. doi:10.1108/17554211211198552

Laursen, F. (2010). *Comparative Regional Integration. Europe and Beyond*. London: Ashgate Publications.

Levenson, R. W., Ekman, P., & Friesen, W. W. (1990). Voluntary facial expression generates emotions-specific nervous system activity. *Psychophysiology*, *27*, 363–384. doi:10.1111/j.1469-8986.1990.tb02330.x PMID:2236440

Locke, J. (1993). *Two Treatises of Government*. Cambridge, UK: Cambridge University Press.

Luhmann, N. (2006). *Sociología del Riesgo* [The sociology of Risk]. México City: Universidad Iberoamericana.

McCauley, C., & Moskalenko, S. (2008). Mechanisms of political radicalization: Pathways toward terrorism. *Terrorism and Political Violence*, *20*(3), 415–433. doi:10.1080/09546550802073367

Oliver-Smith, A. (2002). Theorizing disasters, nature, power and culture? In S. Hoffman & A. Oliver-Smith (Eds.), *Catasthrope and culture: the anthropology of disasters* (pp. 23–47). Oxford, UK: School of American Research James Currey.

Panksepp, J. (1982). Toward a General psychobiological theory of Emotions. *Behavioral and Brain Sciences*, *5*(03), 407–467. doi:10.1017/S0140525X00012759

Piazza, J. A. (2007). Draining the swamp: Democracy promotion, state failure, and terrorism in 19 Middle Eastern countries. *Studies in Conflict and Terrorism*, *30*(6), 521–539. doi:10.1080/10576100701329576

Piazza, J. A. (2008). Incubators of terror: Do failed and failing states promote transnational terrorism? *International Studies Quarterly*, *52*(3), 469–488. doi:10.1111/j.1468-2478.2008.00511.x

Richardson, J. (2010). The certainty of uncertainty: Risk management revisited. *Foresight*, *12*(4), 47–64. doi:10.1108/14636681011062997

Schmitt, R. (1995). *Beyond Separateness*. Boulder, CO: Westview Press.

Sennett, R. (2011). *The corrosion of character: The personal consequences of work in the new capitalism*. New York: WW Norton & Company.

Skinner, Q. (1964). Hobbess Leviathan. *Historical Journal (Cambridge, England)*, 7(02), 321–333. doi:10.1017/S0018246X00025498

Skoll, G. R. (2007). Meanings of terrorism. *International Journal for the Semiotics of Law-Revue internationale de Sémiotique juridique, 20*(2), 107-127.

Skoll, G. R., & Korstanje, M. E. (2012). Risks, totems, and fetishes in Marx and Freud. *Sincronía*, (2), 11-27.

Soyinka, W. (2005). *The Climate of Fear. The quest for dignity in a dehumanized World*. New York: Random House.

Sunstein, C. R. (2002). *Risk and reason: Safety, law, and the environment*. Cambridge, UK: Cambridge University Press.

Sunstein, C. R. (2005). *Laws of fear: Beyond the precautionary principle* (Vol. 6). Cambridge, UK: Cambridge University Press. doi:10.1017/CBO9780511790850

Tierney, K. (1994). *Sociology's Unique Contributions to the Study of Risk*. Disaster Research Center, Preliminary Paper, 204.

Timmermann, F. (2015). *El Gran Terror. Miedo, Emocion y discurso Chile 1973-1980*. Santiago: Ediciones Copygraph.

Urry, J. (2007). *Mobilities*. Cambridge, UK: Polity Press.

Virilio, P. (2010). *The University of Disaster*. Cambridge, UK: Polity Press.

WarLeigh-Lack, A (2010). The EU in comparative perspective: comparing the EU and Nafta. In Comparative Regional Integration. Europe and Beyond. London: Ashgate.

ADDITIONAL READING

Bailyn, B. (1965). *The Origins of American Politics*. New York: Vintage Books.

Bellesisles, M. A. (2010). *1877: America's Year of Living Violently*. New York: The New Press.

Boorstin, D. (1962). From traveler to tourist: The lost art of travel. *The image: Or what happened to the American dream*, 77-117.

Coleman, S. (2013). *"Actors of History? Religion, Politics, and Reality within the Protestant Right in America". Religion, Politics & Globalization. Lindquist, G. & Handelman, D* (pp. 171–188). New York: Berghan Book.

Domínguez, P., Burguette, E., & Bernard, A. (2003). Efectos del 11 de Septiembre en la hotelería Mexicana: Reflexión sobre la mono-dependencia turística. *Estudios y Perspectivas en Turismo, 12*(3-4), 335–348.

Dworking, R. (1996). *Freedom's Law: the moral reading of the American Constitution*. Massachusetts: Harvard University Press.

Elías, N. (1998). *La Civilización de los Padres y otros Ensayos*. Bogotá: Editorial Norma.

Ertuna, C., & Ertuna, Z. I. (2009). The Sensitivity of German and British tourists to new shocks. *Tourism Review., 64*(3), 19–27. doi:10.1108/16605370910988809

Fernández de Castro, R. (2008) "Introducción". (pp 7-16) En ¿Qué Es Estados Unidos? R. de Castro Fernandez & Blackmore H. México, Fondo de Cultura Económica.

Fitzgerald, F. (1986). *Cities on a Hill*. New York: Simon & Schuster.

Floyd, M., ... (2003). "The Effects of Risk Perception on Intention to Travel in the Aftermath of September 11, 2001". In Safety and Security in Tourism: relationships, Management and Marketing, (Eds) Hall, M. Timothy, D. y Duval, T. New York, Haworth Hospitality Press.

Floyd, M., & Pennington-Gray, L. (2004). Profiling Risk: Perception of tourist. *Annals of Tourism Research, 31*(4), 1051–1054. doi:10.1016/j.annals.2004.03.011

Freeland, R. (1985). *The Truman Doctrine and the Origins of McCarthyism: foreign policy, domestic politics, and internal security 1946-1948*. New York: New York University Press.

Fromm, E. (2005). *El Miedo a la libertad*. Buenos Aires: Paidos.

Fuchs, G., & Reichel, A. (2004). Cultural Differences in tourist destination risk perception: An exploratory study. *Tourism., 52*(1), 21–37.

Fuchs, G., Uriely, N., Reichel, A., & Maoz, D. (2012). Vacationing in Terror-Stricken Destinations: Tourists Risk Perception and Rationalizations. *Journal of Travel Research*.

Glassner, B. (1999). *The Culture of Fear: Why Americans Are Afraid of the Wrong Things*. New York: Basic Books.

Greven, P. (1988). *The Protestant Temperament. Pattersn of child-rearing religious experience, and the self in Early America*. Illinois: Chicago University Press.

Gutmann, A. (2001) "Introduction". In Human rights, as politics and idolatry. Michael Ignatieff. New Jersey, Princeton University Press. Pp Vii-xxviii

Ignatieff, M. (2001). *Human Rights. As Politics and Idolatry* (pp. 3–53). New Jersey: Princeton University Press.

Korstanje, M. (2013). "Empire and Democracy, a critical reading of Michael Ignatieff". *Nómadas: Revista crítica de Ciencias Sociales y Jurídicas. Número*, *38*(II), 69–78.

Korstanje, M. E., & Olsen, D. H. (2011). The discourse of risk in horror movies post 9/11: Hospitality and hostility in perspective. *International Journal of Tourism Anthropology*, *1*(3), 304–317. doi:10.1504/IJTA.2011.043712

Krehbiel, K. (2010). *Pivotal politics: A theory of US lawmaking*. University of Chicago Press.

MacCannell, D. (1976). *The tourist: A new theory of the leisure class*. Univ of California Press.

Revel, J. F. (2002). *Anti-americanism*. San Francisco: Encounter Books.

Sackett, H., & Botterill, D. (2006). "Perception of International Travel Risk: an exploratory study of the influence of proximity to terrorist attack". *E-review of tourism Research*. Vol. 4 (2): 44-49. A&M Texas University, Estados Unidos, Material Disponible en www.ertr.tamu.edu. Extraído el 02 de Febrero de 2009.

Skoll, G. (2009). *Contemporary Criminology and Criminal Justice Theory. Evaluating justice Systems in Capitalist Societies*. New York: Macmillan. doi:10.1057/9780230101111

Steiner, C. (2007). Political instability, Transnational Tourist Companies and Destination Recovery in the Middle East after 9/11. *Tourism and Hospitality Planning & Development.*, *4*(3), 169–190. doi:10.1080/14790530701733421

Steiner, C. (2009). "Political risk, violent political unrest and Foreign Direct investment in the Tourism Industry". *Geographishes Institut. Papers of C. Steiner*. Disponible en http://www.staff.uni-mainz.de/steinec/Publikationen/Steiner_VPU-Tourism-FDI_1009_WP.pdf

Temple, R. C. (1961). *Americans Abroad*. New York: Sterling Publishing.

Weber, M. (1958). *Essays in Sociology*. New York: Oxford University Press.

Weber, M. (1964). *Economía y sociedad: esbozo de sociología comprensiva*. México: FCE.

Weber, M. (1995). *La Ética Protestante y el Espíritu del Capitalismo*. Barcelona: Península.

Wildman, S. M. (1996). *Privilege revealed: How invisible preference undermines America*. New York: NYU Press.

Wong, J. Y., & Yeh, C. (2009). Tourist Hesitation in Destination decision Making. *Annals of Tourism Research*, *36*(1), 6–23. doi:10.1016/j.annals.2008.09.005

Woods, J. et al.. (2008). Terrorism Risk Perception and Proximity to Primary Terrorist Targets: How close is too close? *Research in Human Ecology.*, *15*(1), 63–70.

KEY TERMS AND DEFINITIONS

Capitalism: Cultural and economic project centered on the hegemony of capital over other cultural values.

Excemptionalism: Political doctrine oriented to highlight the own supremacy or exception respecting to laws.

Exploitation: The usufruct of local resources towards profit-oriented goals.

Fear: Sentiment associated to the needs of protection.

Prevention: Act of identifying risks to prevent their affects.

State of Nature: State of wilderness where predominates the war of all against all.

Terrorism: Signals to the illegal use of force or violence to vulnerates the rights of weaker ones.

Compilation of References

@ Refilwel. (2015, 21 October). Tweet. Retrieved from https://twitter.com/RefilweI/status/656694365604204544?lang=en

Abeles, M. (2006). Globalization, Power and Survival: An Anthropological Perspective. *Anthropological Quarterly*, *79*(3), 483–508. doi:10.1353/anq.2006.0030

Achcar, G. (2006). *The Clash of Barbarisms. The Making of the New World Disorder.* London: Saqi.

Agarwal, H., Bommadevara, N., & Weinberg, A. (2013). *Using a Plan-Build-Run Organisational Model to Drive IT Infrastructure Objectives*. Retrieved from http://www.mckinsey.com/~/media/mckinsey/dotcom/client_service/BTO/ PDF/MOBT32_28-36_Plan-Build-Run_R5.ash

Altheide, D. L. (2006). Terrorism and the Politics of Fear. *Cultural Studies Critical Methodologies, 6*(4), 415-439.

Altheide, D. L. (2004). Consuming terrorism. *Symbolic Interaction*, *27*(3), 289–308. doi:10.1525/si.2004.27.3.289

Anderson, A. (2013). The Computer Fraud and Abuse Act: Hacking into the Authorization Debate. *Jurimetrics*, *53*(4), 447–466.

Andress, J., & Winterfield, S. (2011). *Cyber Warfare: Techniques, Tactics and Tools for Security Practitioners*. Waltham: Elsevier. doi:10.1016/B978-1-59749-637-7.00001-0

Anon. (2015). *What is terrorism?* Retrieved July 2, 2015, from http://http://www.terrorism-research.com/

Anonvnops. (2012, September 24). *We are Legion – The Story of the Hacktivists* [Video file]. Retrieved from http://www.youtube.com/watch?v=8TMtbNJ81X0

AnonymousPrime99. (2012, May 12). *We Are Legion: Digital (R)Evolution* [Video file]. Retrieved from http://www.youtube.com/watch?v=bLdhRP8iB6E

Armerding, T. (2012, September 13). *Anonymous Had Bad Month, But No Less "Reliable"*. Retrieved from http://www.csoonline.com/article/716092/anonymous-had-bad-month-but-no-less-reliable-

Arquilla, J., & Ronfeldt, D. (2001). *Networks and Netwars: The Future of Terror*. Crime, and Militancy.

Ashford, W. (2013, October 28). Cyber attack shuts down Israeli toll road tunnel. *Computer Weekly*. Retrieved from http://www.computerweekly.com/news/2240207924/Cyber-attack-shuts-down-Israeli-toll-road-tunnel

AV-TEST. (2015). *Malware*. Retrieved from AV Test The Independant IT Security Institute: https://www.av-test.org/en/statistics/malware/

Awad, N. F., & Krishnan, M. S. (2006). The Personalization Privacy Paradox: An Empirical Evaluation of Information Transparency and the Willingness to Be Profiled Online for Personalization. *Management Information Systems Quarterly*, *30*(1), 13–28.

Ayoob, M. (2009). State making, state breaking, and state failure. In C. Crocker, F. O. Hampson, & P. Aall (Eds.), *Leashing the Dogs of War, conflict management in a divided world* (pp. 83–114). Washington, DC: United States Institute of Peace press.

Ayres, N., Maglaras, L. A., Janicke, H., Smith, R., & He, Y. (2016). The mimetic virus: A vector for cyber security. *International Journal of Business Continuity and Risk Management*.

Ayres, N., & Maglaras, L. A. (2016). *Cyberterrorism targeting general public through social media*. Security and Communication Networks.

BAE Systems. (2015). *Intelligence-Led Threat Mitigation*. Retrieved from www.baesystems.com/ai

Balasundaram, I., & Ramaraj, E. (2012). An efficient technique for detection and prevention of SQL injection attack using ASCII based string matching. *Procedia Engineering*, *30*(2011), 183–190. http://doi.org/10.1016/j.proeng.2012.01.850

Balduzzi, M., Wihoit, K., & Pasta, A. (2013). Hey Captain, where's your ship? Attacking vessel tracking systems for fun and profit. *11th Annual Hack in the Box (HITB) Security Conference in Asia*. Retrieved from https://conference.hitb.org/hitbsecconf2013kul/materials/D1T1%20-%20Marco%20Balduzzi,%20Kyle%20Wilhoit%20Alessandro%20Pasta%20-%20Attacking%20Vessel%20Tracking%20Systems%20for%20Fun%20and%20Profit.pdf

Ball, E. (2006). *Getting to Know ITIL*. Retrieved from http://www.theiia.org/intAuditor/itaudit/archives/ 2006/june/getting-to-know-itil/

Bannon, L. J. (1991). From Human Factors to Human Actors: The Role of Psychology and Human Computer Interaction Studies in System Design. In J. Greenbaum & M. Kyng (Eds.), *Design At Work: Cooperative Design of Computer Systems* (pp. 25–44). Lawrence Erlbaum Associates, Inc. Publishers.

Barber, B. R. (1984). *Strong Democracy: Participatory Politics for a New Age*. Berkeley, CA: University of California Press.

Bassham, G. (2005). *The Religion and the Matrix and the problem of pluralism*. Chicago: Open Court.

Bath, C. (2015). *Twitter Bombarded By #Feesmustfall Tweets*. Retrieved 19 August 2016 from http://ewn.co.za/2015/10/23/More-than-140000-FeesMustFall-tweets-overnight

Baudrillard, J. (2006). Virtuality and Events: the hell of power. *Baudrillard Studies, 3*(2). Available at http://www.ubishops.ca/BaudrillardStudies/

Baudrillard, J. (1983). *Simulations*. New York: Semiotexte.

Baudrillard, J. (1989). *La Transparencias del Mal*. Barcelona: Anagrama.

Baudrillard, J. (1989). *The Transparency of Evil: essays on extreme phenomena*. Barcelona: Anagrama.

Baudrillard, J. (1993). *The Transparency of Evil: essays of extreme phenomena*. New York: Verso.

Baudrillard, J. (1994). *Simulacra & Simulation*. The University of Michigan Press.

Baudrillard, J. (1995a). *The systems of the objects*. Siglo XXI.

Baudrillard, J. (1995b). *The Gulf War Did Not Take Place*. Sydney: Power Publications.

Baudrillard, J. (2002). *Cool Memories IV*. New York: Verso.

Baudrillard, J. (2003). *The Spirit of Terrorism*. New York: Verso.

Baudrillard, J. (2006). *Virtuality and Events: the hell of power* (Vol. 3). Baudrillard Studies.

Baughman, H. M., Dearing, S., Giammarco, E., & Vernon, P. A. (2012). Relationships between Bullying Behaviours and the Dark Triad: A Study with Adults. *Personality and Individual Differences, 52*(5), 571–575. doi:10.1016/j.paid.2011.11.020

Bauman, Z. (2011). *Society under Siege*. Buenos Aires: Fondo de Cultura Económica.

Bauman, Z. (2013). *Consuming life*. New York: John Wiley & Sons.

Bauman, Z., & Lyon, D. (2013). *Liquid surveillance: A conversation*. Cambridge, UK: Polity Press.

Bay, A. (2011, January 18). *Tunisia's remarkable revolt*. Retrieved from http://www.strategypage.com/on_point/20110118224752.aspx

BBC. (2015). *Macedonia corruption: Fourth night of protests as snap election called*. Retrieved from http://www.bbc.com/news/world-europe-36058173

Beauchamp, Z. (2015). *South Africa's huge student protests*. Retrieved from http://www.vox.com/2015/10/24/9607236/south-africas-huge-student-protests-explained

Beck, U. (1998). *La Invención de lo Social*. FCE.

Beck, U. (2006). *La Sociedad del riesgo, hacia una nueva modernidad*. Buenos Aires: Paidos.

Beck, U. (2006). *Risk Society. Towards a new modernity: hacia una nueva modernidad*. Buenos Aires: Paidos.

Beck, U. (2011). Convivir con el Riesgo Global. In *La Humanidad Amenazada: gobernar los riesgos globales* (pp. 21–32). Madrid: Paidos.

Beike, D., & Wirth-Beaumont, E. (2005). Psychological Closure as a Memory Phenomenon. *Memory (Hove, England), 13*(6), 574–593. doi:10.1080/09658210444000241 PMID:16076673

Bellamy, A. J. (2005). Is the war on terror just? *International Relations, 19*(3), 275–296. doi:10.1177/0047117805055407

Bellamy, C., & Taylor, J. A. (1998). *Governing in the Information Age*. Maidenhead, UK: Open University Press.

Bell, D. (1974). *The Coming of postindustrial society. A venture in social forecasting*. London: Heinemann.

Bellovin, S. M. (2008). The Insider Attack Problem Nature and Scope. In S. J. Stolfo, S. M. Bellovin, S. Hershkop, A. Keromytis, S. Sinclair, & S. W. Smith (Eds.), *Insider Attack and Cyber Security: Beyond the Hacker* (pp. 1–4). New York, N.Y.: Springer Link. doi:10.1007/978-0-387-77322-3_1

Bicking, M., & Wimmer, M. A. 2010. Tools and Technologies in eParticipation: Insights from Project Evaluation. In F. De Cindio, A. Machintosh, & C. Peraboni (Eds.), *Online Deliberation:* Proceedings of the Fourth International Conference (pp. 75 – 86).

Bimber, B. (1998). The Internet and Political Transformation: Populism, Community and Accelerated Pluralism. *Polity, 31*(1), 133–160.

BIMCO. CLIA, IS, INTERCARGO & INTERTANKO. (2016, January). *The Guidelines on Cyber Security Onboard Ships*. Retrieved from www.bimco.org

Binns, A. (2012). Dont Feed the Trolls! *Journalism Practice, 6*(4), 547–562. doi:10.1080/17512786.2011.648988

Bishop, P., & Anderson, L. (2004). *E-Government to E-Democracy: 'High-Tech' Solutions to 'No Tech' Problems.* Paper presented to the Australian Electronic Governance Conference, Melbourne, Australia.

Bishop, J. (2012). The Psychology of Trolling and Lurking: The Role of Defriending and Gamification for Increasing Participation in Online Communities using Seductive Narratives. In H. Li (Ed.), *Virtual Community Participation and Motivation: Cross- Disciplinary Theories* (pp. 160–176). Hershey, PA: IGI Global. doi:10.4018/978-1-4666-0312-7.ch010

Bishop, J. (2014a). My Click is My Bond: The Role of Contracts, Social Proof, and Gamification for Sysops to Reduce Pseudo-Activism and Internet Trolling. In J. Bishop (Ed.), *Gamification for Human Factors Integration: Social, Education and Psychological Issues* (pp. 1–16). Hershey, PA: IGI Global. doi:10.4018/978-1-4666-5071-8.ch001

Bishop, J. (2014b). Representations of Trolls in Mass Media Communication: A Review of Media-Texts and Moral Panics Relating to Internet Trolling. *International Journal of Web Based Communities, 10*(1), 7–24. doi:10.1504/IJWBC.2014.058384

Black, I., & Chulov, M. (2011, February 18). *Bahrain, Libya and Yemen try to crush protests with violence*. Retrieved from http://www.guardian.co.uk/world/2011/feb/18/bahrain-libya-yemen-protests-violence

Block, J. (2008). Issues for DSM-V: Internet addiction. *The American Journal of Psychiatry*, *165*(3), 306–307. doi:10.1176/appi.ajp.2007.07101556 PMID:18316427

Blumler, J. C., & Coleman, S. (2001). Realising Democracy Online: A Civic Commons in Cyberspace. IPPR Citizens Online Research Publication No.2.

Brady, L. (2015). Tweet. Retrieved from https://twitter.com/BradyLeanne/status/657421743892615168/photo/1?ref_src=twsrc%5Etfw

Brafman, O., & Beckstrom, R. A. (2006). *The Starfish and the Spider: The Unstoppable Power of Leaderless Organizations*. New York: Penguin Publishing.

British Broadcasting Corporation (BBC). (2015, June 21). *Polish LOT aeroplanes grounded by computer hack*. Retrieved from http://www.bbc.com/news/world-europe-33219276

Brook, C. (2015, June 22). Polish planes grounded after airline hit with DDoS attack. *Threatpost*. Retrieved from https://threatpost.com/polish-planes-grounded-after-airline-hit-with-ddos-attack/113412

Brown, M. (2009). New Global Dangers. In C. Crocker, F. O. Hampson, & P. Aall (Eds.), *Leashing the Dogs of War, conflict management in a divided world* (pp. 39–52). Washington, DC: United States Institute of Peace press.

Buckels, E. E., Trapnell, P. D., & Paulhus, D. L. (2014). Trolls Just Want to Have Fun. *Personality and Individual Differences*, *67*, 97–102. doi:10.1016/j.paid.2014.01.016

Bunt, G. R. (2003). Islam in the Digital Age: E-Jihad, Online Fatwas and Cyber Islamic Environments. London: Pluto Press. Retrieved from http://www.bandung2.co.uk/books/Files/Religion/Islam%20in%20the%20Digital%20Age%20(E-Jihad,%20Online%20Fatwas%20and%20Cyber%20Islamic%20Environments)%20-%20Gary%20R.%20Bunt.pdf

C, K. (2009). *Goodtimes Virus*. Retrieved from Know Your Meme: http://knowyourmeme.com/memes/goodtimes-virus

Campbell, M. A. (2005). Cyber Bullying: An Old Problem in a New Guise? *Australian Journal of Guidance & Counselling*, *15*(1), 68–76. doi:10.1375/ajgc.15.1.68

Carayon, P., Hoonakker, P., & Smith, M. J. (2012). Human Factors in Organizational Design and Management. In G. Salvendy (Ed.), *Handbook of Human Factors and Ergonomics* – (4th ed.; pp. 534–552). John Wiley and Sons. doi:10.1002/9781118131350.ch18

Carnegie Mellon University SEI - CERT Coordination Centre. (2003). *Handbook for Computer Security Incident Response Teams (CSIRTs)*. Retrieved from http://www.sei.cmu.edu/library/abstracts/reports/ 03hb002.cfm

Carr, R. (2015, November 2). Empowering pipeline SCADA cybersecurity. *Tripwire*. Retrieved from http://www.tripwire.com/state-of-security/featured/empowering-pipeline-scada-cybersecurity/

Carroll, T., Edgar, T., Manz, D., & Greitzer, F. L. (2012). *Realizing Scientific Methods for Cyber Security*. Research paper presented at LASER '12.

Cassim, B. (2015, October 24). *10 unforgettable moments of solidarity from #FeesMustFall*. The Daily Vox. Retrieved from http://www.thedailyvox.co.za/10-unforgettable-moments-of-solidarity-from-feesmustfall/

Castel, R. (1983). From dangerousness to risk. *Actes de la recherche en sciences sociales*, *47*(4), 119-127.

Catinat, M., & Vedel, T. (2000). Public Policies for Digital Democracy. In K. L. Hacker & J. van Dijk (Eds.), *Digital Democracy: Issues of Theory and Practice* (pp. 184–208). London: Sage Publications. doi:10.4135/9781446218891.n11

Caverlee, J., Cheng, Z., Sui, D. Z., & Kamath, K. Y. (2013). Towards Geo-Social Intelligence: Mining, Analyzing, and Leveraging Geospatial Footprints in Social Media. *IEEE Data Eng. Bull.*, *36*(3), 33–41.

Cavoukian, A., & Jonas, J. (2012). *Privacy by Design in the Age of Big Data*. Ontario: Information and Privacy Commissioner of Ontario, Canada.

Center for the Protection of National Infrastructure (CPNI). (2008). *Process control and SCADA security, Good practice guide 1: Understand the business risk*. Retrieved from http://www.cpni.gov.uk/Documents/Publications/2008/2008024-GPG_SCADA_Business_Risk.pdf

Chabrow, E. (2014, July 13). Details emerge of Boeing hack. *GovInfoSecurity*. Retrieved from http://www.bankinfosecurity.com/details-emerge-boeing-hack-a-7053/op-1

Chan, G. Y., Lee, C. S., & Heng, S. H. (2013). Discovering fuzzy association rule patterns and increasing sensitivity analysis of XML-related attacks. *Journal of Network and Computer Applications*, *36*(2), 829–842. doi:10.1016/j.jnca.2012.11.006

Chan, G. Y., Lee, C. S., & Heng, S. H. (2014). Defending against XML-related attacks in e-commerce applications with predictive fuzzy associative rules. *Applied Soft Computing*, *24*, 142–157. doi:10.1016/j.asoc.2014.06.053

Chavez, N. (2008). *Cuando Los Mundos Convergen, terrorismo, narcotráfico y migración post 9/11* [When the worlds converge, terrorism, narco-traffic and migration post 9/11]. Quito: FLACSO Ediciones.

Chen, Y., Harper, F. M., Konstan, J., & Li, S. X. (2009). Group Identity and Social Preferences. *The American Economic Review*, *99*(1), 431–457. doi:10.1257/aer.99.1.431

Chismon, D., & Ruks, M. (2015). *Threat Intelligence: Collecting, Analysing, Evaluating, MWR InfoSecurity, Centre for the Protection of National Infrastructure, and CERT-UK*. Retrieved from http://CPNI.gov.uk

Chiu, C. M., Hsu, M. H., & Wang, E. T. (2006). Understanding Knowledge Sharing in Virtual Communities: An Integration of Social Capital and Social Cognitive Theories. *Decision Support Systems*, *42*(3), 1872–1888.

Christey, S., & Martin, R. A. (n.d.). *CWE - Vulnerability Type Distributions in CVE*. Retrieved April 17, 2015, from http://cwe.mitre.org/documents/vuln-trends/index.html

Chung, Y., Wu, M., Chen, Y., & Chang, W. (2012). *A Hot Query Bank approach to improve detection performance against SQL injection attacks. In Computers & Security*. Elsevier Ltd. doi:10.1016/j.cose.2011.11.007

Chun, S., Shulman, S., Sandoval, R., & Hovy, E. (2010). Government 2.0: Making Connections Between Citizens, Data and Government. *Information Polity Journal*, *15*(1-2), 1–9.

Cilluffo, F., Cardash, S., & Salmoiraghi, G. (2012). A Blueprint for Cyber Deterrence: Building Stability through Strength. *Military and Strategic Affairs*, *4*(3), 11.

Claburn, T. (2009, May 7). Air traffic control system repeatedly hacked. *Dark Reading*. Retrieved from http://www.darkreading.com/risk-management/air-traffic-control-system-repeatedly-hacked/d/d-id/1079364

Clayton, M. (2013, April 19). Exclusive: Cyberattack leaves natural gas pipelines vulnerable to sabotage. *Christian Science Monitor*. Retrieved from http://www.csmonitor.com/Environment/2013/0227/Exclusive-Cyberattack-leaves-natural-gas-pipelines-vulnerable-to-sabotage

Cluley, G. (2016, January 21). Hackers steal $55 million from Boeing supplier. *Tripwire*. Retrieved from http://www.tripwire.com/state-of-security/security-data-protection/boeing-supplier-hacked-claims-55-million-worth-of-damage-as-stock-price-falls/

Coleman, S. (2005). 'Just How Risky is Online Voting? *Information Polity*, *10*(1), 95–104.

Coleman, S. (2013). *Actors of History? Religion, Politics, and Reality within the Protestant Right in America. In Religion, Politics & Globalization* (pp. 171–188). New York: Berghan Book.

Coleman, S., & Blumler, J. G. (2011). The Wisdom of Which Crowd? On the Pathology of a Listening Government. *The Political Quarterly*, *82*(3), 355–364. doi:10.1111/j.1467-923X.2011.02210.x

Coleman, S., & Gotze, J. (2001). *Bowling Together: Online Public Engagement in Policy Deliberation*. London: Hansard Society.

Collin, B. (1997). *The Future of Cyberterrorism. In Criminal & Justice International* (pp. 15–18). Retrieved from The Future of Cyberterrorism.

Coulter, G. (n.d.). *Jean Baudrillard: from the Ocean to the desert or the Poetics of Radicality*. New Smyrna Beach, FL: Intertheory Press.

Cowen, N. (2015). *High Court Issues Interdiction against A Hashtag In #Feesmustfall Doc, 20 October*. Retrieved 11 April from http://www.htxt.co.za/2015/10/20/high-court-issues-interdiction-against-a-hashtag-in-feesmustfall-doc/

Cranor, L.F., Reagle, J., & Ackerman, M.S. (1999, August 17). Beyond Concern: Understanding Net Users' Attitudes About Online Privacy. *Privacy Study.*

Crosston, M. (2012). Virtual Patriots and a New American Cyber Strategy: Changing the Zero-Sum Game. Strategic Studies Quarterly, 6(4).

Crosston, M. (2013). Duqu's Dilemma: The Ambiguity Assertion and the Futility of Sanitized Cyber War. *Journal of Military and Strategic Affairs, 5*(1).

Crosston, M. D. (2011, Spring). World Gone Cyber MAD: How "Mutually Assured Debilitation" Is the Best Hope for Cyber Deterrence. *Strategic Studies Quarterly.*

Cruickshank, P., Edelmann, N., & Smith, C. (2010). Signing an e-Petition as a Transition from Lurking to Participation. In H. J. Scholl & M. Janssen (Eds.), *Electronic Government and Electronic Participation: Joint Proceedings of Ongoing Research and Projects of IFIP EGOV and IFIP ePart 2010* (pp. 275–282). Linz, Austria: Trauner Verlag.

CSA Big Data Working Group. (2013, September). *Big Data Analytics for Security Intelligence, Cloud Security Alliance.* Retrieved from http://www.cloudsecurityalliance.org/research/big-data

CyberKeel. (2014, October 15). *Maritime cyber-risks.* Retrieved from www.cyberkeel.com

Czaja, S. J., & Nair, S. N. (2012). Human Factors Engineering and Systems Design. In G. Salvendy (Ed.), *Handbook of Human Factors and Ergonomics* – (4th ed.; pp. 38–56). John Wiley and Sons. doi:10.1002/9781118131350.ch2

Dalziel, H. (2015). *How to Define and Build an Effective Cyber Threat Intelligence Capability.* Waltham, MA: Syngress.

Dammert, L., & Arias, P. (2007). El desafío de la delincuencia en América Latina: diagnóstico y respuestas de política. *Seguridad y violencia: desafíos para la ciudadanía. FLACSO*, 21-66.

Dark Reading. (2016, June 14). *Preliminary Agreement Reached On Airline Cybersecurity.* Retrieved from http://www.darkreading.com/risk/preliminary-agreement-reached-on-airline-cybersecurity/d/d-id/1325904

Denning, D. E. (2000). *Cyberterrorism Testimony before the Special Oversight Panel on Terrorism.* Georgetown University.

Department of Justice and Correctional Services. (2015). *Cybercrimes and Cybersecurity Bill.* Pretoria: Government of South Africa.

di Gennaro, C., & Dutton, W. H. (2006). The Internet and the Public: Online and Offline Political Participation in the UK. *Parliamentary Affairs, 59*(2), 299–313. doi:10.1093/pa/gsl004

Dickerson, C., & Mavris, D. N. (2010). *Architecture and Principles of Systems Engineering.* CRC Press.

Diken, B. (2011). Militarization as comedy of t(error). *Economia Autonoma, 4*(7), 72–92.

Diken, B., & Laustsen, C. B. (2002). Zones of indistinction security, terror, and bare life. *Space and Culture*, *5*(3), 290–307. doi:10.1177/1206331202005003009

Dodds, E. R. (1997). *Los Griegos y lo Irracional*. Madrid: Alianza.

Donath, J. S. (1999). Identity and Deception in the Virtual Community. In M. A. Smith & P. Kollock (Eds.), Communities in Cyberspace (pp. 29-59). London: Routledge Press.

Douglas, M. (1992). *Risk and Blame*. Routledge.

Douglas, M. (1992). *Risk and Blame*. New York: Routledge. doi:10.4324/9780203430866

Douglas, M. (2007). *Pureza y Peligro: un análisis de los conceptos de contaminación y tabú*. Buenos Aires: Nueva Visión.

Douglas, M., & Wildawsky, A. (1983). *Risk and Culture: an essay on the selection of technological and environmental Dangers*. Los Angeles, CA: University of California Press.

Duclos, D. (1987). *Le Risque: une construction sociale?. In La Societeé Vulnerable* (pp. 91–92). Paris: Ecole Normale Superieure.

Duda, R. O., Hart, P. E., & Stork, D. G. (2012). *Pattern classification*. John Wiley & Sons.

Duell, M., & Glanfield, E. (2015, November 4). Scuffles break march against university tuition fees shadow chancellor John McDonnell accuses Government betraying students. *The Daily Mail*. Retrieved from http://www.dailymail.co.uk/news/article-3303774/Scuffles-break-march-against-university-tuition-fees-shadow-chancellor-John-McDonnell-accuses-Government-betraying-students.html

Dunn, J. E. (2013, October 16). Hackers planted remote devices to smuggle drugs through Antwerp port, Europol reveals. *Techworld*. Retrieved from http://news.techworld.com/security/3474018/hackers-planted-remote-devices-to-smuggle-drugs-through-antwerp-port-europol-reveals/

Duran, M. S. (2002). Somebody else's civil war. *Foreign Affairs, 81*(1), 22-42.

Durkheim, E. (1938). The Rules of the Sociological Method (S. A. Solovay & J. H. Mueller, Trans.; G. E. G. Catlin, Ed.). Glencoe, IL: The Free Press.

Durkheim, E. (1984). The Division of Labor in Society (W. D. Halls, Trans.). New York, NY: The Free Press.

Durkheim, E. (1915). *The Elementary Forms of the Religious Life* (J. W. Swain, Trans.). New York, NY: The Free Press.

Dutta, S., Dutton, W. H., & Law, G. (2011). *The New Internet World: A Global Perspective on Freedom of Expression, Privacy, Trust and Security Online*. INSEAD Working Paper No. 2011/89/TOM.

Dutton, W. H., & Peltu, M. (2007). *Reconfiguring Government Public Engagements: Enhancing the Communicative Power of Citizens*. Oxford Internet Institute Forum Discussion Paper No. 9.

Dutton, W. (1998). *Society on the Line: information politics in the digital Age*. Oxford, UK: Oxford University Press.

Dutton, W. H., & Shepherd, A. (2006). Trust in the Internet as an Experience Technology. *Information Communication and Society*, *9*(4), 433–451. doi:10.1080/13691180600858606

Dwan, B. (2001). Cyber-terrorism: Virtual for Who? *Computer Fraud & Security*, *11*, 12–14.

Dyryavyy, Y. (2014). *Preparing for cyber Battleships – electronic chart display and information system security*. NCC Group. Retrieved from https://www.nccgroup.com/media/481230/2014-03-03_-_ncc_group_-_whitepaper_-_cyber_battle_ship_v1-0.pdf

Earl, J., & Kimport, K. (2011). *Digitally Enabled Social Change: Activism in the Internet Age*. Cambridge, MA: The MIT Press. doi:10.7551/mitpress/9780262015103.001.0001

Eggers, W. D. (2005). *Government 2.0: Using Technology to Improve Education, Cut Red Tape, Reduce Gridlock and Enhance Democracy*. Lanham, MD: Rowman and Littlefield Publishers Inc.

Eid, M. (Ed.). (2014). *Exchanging terrorism oxygen for media airwaves: The age of terroredia*. Hershey, PA: IGI Global. doi:10.4018/978-1-4666-5776-2

Eishenhauer, J., Donnelly, P., Ellis, M. & O'Brien, M. (2006, January). *Roadmap to Secure Control System in the Energy Sector*. Columbia, MD: Energetics Incorporated.

Emspak, J. (2012, May 7). GPS vulnerable to dangerous hacks and spamming. *Discovery Channel*. Retrieved from http://www.msnbc.msn.com/id/47324401/ns/technology_and_science-security/#.T7BXxoGvPLU

ENISA. (2006). *A Step-By-Step Approach On How To Set Up a CSIRT*. Deliverable WP2006/5.4 (CERT-D1/D2).

ENISA. (2015). *CSIRT Structure*. Retrieved from https://www.enisa.europa.eu/activities/cert/support/ guide2/internal-management/structure

Ermoliev, Y., Makowski, M., & Marti, K. (2012). *Managing Safety of Heterogeneous Systems*. Heidelberg, Germany: Springer. doi:10.1007/978-3-642-22884-1

European Network and Information Security Agency (ENISA). (2011, November). *Analysis of cyber security aspects in the maritime sector*. Retrieved from http://www.enisa.europa.eu/activities/Resilience-and-CIIP/critical-infrastructure-and-services/dependencies-of-maritime-transport-to-icts/cyber-security-aspects-in-the-maritime-sector-1

Farnsworth, T. (2012). US Officials Detail Cyber Policy. *Arms Control Today*, *42*(9), 32.

Fearn-Banks, K. (2007). *Crisis Communications: A Casebook Approach*. Mahwah, NJ: Lawrence Erlbaum Associates.

FedSolutions. (2012, May 24). *Are You Paying Too Much For DDoS Attacks?* Retrieved from http://www.fedsolutions.com/2012/05/cost-of-ddos-attacks/

Feni, M., & Mama, S. (2012). *Student protests turn violent.* Available at: http://www.iol.co.za/news/south-africa/western-cape/student-protests-turn-violent-1.1214650#.VXAgO8-qqko

Ferguson, R. (2006). *Digital Dialogues: Interim Report, December 2005 – August 2006.* London: The Hansard Society.

Fieldhouse, E., Tranmer, M., & Russel, A. (2007). Something about Young People or Something about Elections? Electoral Participation of Young People in Europe: Evidence from a Multilevel Analysis of the European Social Survey. *European Journal of Political Research, 46*(6), 797–822. doi:10.1111/j.1475-6765.2007.00713.x

Fink, G., Best, D., Manz, D., Popovsky, V., & Endicott-Popovsky, B. (2013). Gamification for Measuring Cyber Security Situational Awareness. In D. D. Schmorrow & C. M. Fidopiastis (Eds.), *Foundations of Augmented Cognition: Advancing Human Performance and Decision-Making Through Adaptive Systems* (pp. 656–665). Berlin: Springer-Verlag. doi:10.1007/978-3-642-39454-6_70

Fishkin, J. S. (1995). *The Voice of the People: Public Opinion and Democracy.* New York: Yale University Press.

Fitri, N. (2011, April). *Democracy Discourses Through the Internet Communication: Understanding the Hacktivism for the Global Changing.* Retrieved from http://www.academia.edu/1438665/Hacktivism_for_Global_Changing_Nofia_Fitri_ojcmt.net_

Flaherty, M. P., Samenow, J., & Rein, L. (2014, November 12). Chinese hack U.S. weather systems, satellite network. *The Washington Post.* Retrieved from http://www.washingtonpost.com/local/chinese-hack-us-weather-systems-satellite-network/2014/11/12/bef1206a-68e9-11e4-b053-65cea7903f2e_story.html

Flaxman, G., & Rogerson, B. (2011). The Economy of fear. *Symploke, 18*(1-2), 333–336.

Foucault, M. (1992). *Microfísica del Poder.* Madrid: La Piqueta.

Foucault, M. (2000). *Los Anormales. Curso en el College de France. 1974-1975.* Buenos Aires: Fondo de Cultura Económica.

Foucault, M. (2003). *Society must be defended: Lectures at the Collège de France* (D. Macey, Trans.). New York: Picador.

Foucault, M. (2006). *Seguridad, Territorio y Población.* Buenos Aires: Fondo de Cultura Económica.

Foucault, M. (2008). *El Nacimiento de la Clínica.* Buenos Aires: Siglo XXI.

Fraisse, P. (1973). *Las Emociones. In Motivación, Emoción y Personalidad* (pp. 106–187). Buenos Aires: Editorial Paidos.

Fraud Alert. (2013, December 18). *SAA warns against email phishing scam.* Retrieved from http://fraudalert.co.za/saa-warns-against-email-phishing-scam/

Fromm, E. (2005). *El Miedo a la libertad.* Buenos Aires: Paidos.

Furlong, A. (2015, October 19). *Students shut down UCT, Ground Up*. Retrieved from http://groundup.org.za/article/students-shut-down-uct_3412

Gallego, R. (2015). *Security Intelligence Is Useful Only If Isolated From The Noise, Computer Weekly*. Retrieved from http://www.computerweekly.com/news/4500255079/Security-Think-Tank-Security-intelligence-useful-only-if-isolated-from-the-noise

Gandotra, V., Singhal, A., & Bedi, P. (2009). Threat Mitigation, Monitoring and Management Plan - A New Approach in Risk Management. *2009 International Conference on Advances in Recent Technologies in Communication and Computing*. Retrieved from http://ieeexplore.ieee.org/stamp/stamp.jsp? tp=&arnumber=5328897

García, C., Chauveau, P., Ledezma, J., & Pinto, M. (2013). What can social media teach us about protests? Analyzing the Chilean 2011-12 student movement's network evolution through twitter data. In *Proceedings of the 4th International Conference on Collaborative Innovation Networks COINs13*.

Gartner. (2013). *Managed Security Service Provider (MSSP)*. Retrieved from http://www.gartner.com/it-glossary/mssp-managed-security-service-provider

Gaudin, S. (2011, March 11). *Twitter, Facebook become lifelines after Japan Quake, Computerworld*. Retrieved from http://www.computerworld.com/s/article/9214213/Twitter_Facebook_become_lifelines_after_Japan_quake

Geach, C. (2015). *FeesMustFall: 10 powerful placards, Independent Online*. http://www.iol.co.za/news/south-africa/feesmustfall-10-powerful-placards-1933343

Gebauer, G., & Wulf, C. (1995). *Mimesis: culture, art, society*. University of California Press.

Geman, S., & Geman, D. (1993). Stochastic relaxation, Gibbs distributions and the Bayesian restoration of images. *Journal of Applied Statistics*, *20*(5-6), 25–62. doi:10.1080/02664769300000058

Geoghegan, T. (2007). The Petition, the 'Prat' and the Political Ideal. *BBC News – Magazine*. Available at: http://news.bbc.co.uk/1/hi/magazine/6354735.stm

Ghena, B., Beyer, W., Hillaker, A., Pevarnek, J., & Halderman, J. A. (2014, August 19). Green lights forever: analyzing the security of traffic infrastructure. *Proceedings of the 8th USENIX Workshop on Offensive Technologies (WOOT '14)*. Retrieved from https://www.usenix.org/conference/woot14/workshop-program/presentation/ghena

Giddens, A. (1991). *Modernity and Self-Identity: Self and Society in the Late Modern Age*. Stanford University Press.

Giddens, A. (1991). *Modernity and self-identity: self and society in the late modern age*. Stanford, CA: Stanford University Press.

Giddens, A. (1997). *The Transformation of Intimacy*. Oxford, UK: Polity Press.

Giddens, A. (1999). *Consecuencias de la Modernidad*. Madrid: Alianza Editorial.

Gjelten, T. (2011, July 20). *FBI Tries To Send Message With Hacker Arrests*. Retrieved from http://m.npr.org/news/Technology/138555799

Gollmann, D. (2008). Securing Web applications. *Information Security Technical Report*, *13*(1), 1–9. doi:10.1016/j.istr.2008.02.002

Goodhart, M. (2007). Europes Democratic Deficits through the Looking Glass: The European Union as a Challenge for Democracy. *Perspectives on Politics*, *5*(3), 567–584. doi:10.1017/S1537592707071551

Gordon, S., & Ford, R. (2002). Cyberterrorism? *Computers & Security*, *21*(7), 636–647. doi:10.1016/S0167-4048(02)01116-1

Government Accountability Office (GAO). (2014, June). *Maritime critical infrastructure protection: DHS needs to better address port cybersecurity*. Retrieved from www.gao.gov/assets/670/663828.pdf

Government Accountability Office (GAO). (2015a, January). *FAA Needs to Address Weaknesses in Air Traffic Control Systems*. Retrieved from http://www.gao.gov/assets/670/668169.pdf

Government Accountability Office (GAO). (2015b, April). *FAA Needs a More Comprehensive Approach to Address Cybersecurity As Agency Transitions to NextGen*. Retrieved from http://www.gao.gov/assets/670/669627.pdf

Graham, G. (2002). *Genes: a philosophical inquiry*. Routledge.

Graves, T. (2012). *Service, Function and Capability*. Retrieved from http://weblog.tetradian.com/2012/09/ 22/service-function-capability-again/

Gray, C. S. (2007). *The implications of preemptive and preventive war doctrines: A Reconsideration*. Army War Coll Strategic Studies Inst Carlisle Barracks PA.

Greenberg, A. (2015, July 21). Hacker remotely kill a jeep on the highway. *Wired*. Retrieved from http://www.wired.com/2015/07/hackers-remotely-kill-jeep-highway/

Greenberg, A. (2012). *This Machine Kills Secrets: How WikiLeakers, Cypherpunks, and Hacktivists Aim to Free the World's Information*. New York: Penguin Publishing.

Greenburg, A. (2011). *Tech*. Retrieved 8 25, 2015, from Forbes: http://www.forbes.com/sites/andygreenberg/2011/07/22/apple-laptops-vulnerable-to-hack-that-kills-or-corrupts-batteries/

Green, J. (2002). *The Myth of Cyberterrorism. Washington Monthly*.

Gregory, P. R. (2014b). You Tube shatters Russian lies about troops in Ukraine: Putin denies truth to Obama. *Forbes*, *14*(April). Retrieved from http://www.forbes.com/sites/paulroderick-gregory/2014/04/14/you-tube-shatters-russian-lies-about-troops-in-ukraine/

Gresty, D., Taylor, M., & Lunn, J. (2005). Legal Considerations for Cyber-Activism. *Systems Research and Behavioral Science*, *22*(6), 566–567. doi:10.1002/sres.625

Greven, P. (1988). *The Protestant Temperament. Pattersn of child-rearing religious experience, and the self in Early America*. Chicago University Press.

Griffiths, M. D. (1995). Technological Addictions. *Clinical Psychology Forum, 76*, 14-19.

Griffiths, M. D. (1998). Internet Addiction: Does It Really Exist? In J. Gackenbach (Ed.), *Psychology and the Internet: Intrapersonal, Interpersonal, and Transpersonal Implications* (pp. 61–75). New York, NY: Academic Press.

Gruzd, A. (2015). *The Use of Social Media During the 2014 Crisis in Ukraine*. University of Pennsylvania. Retrieved from https://www.asc.upenn.edu/news-events/events/use-social-media-during-2014-crisis-ukraine-anatoliy-gruzd

Hakar, V. (2003). The Right to Civil Disobedience. *Osgoode Hall Law Journal, 41*(2/3), 411.

Halfond, W. G. J., & Orso, A. (2005). AMNESIA: Analysis and Monitoring for NEutralizing SQL-injection Attacks. In *Proceedings of the 20th IEEE/ACM International Conference on Automated Software Engineering* (pp. 174–183). http://doi.org/ doi:10.1145/1101908.1101935

Hampson, N. (2012). Hacktivism: A New Breed of Protest in a Networked World. *Boston College International and Comparative Law Review, 35*(2), 531–532.

Hansen, S. (2002). Excessive Internet Usage or Internet Addiction? The Implications of Diagnostic Categories for Student Users. *Journal of Computer Assisted Learning, 18*(2), 232–236. doi:10.1046/j.1365-2729.2002.t01-2-00230.x

Hardaker, C. (2010). Trolling in Asynchronous Computer-Mediated Communication: From User Discussions to Academic Definitions. *Journal of Politeness Research, 6*(2), 215–242. doi:10.1515/jplr.2010.011

Harrison, V., & Pagliery, J. (2015, April 14). *Nearly 1 million new malware threats released every day*. Retrieved from CNN Money: http://money.cnn.com/2015/04/14/technology/security/cyber-attack-hacks-security/

Heinzelman, J., & Waters, C. (2010). *Crowdsourcing Crisis Information in Disaster-Affected Haiti, Special Report 252*. United States Institute of Peace.

Hendawi, H. (2011, January 27). *Egypt protests pose threat for regime*. Retrieved from http://news.yahoo.com/s/ap/20110127/ap_on_re_mi_ea/ml_egypt_protest_66

Hewlett-Packard. (2013). *5G/SOC: SOC Generations*. Retrieved from http://www8.hp.com/us/en/software-solutions/software.html?compURI=1343719#!

Himma, K. E., & Tavani, H. T. (2008). *The Handbook of Information and Computer Ethics*. Hoboken, NJ: John Wiley & Sons, Inc. Retrieved from http://www.e-reading.org.ua/bookreader.php/141465/The_Handbook_of_Information_and_Computer_Ethics.pdf

Hirayama, L. (2015, 22 October). Tweet. Retrieved from https://twitter.com/Lalla_Hirayama/status/657216095347200000?ref_src=twsrc^tfw

Hobbes, T. (2004). Leviatán. Buenos Aires: Ediciones el Libertador.

Hobbes, T. (1994). *Leviathan: with selected variants from the Latin edition of 1668* (Vol. 2). New York: Hackett Publishing.

Hoff, J. (2006). The Shaping of Digital Political Communication – Creating e-Democracy in a Danish Municipality: Intentions and Realities. In H. K. Hansen & J. Hoff (Eds.), *Digital Governance: Networked Societies. Creating Authority, Community and Identity in a Globalized World* (pp. 261–299). Copenhagen: Samfundslitteratur Press/NORDICOM.

Hoffman, B. (2013). *Inside terrorism*. Columbia University Press.

House homeland security subcommittee on oversight, investigations, and management hearing. (2012). Lanham, MD: Federal Information & News Dispatch, Inc.

Howie L (2012) *Witnesses to Terror: Understanding the Meanings and Consequences of Terrorism*. Hampshire, UK: Palgrave-Macmillan.

Howie, L. (2011). *Terror on the screen: Witnesses and the re-animation of 9/11 as image-event, popular culture*. Washington, DC: New Academia Publishing, LLC.

Ho, Y. C., & Kashyap, R. L. (1965). An algorithm for linear inequalities and its applications. *IEEE Transactions on Electronic Computers*, *5*(5), 683–688. doi:10.1109/PGEC.1965.264207

Hsu, F. D., & Marinucci, D. (2013). *Advances in Cyber Security: Technology, Operation, and Experiences*. Fordham University Press. doi:10.5422/fordham/9780823244560.001.0001

Huysman, M., & Wulf, V. (2004). *Social Capital and Information Technology*. Hong Kong: M.I.T. Press.

Iasiello, E. (2014). Hacking Back: Not the Right Solution. *Parameters*, *44*(3), 113.

IBM. (2013). *Protecting Your Critical Data With Integrated Security Intelligence, Software Thought Leadership White Paper*. Retrieved from http://www-01.ibm.com/common/ssi/cgi-bin/ssialias? htmlfid=WGW03023USEN

ICS-CERT. (2016). *NCCIC/ICS-CERT Year in Review: FY2015*. Washington, DC: Department of Homeland Security.

Ignatieff, M. (2013). *The lesser evil: Political ethics in an age of terror*. Princeton University Press.

Ilcan, S., & Lacey, A. (2013). Networks of Social Justice: Transnational Activism and Social Change. *Studies in Social Justice*, *7*(1).

Independent Online. (2015, October 22). *Giving #FeesMustFall a voice*. Retrieved from: http://www.iol.co.za/capeargus/giving-feesmustfall-a-voice-1.1934338#.VioK634rLRY

INGAA. (2011, January 31). *Control Systems Cyber Security Guidelines for the Natural Gas Pipeline Industry*. Washington, DC: INGAA.

INSA Cyber Intelligence Task Force. (2014). *Operational Cyber Intelligence*. Arlington, VA: Intelligence and National Security Alliance. Retrieved from www.insaonline.org

INSA Cyber Intelligence Task Force. (2015). *Tactical Cyber Intelligence*. Arlington, VA: Intelligence and National Security Alliance. Retrieved from www.insaonline.org

Irwin, W. (2005). Computers, Caves and Oracles: neo and Socrates. In W. Irwin (Ed.), *The Matrix and Philosophy* (pp. 5–15). Chicago: Open Court.

ISACA. (2012). *COBIT 5 – A Business Framework for the Governance and Management of Enterprise IT*. Retrieved from http://www.isaca.org/cobit/pages/cobit-5-framework-product-page.aspx

Ismail, S., Sitnikova, E., & Slay, J. (2015). SCADA systems cyber security for critical infrastructures: case Studies in the transport sector. In *Proceedings of the 10th International Conference on Cyber Warfare and Security (ICCWS 2015)*, (pp. 425-433).

ISO/IEC. (2009). *An Introduction to ISO 27001, ISO 27002.... ISO 27008*. Retrieved from http://www.27000.org/

Israeli Homeland Security.com. (2013, August 7). *$80 million yacht hijacked by students spoofing GPS signals*. Retrieved from http://i-hls.com/2013/08/80-million-yacht-hijacked-by-students-spoofing-gps-signals/

ITWeb. (2015). *Vodacom talks cyber hunting*. Retrieved from http://www.itweb.co.za/index.php?option=com_content&view=article&id=143114:Vodacom-talks-cyberhunting&catid=234

Jacko, J. A., Yi, J. S., Sainfort, F., & McClellan, M. (2012). Human Factors and Ergonomic Methods. In G. Salvendy (Ed.), *Handbook of Human Factors and Ergonomics* (4th ed.; pp. 289–329). John Wiley and Sons. doi:10.1002/9781118131350.ch10

Jacobs, P. (2015). *Towards a Framework For Building Security Operation Centres*. Rhodes University. Retrieved from http://contentpro.seals.ac.za/iii/cpro/DigitalItemViewPage.external?lang=eng&sp=1017932&sp=T&suite=def

Jacobs, P., Von Solms, S. H., & Grobler, M. M. (2016). E-CMIRC – Towards a Model For The Integration of Services Between SOCs and CSIRTs. *Proceedings of the 15th European Conference on Cyber Warfare and Security (ECCWS) 2016.*

Jacobs, J., & Rudis, B. (2014). *Data Driven Security*. Indianapolis, IN: Wiley.

Jang, Y.-S., & Choi, J.-Y. (2014). Detecting SQL injection attacks using query result size. *Computers & Security*, *44*, 104–118. doi:10.1016/j.cose.2014.04.007

Jansen, E., & James, V. (2002). *NetLingo: The Internet Dictionary*. Oxnard, CA: Netlingo Inc.

Janssen, D., & Kies, R. (2004). *Online Forums and Deliberative Democracy: Hypotheses, Variables and Methodologies*. e-Democracy Centre e-Working Papers No. 1.

Jiejin, H. (2009). *A Practical Approach to the Operation of Telecommunication Services Driven by the TMF eTOM Framework* (Master's Thesis). Universitat Poliècnica de Catalunya.

Jones, A. (2005). Cyberterrorism: Fact or Fiction. *Computer Fraud & Security*, *6*, 4–7.

Jones, D. A. (2005). Cyber Terrorism: Fact or Fiction. *Computer Fraud & Security*, 4.

Kamis, B., & Thiel, T. (2013). *The Original Battle Trolls: How States Represent the Internet as a Violent Place.* Paper presented at the ECPR General Conference, Bordeaux, France.

Kamzi, A. (2011, September 27). How Anonymous Emerged to Occupy Wall Street. *The Guardian*. Retrieved from http://www.guardian.co.uk/commentisfree/cifamerica/2011/sep/27/occupy-wall-street-anonymous

Kaplan, E. (2009). *Terrorists and the Internet.* Retrieved July 3, 2015, from http://http://www.cfr.org

Karwowski, W. (2012). The Discipline of Human Factors and Ergonomics. In G. Salvendy (Ed.), *Handbook of Human Factors and Ergonomics* (4th ed.; pp. 3–37). John Wiley and Sons. doi:10.1002/9781118131350.ch1

Kaspersky Lab. (2013). *The 'Icefog' APT: a tale of cloak and three daggers*. Retrieved from http://kasperskycontenthub.com/wp-content/uploads/sites/43/vlpdfs/icefog.pdf

Keepers, G. A. (1990). Pathological Preoccupation with Video Games. *Journal of the American Academy of Child and Adolescent Psychiatry*, *29*(1), 49–50. doi:10.1097/00004583-199001000-00009 PMID:2295578

Kellner, D. (2005). Baudrillard, Globalization and Terrorism: some comments in recent adventures of the Image and Spectable on the occasion of Baudrillard's 75th birthday. Bishop's University. Retrieved from http://www.ubishops.ca/BaudrillardStudies/

Kenney, M. (2015). *Cyber-Terrorism in a Post-Stuxnet World.* Foreign Policy Research Institute.

Kies, R., & Wojcik, S. (2010). European Web-Deliberation: Lessons from the European Citizens Consultation. In F. De Cindio, A. Machintosh, & C. Peraboni (Eds.), *Online Deliberation: Proceedings of the Fourth International Conference - OD2010* (pp. 198-211).

Kim, M. Y., & Lee, D. H. (2014). Data-mining based SQL injection attack detection using internal query trees. *Expert Systems with Applications*, *41*(11), 5416–5430. doi:10.1016/j.eswa.2014.02.041

King, R. (2014, July 9). Hacking groups target shipping ports. *Wall Street Journal*. Retrieved from http://mobile.blogs.wsj.com/cio/2014/07/09/hacking-groups-target-shipping-ports/?mg=blogs-wsj

Kirk, J. (2009, January 19). *Virus attacks Ministry of Defence*. Retrieved from http://www.cio.co.uk/news/3460/virus-attacks-ministry-of-defence/

Klein, N. (2007). *The shock doctrine: The rise of disaster capitalism.* New York: Macmillan.

Knoblauch, M. (2014). *Internet Users Send 204 Million Emails Per Minute*. Retrieved 8 25, 2015, from MashableUK: http://mashable.com/2014/04/23/data-online-every-minute/

Knox, J. (2015, June 15). *Coast Guard Commandant on Cyber in the maritime domain, US Coast Guard.* Retrieved from http://mariners.coastguard.dodlive.mil/2015/06/15/6152015-coast-guard-commandant-on-cyber-in-the-maritime-domain/

Kopstein, J. (2013, June 26). UK police secretly monitoring 9,000 political campaigners using social media surveillance. *The Verge*. Retrieved from http://www.theverge.com/2013/6/26/4467056/uk-police-monitoring-9000-political-activists-using-social-media

Korstanje, M. E. (2011). Why risk why now? Conceptual problems around the risk perception in tourism industry. *Revista Brasileira de Pesquisa em turismo, 5*(1), 4-22.

Korstanje, M. E. (2015). El papel del riesgo en la configuración de las sociedades anglosajonas. *Estudios Sociales: Revista de investigación científica, 23*(45), 283-307.

Korstanje, M. (2012). Aportaciones de la Mitología Nórdica para comprender la modernidad, revisión de Max Weber. *Historia Actual., 28*(1), 175–190.

Korstanje, M. E. (2014a). Why Risk-Research is More Prominent in English Speaking Countries in the Digital Society. *International Journal of Cyber Warfare & Terrorism, 4*(1), 8–18. doi:10.4018/ijcwt.2014010102

Korstanje, M. E. (2014b). Chile helps Chile: Exploring the effects of earthquake Chile 2010. *International Journal of Disaster Resilience in the Built Environment, 5*(4), 380–390. doi:10.1108/IJDRBE-01-2012-0002

Korstanje, M. E. (2015). *A difficult world, examining the roots of capitalism.* New York: Nova Science Pubs.

Korstanje, M. E. (2015). *A Difficult World, examining the roots of Capitalism.* New York: Nova Science.

Korstanje, M. E. (2016a). *The rise of Thana capitalism and tourism.* Abingdon, UK: Routledge.

Korstanje, M. E. (2016b). *Terrorism in the global village.* New York: Nova Science Pubs.

Korstanje, M. E., & Clayton, A. (2012). Tourism and terrorism: Conflicts and commonalities. *Worldwide Hospitality and Tourism Themes, 4*(1), 8–25. doi:10.1108/17554211211198552

Kostyuchenko, Yu. V. (2015). Geostatistics and remote sensing for extremes forecasting and disaster risk multiscale analysis. In S. Kadry & A. El Hami (Eds.), Numerical Methods for Reliability and Safety Assessment: Multiscale and Multiphysics Systems. Springer International Publishing. doi:10.1007/978-3-319-07167-1_16

Kostyuchenko, Yu. V. (2016). Risk Perception Based Approach to Analysis of Social Vulnerability. In Risk Perception: Theories and Approaches. Nova Publ.

Kostyuchenko, Yu. V., & Movchan, D. (2015). Quantitative parameter of risk perception: can we measure a geoethic and socio-economic component in disaster vulnerability? In Geoethics: the Role and Responsibility of Geoscientists. Geological Society. doi:10.1144/SP419.10

Kostyuchenko, Yu. V., Movchan, D., Kopachevsky, I., & Bilous, Y. (2015). Robust Algorithm of Multi-Source Data Analysis for Evaluation of Social Vulnerability in Risk Assessment Tasks. In *Proc. of SAI IntelSys*. doi:10.1109/IntelliSys.2015.7361256

Kotsiopoulos, I., Hagen, K., Watson, H., & Kalemaki, E. (2014). *Towards Modelling Crisis Concepts and Dynamics*, Deliverable D1.4.1, COSMIC. Retrieved from www.cosmic-project.eu

Kovacs, E. (2014, June 6). Default password exposes digital highway signs to hacker attacks. *Security Week*. Retrieved from http://www.securityweek.com/default-password-exposes-digital-highway-signs-hacker-attacks

Kovacs, E. (2015, May 18). Nigerian fraudsters attack oil logistics firms without using malware: report. *Security Week*. Retrieved from http://www.securityweek.com/nigerian-fraudsters-attack-oil-logistics-firms-without-using-malware-report

KPMG. (2012). *Security Operations Centre (SOC) Globalization*. Retrieved from http://www.kpmg.com/SG/en/IssuesAndInsights/ArticlesPublications/Documents/Advisory-CS-Security-Operations-Center-SOC-Globalization.pdf

KPMG. (2014). *Next Generation IT Operating Models*. Retrieved from http://www.kpmg.com/BE/en/IssuesAndInsights/ArticlesPublications/Documents/Next-Generation-IT-Delivery-Models.pdf

Kramek, J. (2013, July 3). *The critical infrastructure gap: U.S. port facilities and cyber vulnerabilities, Center for 21st Century Security and Intelligence*. Brookings Institute. Retrieved from http://www.brookings.edu/~/media/research/files/papers/2013/07/02-cyber-port-security-kramek/03-cyber-port-security-kramek.pdf

Krause, J., Croft, D. P., & James, R. (2007). Social network theory in the behavioural sciences: Potential applications. *Behavioral Ecology and Sociobiology*, *62*(1), 15–27. doi:10.1007/s00265-007-0445-8

Kravets, D. (2009, March 18). Feds: hacker disabled offshore oil platforms' leak-detection system. *Wired*. Retrieved from http://www.wired.com/2009/03/feds-hacker-dis/

Kuutti, K. (1996). Activity Theory as a Potential Framework for Human Computer Interaction Research. In B. A. Nardi (Ed.), Context and Consciousness: Activity Theory and Human Computer Interaction (pp. 17-44). Boston, MA: M.I.T. Press.

Lasch, C. (1991). *The culture of narcissism: American life in an age of diminishing expectations*. New York: WW Norton & Company.

Laursen, F. (2010). *Comparative Regional Integration. Europe and Beyond*. London: Ashgate Publications.

Law, D. M., Shapka, J. D., Hymel, S., Olson, B. F., & Waterhouse, T. (2012). The Changing Face of Bullying: An Empirical Comparison Between Traditional and Internet Bullying and Victimization. *Computers in Human Behavior*, *28*(1), 226–232. doi:10.1016/j.chb.2011.09.004

Lawler, J. (2005). We are the one: Kant explains how to manipulate the Matrix. In W. Irwin (Ed.), *The Matrix and Philosophy* (pp. 138–152). Chicago: Open Court.

Lee, D. (2014, October 30). The role of social media in Occupy protests, on the ground and around the world. *The South China Morning Post.* Retrieved from http://www.scmp.com/news/hong-kong/article/1628305/role-social-media-occupy-protests-ground-and-around-world

Lee, R. M., Assante, M. J., & Conway, T. (2014, December 20). *ICS CP/PE (Cyber-to-physical or process effects) case study paper – media report of the Baku-Tbilisi-Ceyhan (BTC) pipeline cyber attack.* SANS Institute. Retrieved from https://ics.sans.org/media/Media-report-of-the-BTC-pipeline-Cyber-Attack.pdf

Lee, I., Jeong, S., Yeo, S., & Moon, J. (2012). A novel method for SQL injection attack detection based on removing SQL query attribute values. *Mathematical and Computer Modelling*, *55*(1-2), 58–68. doi:10.1016/j.mcm.2011.01.050

Leithauser, T. (2014, November 10). Elections Muddle Prospects for Cyber, Surveillance Bills. *Cybersecurity Policy Report.*

Leopoldi, R. (2005). *The Role of ITIL in IT Governance - ITSM.* Retrieved from www.itsm.info/ITSM ITIL and IT Governance.pdf

Lerbinger, O. (2012). *The crisis manager.* Routledge.

Lerman, K. (2013). Social Informatics: Using Big Data to Understand Social Behavior. In Handbook of Human Computation (pp. 751-759). Springer New York.

Lesieur, H. R., & Blume, S. B. (1993). Pathological Gambling, Eating Disorders, and the Psychoactive Substance Use Disorders. *Journal of Addictive Diseases*, *12*(3), 89–102. doi:10.1300/J069v12n03_08 PMID:8251548

Levenson, R. W., Ekman, P., & Friesen, W. W. (1990). Voluntary facial expression generates emotions-specific nervous system activity. *Psychophysiology*, *27*, 363–384. doi:10.1111/j.1469-8986.1990.tb02330.x PMID:2236440

Levy, S. (1996). Breathing is Also Addictive. *Newsweek*, *128*, 52-53.

Leyden, J. (2008, January 11). Polish teen derails tram after hacking train network. *The Register.* Retrieved from http://www.theregister.co.uk/2008/01/11/tram_hack/

Leyden, J. (2016, April 12). Sweden 'secretly blames' hackers – not solar flares – for taking out air traffic control. *The Register.* Retrieved from http://www.theregister.co.uk/2016/04/12/sweden_suspects_russian_hackers_hit_air_traffic_control/

Libicki, M. C. (2007). Conquest. In *Cyberspace: National Security and Information Warfare.* New York, NY: Cambridge University Press.

Li, Q., & Clark, G. (2015). *Security Intelligence: A Practitioner's Guide to Solving Enterprise Security Challenges.* Indianapolis, IN: Wiley. doi:10.1002/9781119183617

Locke, J. (1993). *Two Treatises of Government.* Cambridge, UK: Cambridge University Press.

Loukis, E., & Wimmer, M. A. (2010). Analysing Different Models of Structured Electronic Consultation on Legislation Under Formation. In F. De Cindio, A. Macintosh & C. Peraboni (Eds.), *Online Deliberation: Proceedings of the Fourth International Conference – OD2010* (pp. 14–26).

Luhmann, N. (2006). *Sociología del Riesgo* [The sociology of Risk]. México City: Universidad Iberoamericana.

Macintosh, A., Malina, A., & Farrell, S. (2002). Digital Democracy through Electronic Petitioning. In W. J. McIver Jr & A. K. Elmagarmid (Eds.), *Advances in Digital Government: Technology, Human Factors and Policy* (pp. 137–162). Dordrecht: Kluwer Academic Publishing.

Maffessoli, M. (2003). *L'instant éternel. Le retour du tragique dans les sociétés postmodernes.* Paris: La Table Ronde.

Makhati, T. (2015, October 30). *19 of our favourite posters from the #FeesMustFall protests, Live Mag*. Retrieved from http://livemag.co.za/feesmustfall/19-of-our-favourite-posters-from-the-feesmustfall-protests/

Mann, A. (2008). Spaces for talk: Communication technologies (ICTs) and genuine dialogue in an International Advocacy Movement. *Asian Social Science*, *4*(10), 3–13.

Maritime Executive. (2016, June 9). *Hackers attempt to track Indian aircraft carriers.* Retrieved from http://maritime-executive.com/article/hackers-attempt-to-track-indian-aircraft-carriers

Markey, E. (2016). *Cybersecurity Standards for Aircraft to Improve Resilience Act of 2016.* Retrieved from http://www.markey.senate.gov/imo/media/doc/Cyber%20AIR%20Act%204-7-16.pdf

Marks, I. (1990). Behavioural (non-chemical) Addictions. *British Journal of Addiction*, *85*(11), 1389–1394. doi:10.1111/j.1360-0443.1990.tb01618.x PMID:2285832

Martin, R. (2015). Rawls on International Economic Justice in the Law of Peoples. *Journal of Business Ethics*, *127*(4), 743–759. doi:10.1007/s10551-014-2184-x

McCarthy, P., Rylance, J., Bennett, R., & Zimmermann, H. (2001). *Bullying: From Backyard to Boardroom* (2nd ed.). Sydney, Australia: The Federation Press.

McCauley, C., & Moskalenko, S. (2008). Mechanisms of political radicalization: Pathways toward terrorism. *Terrorism and Political Violence*, *20*(3), 415–433. doi:10.1080/09546550802073367

McCue, A. (2003, October 7). 'Revenge' hack downed US port systems. *ZDNet*. Retrieved from http://www.zdnet.com/article/revenge-hack-downed-us-port-systems/

McGhee, W. E. (2008). *Information technology governance: An exploratory study of the impact of organizational information technology security planning* (Doctoral Thesis, Capella University). ProQuest Dissertations and Theses.

McVeigh, T. (2012). *Text messaging turns 20.* Retrieved 8 25Text, 2015, from The Observer: http://www.theguardian.com/technology/2012/dec/01/text-messaging-20-years

Mehroof, M., & Griffiths, M. D. (2010). Online Gaming Addiction: The Role of Sensation Seeking, Self-control, Neuroticism, Aggression, State Anxiety, and Trait Anxiety. *Cyberpsychology, Behavior, and Social Networking*, *13*(3), 313–316. doi:10.1089/cyber.2009.0229 PMID:20557251

Menemencioğlu, O., Sonuç, E., Karaş, İ. R., & Orak, İ. M. (2013). Academic Curriculum Vitae based Faculty Information System. In XV (pp. 1123–1127). Antalya, Turkey: Akademik Bilişim Conference Proceedings. Retrieved from http://ab.org.tr/ab13/kitap/menemencioglu_sonuc_AB13.pdf

Miller, B., & Rowe, D. C. (2012, October 11-13). A survey of SCADA and critical infrastructure incidents. *ACM Special Interest Group on Information Technology Education (SIGITE) Research in IT Conference.*

MITRE. (2016). *Risk Mitigation Planning, Implementation, and Progress Monitoring, Systems Engineering Guide*. Retrieved from https://www.mitre.org/publications/systems-engineering-guide/ acquisition-systems-engineering/risk-management/risk-mitigation-planning-implementation-and-progress-monitoring

Mkwanazi, N. (2015, October 26). *A One-Sided Story: My Thoughts on the #FeesMustFall Movement, exPress impress*. Retrieved from http://www.expressimpress.org/2015/10/26/a-one-sided-story-my-thoughts-on-the-feesmustfall-movement/

Mlambo, S. (2015, November 6). The Young & the Restless. *The Daily News*, p. 13.

Morahan-Martin, J. (2007). Internet Use and Abuse and Psychological Problems. In *Oxford Handbook of Internet Psychology* (pp. 331–345). Oxford, UK: Oxford University Press.

Morgan, J. (2015, November 4). Student fees and grants protest: Jeremy Corbyn gives support. *The Telegraph*. Retrieved 22 September 2016 from https://www.timeshighereducation.com/blog/student-fees-and-grants-protest-jeremy-corbyn-gives-support

Morgan, W. (1979). Negative Addiction in Runners. *The Physician and Sportsmedicine*, *7*, 56–69.

Moskalenko, S., & McCauley, C. (2011). The psychology of lone-wolf terrorism. *Counselling Psychology Quarterly*, *24*(2), 115–126. doi:10.1080/09515070.2011.581835

Muthuprasanna, M., Ke, W., & Kothari, S. (2006). Eliminating SQL injection attacks - A transparent defense mechanism. *Proceedings of the Eighth IEEE International Symposium on Web Site Evolution, WSE 2006*, (pp. 22–30). http://doi.org/ doi:10.1109/WSE.2006.9

Naisbitt, J. (1984). *Megatrends: Ten new directions transforming our lives (No. 04; HN59. 2, N3.)*. New York: Warner Books.

Natarajan, K., & Subramani, S. (2012). Generation of Sql-injection Free Secure Algorithm to Detect and Prevent Sql-Injection Attacks. *Procedia Technology*, *4*, 790–796. doi:10.1016/j.protcy.2012.05.129

National Institute of Standards and Technology (NIST). (2014, February 12). *Framework for Improving Critical Infrastructure Cybersecurity*. Retrieved from http://www.nist.gov/cyberframework/

National Research Council. (1990). *Computers at Risk Safe Computing in the Information Age*. National Academies Press.

Neuhaus, R. J. (1989). The Naked Public Square: Religion and Democracy in America, Second Edition. Grand Rapids, MI: William B. Eedermans Publishing.

Nigg, J. (1995). *Disaster Recovery as a social process*. Preliminary Paper #219. Disaster Research Center, Universidad of Delaware.

Nigg, J. (1996). *Policy Issues for post disaster mitigation the need for a process*. Preliminary Paper # 244. Disaster Research Center, Universidad of Delaware.

NIST. (2012). *NIST Publishes Draft Implementation Guidance for Continuously Monitoring an Organization's IT System Security*. Retrieved from http://www.nist.gov/itl/csd/monitoring-012412.cfm

NIST. (n.d.). *Web Application Vulnerability Scanners*. Retrieved June 11, 2013, from http://samate.nist.gov/index.php/Web_Application_Vulnerability_Scanners.html

Nixon, D. (2005). The Matrix Possibility. In W. Irwin (Ed.), *The Matrix and Philosophy* (pp. 28–40). Chicago: Open Court.

Northcutt, S. (2009). *Security Controls*. Retrieved from http://www.sans.edu/research/security-laboratory/article/security-controls

Nyamnjoh, F., Nkwi, W. G., & Konings, P. (2007). *University Crisis and Student Protests in Africa*. Retrieved from http://www.africanbookscollective.com/books/university-crisis-and-student-protests-in-africa

Nye, J. Jr. (2013). From Bombs to Bytes. *The Bulletin of the Atomic Scientists*, *69*(5), 12. doi:10.1177/0096340213501338

O'Donnell, C. (2011). *New study quantifies use of social media in Arab Spring*. Retrieved 11 April 2016 from http://www.washington.edu/news/2011/09/12/new-study-quantifies-use-of-social-media-in-arab-spring/

Ogheneovo, E., & Asagba, P. (2013). A Parse Tree Model for Analyzing And Detecting SQL Injection Vulnerabilities. *West African Journal of Industrial and Academic Research*, *6*(1), 33–49. Retrieved from http://www.ajol.info/index.php/wajiar/article/view/87436

Oliver-Smith, A. (2002). Theorizing disasters, nature, power and culture? In S. Hoffman & A. Oliver-Smith (Eds.), *Catasthrope and culture: the anthropology of disasters* (pp. 23–47). Oxford, UK: School of American Research James Currey.

Olson, P. (2012). *We Are Anonymous: Inside the Hacker World of LulzSec, Anonymous, and the Global Cyber Insurgency*. New York: Little, Brown and Company.

Olweus, D. (1993). *Bullying at School: What We Know and What We Can Do*. London: John Wiley.

Onuch, O. (2014, January 2). Social networks and social media in Ukrainian "Euromaidan" protests. *Washington Post Monkey Cage Blog*. Retrieved from http://www.washingtonpost.com/blogs/monkey-cage/wp/2014/01/02/social-networks-and-social-media-in-ukrainian-euromaidan-protests-2/

OWASP. (n.d.). *Category:OWASP Top Ten Project - OWASP*. Retrieved April 17, 2015, from https://www.owasp.org/index.php/Top_Ten

Paganini, P. (2012, August 8). Anonymous Ethical Code. Back to the Origins. *Security Affairs*. Available at http://securityaffairs.co/wordpress/7986/cyber-crime/anonymous-ethical-code-back-to-the-origins.htm

Panetta, L. E. (2012, October 11). *Remarks by Secretary Panetta on Cybersecurity to the Business Executives for National Security*. Retrieved July 8, 2015, from US Department of Defense: http://www.defense.go

Panksepp, J. (1982). Toward a General psychobiological theory of Emotions. *Behavioral and Brain Sciences*, *5*(03), 407–467. doi:10.1017/S0140525X00012759

Paradia, S. (2012). *Enterprise Service Operation Center*. Retrieved from http://www.slideshare.net/sameer paradia/it-enterprise-service-operation-center

Parker, D. B. (1976). *Crime by Computer*. New York, NY: Scribner's.

Pauli, D. (2016, January 4). Irked train hackers talk derailment flaws, drop SCADA password list. *The Register*. Retrieved from http://www.theregister.co.uk/2016/01/04/irked_train_hackers_talk_derailment_flaws_drop_scada_password_list/

Peele, S., & Brodsky, A. (1979). *Love and Addiction*. Scarborough, Ontario: New American Library of Canada.

Pfleeger, C. P. (2008). Reflections on the Insider Threat. In S. J. Stolfo, S. M. Bellovin, S. Hershkop, A. Keromytis, S. Sinclair, & S. W. Smith (Eds.), *Insider Attack and Cyber Security: Beyond the Hacker* (pp. 6–15). New York, NY: Springer Link. doi:10.1007/978-0-387-77322-3_2

Phillips, W. (2011). LOLing At Tragedy: Facebook Trolls, Memorial Pages, and Resistance to Grief Online. *First Monday*, *16*(12), 1–13. doi:10.5210/fm.v16i12.3168

Phillips, W. (2013). The House That Fox Built: Anonymous, Spectacle, and Cycles of Amplification. *Television & New Media*, *14*(6), 494–509. doi:10.1177/1527476412452799

Piazza, J. A. (2007). Draining the swamp: Democracy promotion, state failure, and terrorism in 19 Middle Eastern countries. *Studies in Conflict and Terrorism*, *30*(6), 521–539. doi:10.1080/10576100701329576

Piazza, J. A. (2008). Incubators of terror: Do failed and failing states promote transnational terrorism? *International Studies Quarterly*, *52*(3), 469–488. doi:10.1111/j.1468-2478.2008.00511.x

Picket, K. (2011). *Read Demands of "Occupy Wall Street"...And Try Not to Laugh*. Retrieved from http://nation.foxnews.com/occupy-wall-street/2011/10/04/read-demands-occupy-wall-street-and-try-not-laugh. October 3

Pillay, K. J. (2012). *Do Web 2.0 Social Media Impact Transnational Social Advocacy? A Study of South African Civil Society and Greenpeace* (PhD Thesis). University of KwaZulu-Natal.

Pogue, D. (2016, February 22) Why car hacking is nearly impossible. *Scientific American*. Retrieved from http://www.scientificamerican.com/article/why-car-hacking-is-nearly-impossible/

Pollitt, M. M. (1998). Cyberterrorism: Fact or Fancy? *Computer Fraud & Security*, *1998*(2), 8–10. doi:10.1016/S1361-3723(00)87009-8

Potter, N. (2011, August 9). *London riots 2011: Protestors use Blackberry Messenger; hackers support them*. ABC News. Retrieved from http://abcnews.go.com/Technology/london-riots-2011-protesters-blackberry-messenger-hackers-back/story?id=14264839

Prabhu, V. (2016, March 2). Pirates hack shipping company to find their target ships for raids. *Techworm*. Retrieved from http://www.techworm.net/2016/03/pirates-hacked-shipping-company-gather-intel-target-ships.html

Presence, C. (2015, October 20). *Blade: Student protests not a national crisis*. Retrieved from http://mg.co.za/article/2015-10-20-blade-student-protests-not-a-national-crisis

Prigg, M. (2014, August 20). How to get green lights all the way to work: Hackers reveal how simple it is to control traffic lights in major cities using just a laptop. *Daily Mail*. Retrieved from http://www.dailymail.co.uk/sciencetech/article-2730096/How-green-lights-way-work-Hackers-reveal-simple-control-traffic-lights-major-cities-using-just-laptop.html

Proctor, R. W., & Vu, K.-P. L. (2012). Selection and Control of Action. In G. Salvendy (Ed.), *Handbook of Human Factors and Ergonomics* (4th ed.; pp. 95–116). John Wiley and Sons. doi:10.1002/9781118131350.ch4

Pyzalski, J. (2011). Electronic Aggression Among Adolescents: An Old House with a New Facade (or Even a Number of Houses). In C. Hällgren, E. Dunkels, & G.-M. Frånberg (Eds.), *Youth Culture and Net Culture: Online Social Practices* (pp. 278–295). Hershey, PA: IGI Global. doi:10.4018/978-1-60960-209-3.ch016

Rachlin, H. (1990). Why Do People Gamble and Keep Gambling Despite Heavy Losses? *Psychological Science*, *1*(5), 294–297. doi:10.1111/j.1467-9280.1990.tb00220.x

Ragan, S. (2012, January 25). Railway network disrupted after cyber attack. *Security Week*. Retrieved from http://www.securityweek.com/railway-network-disrupted-after-cyber-attack-report-says

Ratliff, T., & Hall, L. (2014). Practicing the Art of Dissent: Toward a Typology of Protest Activity in the United States. *Humanity & Society*, *38*(3), 271–272. doi:10.1177/0160597614537796

Rawls, J. (1999). *The Law of Peoples*. Cambridge, MA: Harvard University Press.

Reeve, T. (2016, May 16). CEO sacked after aircraft company grounded by whaling attack. *SC Magazine*. Retrieved from http://www.scmagazine.com/ceo-sacked-after-aircraft-company-grounded-by-whaling-attack/article/499258/

Release, Sophos Press. (1999, November 23). *Don't fall for a virus hoax*. Retrieved July 10, 2015, from Sophos: https://www.sophos.com/en-us/press-office/press-releases/1999/11/va_hoaxes.aspx#cost

Reynolds, A. (2015, June 22). Denial is the enemy of Crisis Management. *Public Relations Weekly*. Retrieved 21 September 2016 from http://www.prweek.com/article/1352151/denial-enemy-crisis-management

Rheingold, H. (1993). *The Virtual Community: Homesteading on the Electronic Frontier*. Reading, MA: Addison-Wesley.

Rial, N. (2012). *Mexican Hackers Attack Official Sites*. Retrieved from http://www.neurope.eu/article/mexican-hackers-attack-official-sites

Richardson, J. (2010). The certainty of uncertainty: Risk management revisited. *Foresight*, *12*(4), 47–64. doi:10.1108/14636681011062997

Rifkin, J. (1998). *The biotech Century: a coming age of genetic commerce*. London: Victor Gollancz.

Roadmap to Secure Control System in the Transportation Sector Working Group. (2012). *Roadmap to Secure Control System in the Transportation Sector*. Washington, DC: Department of Homeland Security.

Roberts, P. (2011, October 12). Air Force struggled for weeks with malware in drone fighter systems. *ThreatPost*. Retrieved from http://threatpost.com/en_us/blogs/report-air-force-struggled-weeks-malware-drone-fighter-systems-101211

Robertson, J., & Riley, M. (2014, December 10). Mysterious '08 Turkey pipeline blast opened new cyberwar. *Bloomberg*. Retrieved from http://www.bloomberg.com/news/articles/2014-12-10/mysterious-08-turkey-pipeline-blast-opened-new-cyberwar

Romeo, J. (2012). Hacktivism endures. *SC Magazine*, *23*(3), 24–25.

Ronfeldt, D., & Arquilla, J. (1998). *The Zapatista Social Netwar in Mexico*. Santa Monica, CA: RAND Corporation.

Ross, A. (1990). Hacking Away at the Counterculture. *Postmodern Culture*, *1*(1), 2. doi:10.1353/pmc.1990.0011

Rothke, B. (2012). *Building a Security Operations Center (SOC)*. Retrieved from http://www.rsaconference.com/writable/presentations/file_upload/tech-203.pdf

Rousseau, J. J. (1993). El contrato social. Barcelona: Editorial Altaya.

RSA. (2013). *RSA Advanced Cyber Defense Program for Nextgen SOC, RSA Technical Brief, EMC*[2]. Retrieved from https://www.emc.com/collateral/data-sheet/rsa-advanced-cyber defense-program-nextgen-soc-ds.pdf

RSA. (2014, June). *Building an Intelligence-Driven Security Operations Centre, RSA Technical Brief, EMC*[2]. Retrieved from http://www.rsa.com

Sales, N. (2013). Regulating Cyber-Security. *Northwestern University Law Review*, *107*(4), 1515.

Santamarta, R. (2014). SATCOM terminals hacking by air, sea, and land. *BlackHat USA 2014*. Retrieved from https://www.blackhat.com/docs/us-14/materials/us-14-Santamarta-SATCOM-Terminals-Hacking-By-Air-Sea-And-Land-WP.pdf

Satell, G. (2014, January 18). If you doubt that social media has changed the World, take a look at Ukraine. *Forbes*. Retrieved http://www.forbes.com/sites/gregsatell/2014/01/18/if-you-doubt-that-social-media-has-changed-the-world-take-a-look-at-ukraine/

Sauter, M. (2014). *The Coming Swarm*. New York: Bloomsbury.

Sayed Ally, S. E. (2015). *Tableau Public Data Visualization. #feesmustfall*. Retrieved from https://public.tableau.com/profile/ally5527#!/vizhome/feesmustfall/Story1

Schiller, H. (1981). Who knows: Information in the age of the fortune 500. Norwood.

Schmid, A. P. (2011). *The Routlaedge handbook of terrorism research*. Taylor & Francis.

Schmitt, M. N. (2012). *The 'Use of Force' in Cyberspace: A Reply to Dr. Ziolkowski*. NATO CCD COEP Publications. Retrieved from http://www.google.com/url?sa=t&rct=j&q=michael%20schmitt's%20criteria&source=web&cd=3&cad=rja&ved=0CDwQFjAC&url=http%3A%2F%2Fwww.usnwc.edu%2Fgetattachment%2Fe80ed2a4-675b-4278-8321-840a2c58a27c%2FCyber-ziolkowski-response&ei=RkWgUMS6LeejiAKX0oHgBw&usg=AFQjCNEEUQ2A8ewpyeybf3vyYeyft3-7VA

Schmitt, R. (1995). *Beyond Separateness*. Boulder, CO: Westview Press.

Schurman, K. (2012). *LulzSec: 50 Days of Lulz*. San Francisco, CA: Hyperink.

Schwartz, E. J., Avgerinos, T., & Brumley, D. (2010). All You Ever Wanted to Know About Dynamic Taint Analysis Forward Symbolic Execution (but might have been afraid to ask) A Few Things You Need to Know About Dynamic Taint Analysis Forward Symbolic Execution (but might have been afraid to ask) The Root. *Policy*, 1–5.

Schwartz, M. (2008, August 3). The Trolls Among Us. *The New York Times*.

Sclove, R. E. (1995). *Democracy and Technology*. London: Guilford Press.

Scott, J. (2012). Social network analysis. *Sage (Atlanta, Ga.)*.

SearchSecurity. (2016). *Four Features to Look for in a Threat Intelligence Service, TechTarget.* Retrieved from http://searchsecurity.techtarget.com/ThreatIntelligence/Four-Features-to-Look-for-in-a-Threat-Intelligence-Service

Sekaran, U. (1992). *Research Methods for Business: A Skill-Building Approach* (2nd ed.). New York: John Wiley.

Sennett, R. (2011). *The corrosion of character: The personal consequences of work in the new capitalism.* New York: WW Norton & Company.

Sensage. (2014). *A Practical Guide to Next-Generation SIEM.* Retrieved from http://www.sensage.com

Serracino-Inglott, P. (2013). Is It OK to an Anonymous? *Ethics & Global Politics*, *6*(4), 220. doi:10.3402/egp.v6i4.22527

Seward, M. (2016). *Addressing the New Threat Intelligence Problems of Scale and Relevance, Anomali and Information Week Dark Reading.* Retrieved from http://event.on24.com/r.htm?e=1137629&s=1&k=90E87A5A7DA758E4275E0D35A90B558C

Seymour, G. (1975). *Harry's Game.* London: Collins.

Shachtman, N. (2011, October 8). Computer virus hits US Predator and Reaper drone fleet. *ARS Technica.* Retrieved from http://arstechnica.com/business/news/2011/10/exclusive-computer-virus-hits-drone-fleet

Shah, D. V., Kwak, N., & Holbert, R. L. (2002). Connecting" and" Disconnecting" with Civic Life: Patterns of Internet Use and the Production of Social Capital. *Political Communication*, *18*(2), 141–162.

Ship Technology. (2013, November 28). *Web of intrigue: protecting ports against cyber-terrorism.* Retrieved from http://www.ship-technology.com/features/feature-cybersecurity-port-computer-hackers-us-belgium/

Shirey, R. (2014). *Internet Security Glossary, Version 2.* IETF RFC 4949. Retrieved https://tools.ietf.org/html/rfc4949

Skinner, Q. (1964). Hobbess Leviathan. *Historical Journal (Cambridge, England)*, *7*(02), 321–333. doi:10.1017/S0018246X00025498

Skoll, G. R. (2007). Meanings of terrorism. *International Journal for the Semiotics of Law-Revue internationale de Sémiotique juridique, 20*(2), 107-127.

Skoll, G. R., & Korstanje, M. E. (2012). Risks, totems, and fetishes in Marx and Freud. *Sincronía*, (2), 11-27.

Skoll, G. (2010). *Social Theory of Fear: Terror, Torture, and Death in a Post-Capitalist World.* New York: Palgrave-MacMillan.

Skoll, G. (2016). *Globalization of American Fear Culture: the empire in the Twenty-First Century*. New York: Palgrave-Macmillan.

Skoll, G. R., & Korstanje, M. E. (2013). Constructing an American fear culture from red scares to terrorism. *International Journal of Human Rights and Constitutional Studies*, *1*(4), 341–364. doi:10.1504/IJHRCS.2013.057302

Skoudis, E., & Zeltser, L. (2004). *Malware: Fighting malicious code*. Prentice Hall Professional.

Sky News. (2011). *Middle East: See facts about all the countries in the region*. Retrieved from http://news.sky.com/skynews/Interactive-Graphics/middleeast

SocialMedia. (2013). *Social Networking Reaches Nearly One in Four Around the World - SSocial Networking Reaches Nearly One Four Around World*. Retrieved from eMarketer: http://www.emarketer.com/Article/Social-Networking-Reaches-Nearly-One-Four-Around-World/1009976

Solyom, C. (2012). *Insider Tells Why Anonymous "Might Well Be The Most Powerful Organization On Earth"*. Retrieved from http://news.nationalpost.com/2012/05/12/insider-tells-why-anonymous-might-well-be-the-most-powerful-organization-on-earth/

South African History Online. (2016, July 21) *The June 16 Soweto Youth Uprising*. Retrieved from http://www.sahistory.org.za/topic/june-16-soweto-youth-uprising

Soyinka, W. (2005). *The Climate of Fear. The quest for dignity in a dehumanized World*. New York: Random House.

Sqrrl. (2016). *A Framework for Cyber Threat Hunting*. Retrieved from http://www.sqrrl.com

Staff Writer. (2015, October 23). Political parties hijack #FeesMustFall protest: reports. *Business Tech*. Retrieved from http://businesstech.co.za/news/government/102050/political-parties-hijack-feesmustfall-protest-reports/

State Security Agency. (2015). *The National Cybersecurity Policy Framework*. Pretoria: Government of South Africa.

Stats, I. L. (n.d.). *Internet Users*. Retrieved July 4, 2015, from Internet Live Stats: http://www.internetlivestats.com/internet-users/

Stelter, L. (2015, February 2). Securing the Nation's ports against cyberterrorism. *InHomeland-Security.com*. Retrieved from http://inhomelandsecurity.com/securing-the-nations-ports-against-cyberterrorism/

Sterner, E. (2012, April 23). Hacktivists' Evolution Changes Cyber Security Threat Environment. *World Politics Review*.

Sternstein, A. (2012, January 23). Hackers manipulated railway computers, TSA memo says. *NextGov*. Retrieved from http://www.nextgov.com/cybersecurity/2012/01/hackers-manipulated-railway-computers-tsa-memo-says/50498/

Stormark, K. (2016, April 12). *Sweden issued cyber attack alert as its air traffic reeled.* Retrieved from https://www.aldrimer.no/sweden-issued-cyber-attack-alert-as-its-air-traffic-reeled/

Stouffer, K., Falco, J., & Scarfone, K. (2013). *Guide to Industrial Control Systems (ICS) Security. NIST Special Publication 800-82 Revision 1.* Gaithersburg, MD: NIST.

Strategy&. (2012). *What is a Capability?* Retrieved from http://www.strategyand.pwc.com/global/home/ what-we-think/multimedia/video/mm-video_display/what-is-a-capability

Sultana, M., Paul, P. P., & Gavrilova, M. (2015). Social Behavioral Biometrics: An Emerging Trend. *International Journal of Pattern Recognition and Artificial Intelligence, 29*(08), 1556013. doi:10.1142/S0218001415560133

Sunstein, C. R. (2002). *Risk and reason: Safety, law, and the environment.* Cambridge, UK: Cambridge University Press.

Sunstein, C. R. (2005). *Laws of fear: Beyond the precautionary principle* (Vol. 6). Cambridge, UK: Cambridge University Press. doi:10.1017/CBO9780511790850

Suttner, R. (2015). Op-Ed: Why is the state not a target of student protests? *The Daily Maverick.* Retrieved from http://www.dailymaverick.co.za/article/2015-10-20-op-ed-why-is-the-state-not-a-target-of-student-protests/

Swanbeck, S. (2015, June 22). *Coast Guard Commandant Addresses Cybersecurity Vulnerabilities on Offshore Oil Rigs, CSIS-Tech.org.* Retrieved from http://www.csis-tech.org/blog/2015/6/22/coastguard-commandant-addresses-cybersecurity-vulnerabilities-in-offshore-oil-rigs

Symantec. (2014, April). *2014Internet Security Threat Report, Volume 19, Supplemental Data.* Retrieved from http://www.symantec.com/content/en/us/enterprise/other_resources/b-istr_appendices_v19_221284438.en-us.pdf

Symantec. (2015). *Five Threat Intelligence Traps to Avoid, UBM Tech.* Retrieved from http://www.informationweek.com/whitepaper/risk-management-security/cybersecurity/five-threat-intelligence-traps-to-avoid/364973

Symantec. (2015, April). *2015Internet Security Threat Report, Volume 20, Supplemental Data.* Retrieved from https://www4.symantec.com/mktginfo/whitepaper/ISTR/21347931_GA-internet-security-threat-report-volume-20-2015-appendices.pdf

Symantec. (2016a, April). *2016Internet Security Threat Report, Volume 21.* Retrieved from https://www.symantec.com/security-center/threat-report

Symantec. (2016b, April). *2016Internet Security Threat Report, Volume 21, Supplemental Data.* Retrieved from https://www.symantec.com/security-center/threat-report

Taylor, M. (2015, November 1). Students Protest Our Free Education expected to draw thousands. *The Guardian.* Retrieved from https://www.theguardian.com/world/2015/nov/01/student-protest-attacks-free-education-to-draw-thousands

The 2016 Trafficking in Persons Report. (2016). U.S. Department of State Publications, Office of the Under Secretary of Civilian Security, Democracy and Human Rights.

TheAnonMessage. (2012). *Anonymous: Operation Defense Phase II* [Video file]. Retrieved from http://www.youtube.com/watch?v=Hdv5xR5YqFM&feature=plcp

TheAnonMessage. (2012). *Anonymous: Operation V (Nov. 5th 2012)* [Video file]. Retrieved from http://www.youtube.com/watch?v=jAaCZWQndqA&feature=plcp

Thomas, S. (2015, October 22). *#FeesMustFall: how SA students are using social to subvert traditional media*. Memeburn. Retrieved 21 September 2016 from http://memeburn.com/2015/10/feesmustfall-how-sas-students-are-using-social-to-subvert-traditional-media-narratives/

Thomassen, J., & Schmitt, H. (1999). Introduction: Political Representation and Legitimacy in the European Union. In H. Schmitt & J. Thomassen (Eds.), *Political Representation and Legitimacy in the European Union* (pp. 3–21). Oxford, UK: Oxford University Press. doi:10.1093/0198296614.003.0001

Thoreau, H. D. (1849). *Civil Disobedience*. Available at http://www.gutenberg.org/files/71/71-h/71-h.html

Thruelsen, P. D. (2006). *From Soldier To Civilian: Disarmament Demobilisation Reintegration in Afghanistan*. DIIS Report.

Tierney, K. (1994). *Sociology's Unique Contributions to the Study of Risk*. Disaster Research Center, Preliminary Paper, 204.

Timmermann, F. (2015). *El Gran Terror. Miedo, Emocion y discurso Chile 1973-1980*. Santiago: Ediciones Copygraph.

Tomlinson, K. (2016, January 6). *Even cruise ships can be hacked—and what is being done about it*. Archer Energy Solutions. Retrieved from http://www.archerenergysolutions.com/even-cruise-ships-can-be-hacked-and-what-is-being-done-about-it/

Treanor, J. (2015, April 29). Ryanair confirms hackers stole almost $5m via Chinese bank. *The Guardian*. Retrieved from http://www.theguardian.com/business/2015/apr/29/ryanair-confirms-hackers-stole-almost-5m-via-chinese-bank-electronic-transfer

Tucker, J. A., Metzger, M., & Barbera, P. (2014). *SMaPP Lab Data Report: Ukraine Protests 2013-2014*. Social Media and Political Participation Lab, New York University. Retrieved 31 March 2014 from http://smapp.nyu.edu/reports/Ukraine_Data_Report.pdf

Tucker, J. A., Metzger, M., Penfold-Brown, D., Bonneau, R., Jost, J., & Nagler, J. (2015). *Protest in the Age of Social Media*. Carnegie Corporation of New York. Retrieved from https://medium.com/carnegie-corporation-international-peace-and/protest-in-the-age-of-social-media-7ae9fd940b06#.n857xqa11

Turkle, S. (1995). *Life behind the Screen: Identity in the Age of the Internet*. New York: Simon & Schuster.

Turkle, S. (2011). *Life on the Screen: Identity in the Age of the Internet.* New York, NY: Simon and Schuster.

Twitter. (2015). *Twitter usage / company facts.* Retrieved 8 25, 2015, from Twitter: https://about.twitter.com/company

Urban, G. L., Amyx, C., & Lorenzon, A. (2009). Online Trust: State of the Art, New Frontiers, and Research Potential. *Journal of Interactive Marketing, 23*(2), 179–190. doi:10.1016/j.intmar.2009.03.001

Urry, J. (2007). *Mobilities.* Cambridge, UK: Polity Press.

US Coast Guard. (2015, December 28). *Spoofed business e-mail used to try and defraud maritime organization.* Maritime Cyber Bulletin 002-16.

US Coast Guard. (2016a, January 4). *Maritime organization's websites targeted by actors attempting to exploit Joomla vulnerabilities.* Maritime Cyber Bulletin 003-16.

US Coast Guard. (2016b, January 5). *Shipboard voyage data recorder vulnerabilities.* Maritime Cyber Bulletin 004-16.

Van Niekerk, B., & Jacobs, P. (2015). Towards a Secure Data Centre Model. *ISACA Journal, 3.* Retrieved from http://www.isaca.org/Journal/archives/2015/Volume-3/Pages/toward-a-secure-data-center-model.aspx

van Niekerk, B. (2015). Information Warfare in the 2013-2014 Ukraine Crisis. In J. Richet (Ed.), *Cybersecurity Policies and Strategies for Cyberwarfare Prevention* (pp. 307–339). Hershey, PA: IGI Global. doi:10.4018/978-1-4666-8456-0.ch012

van Niekerk, B., & Maharaj, M. (2013). Social Media and Information Conflict. *International Journal of Communication, 7*, 1162–1184.

van Niekerk, B., Pillay, K., & Maharaj, M. (2011). Analysing the Role of ICTs in the Tunisian and Egyptian Unrest from an Information Warfare Perspective. *International Journal of Communication, 5*, 1406–1416.

Velasquez & LaRose. (2015). Social Media for Social Change: Social Media Political Efficacy and Activism in Student Activist Groups. *Journal of Broadcasting and Electronic Media, 59*(3).

Verizon. (2016a). *Data breach digest. Scenarios from the field.* Retrieved from http://www.verizonenterprise.com/resources/reports/rp_data-breach-digest_xg_en.pdf

Verizon. (2016b). *Data breach investigations report.* Retrieved from http://www.verizonenterprise.com/resources/reports/rp_data-breach-digest_xg_en.pdf

Verton, D., & Brownlow, J. (2003). *Black ice: The invisible threat of cyber-terrorism.* Osborne.

Virilio, P. (2010). *The University of Disaster.* Oxford, UK: Polity Press.

Virkar, S. (2011). *The Politics of Implementing e-Government for Development: The Ecology of Games Shaping Property Tax Administration in Bangalore City* (Unpublished Doctoral Thesis). University of Oxford, Oxford, UK.

Virkar, S. (2007). *(Dis) Connected Citizenship? Exploring Barriers to eConsultation in Europe. Report to the European Commission for the Breaking Barriers to eGovernment: Overcoming Obstacles to Improving European Public Services* Project.

Virkar, S. (2014). Trolls Just Want To Have Fun: Electronic Aggression within the Context of e-Participation and Other Online Political Behaviour in the United Kingdom. *International Journal of E-Politics*, *5*(4), 21–51. doi:10.4018/ijep.2014100102

Wagstaff, J. (2014, April 23). All at sea: global shipping fleet exposed to hacking threat. *Reuters*. Retrieved from http://www.reuters.com/article/2014/04/24/tech-cybersecurity-shipping-graphic-pix-idUSL3N0NG0GP20140424

Walker, D. (2014, July 14). Chinese man charged with hack of Boeing, Lockheed Martin aircraft data. *SC Magazine*. Retrieved from http://www.scmagazine.com/chinese-man-charged-with-hack-of-boeing-lockheed-martin-aircraft-data/article/360786/

Wallace, P. (2001). *The Psychology of the Internet*. Cambridge, UK: Cambridge University Press.

Wallischeck, E. Y. (2013, July). *ICS Security in Maritime Transportation*. Cambridge, MA: John A Volpe National Transportation Systems Center.

Walters, T., Hourizi, R., Moncur, W., & Pitsillides, S. (2011). Does the Internet Change How We Die And Mourn? Overview and Analysis. *Omega*, *64*(4), 275–302. doi:10.2190/OM.64.4.a PMID:22530294

WarLeigh-Lack, A (2010). The EU in comparative perspective: comparing the EU and Nafta. In Comparative Regional Integration. Europe and Beyond. London: Ashgate.

Warnick, B., & Heineman, D. S. (2012). *Rhetoric Online: The Politics of New Media* (2nd ed.). New York: Peter Lang Publishing.

Wassermann, G., & Su, Z. (2007). Sound and precise analysis of web applications for injection vulnerabilities. *ACM SIGPLAN Notices*, *42*(6), 32. doi:10.1145/1273442.1250739

Wasserman, S., & Galaskiewicz, J. (Eds.). (1994). *Advances in social network analysis: Research in the social and behavioral sciences* (Vol. 171). Sage Publications.

Webb, J. (2009, May 11). Somali pirates using London contacts - Spain radio. *Reuters*. Retrieved from http://www.reuters.com/articlePrint?articleId=USLB570114

Weber, M. (1958). *Essays in Sociology*. New York: Oxford University Press.

Weber, M. (1964). *Economía y sociedad: esbozo de sociología comprensiva*. FCE.

Weber, M. (1995). *La Ética Protestante y el Espíritu del Capitalismo*. Barcelona: Península.

Weber, M. (2002). *The Protestant Ethic and the Spirit of Capitalism: and other writings*. New York: Penguin.

Webster, F., & Robins, K. (1986). *Information Technology: a luddite Analysis*. Norwood.

Weimann, G. (2004). *Cyberterrorism, How Real Is the Threat?* Washington, DC: United States Institute of Peace.

Weiss, G. (2008, April 14). *The Farewell dossier: Duping the Soviets, Central Intelligence Agency*. Retrieved from https://www.cia.gov/library/center-for-the-study-of-intelligence/csi-publications/csi-studies/studies/96unclass/farewell.htm

Wesi, T. (2015). *Fees Must Fall protest in Pretoria 'hijacked'*. Retrieved 22 September 2016 from http://citizen.co.za/833478/fees-must-fall-protest-in-pretoria-hijacked/

West-Brown, M., Stikvoort, D., Kossakowski, K. P., Killcrece, G., & Ruefle, R. M. (2003). Handbook for Computer Security Incident Response Teams (CSIRTs) (2nd ed.). CMU/SEI. Retrieved from http://resources.sei.cmu.edu/asset_files/Handbook/2003_002_001_14102.pdf

White House. (2011). *International Strategy for Cyberspace: Prosperity, Security, and Openness in a Networked World*. Retrieved from https://www.whitehouse.gov/sites/default/files/rss_viewer/international_strategy_for_cyberspace.pdf

Widyanto, L., & Griffiths, M. (2011). Internet Addiction: Does It Really Exist? (Revisited). In J. Gackenbach (Ed.), *Psychology and the Internet: Intrapersonal, Interpersonal, and Transpersonal Implications* (pp. 142–166). London: Academic Press.

Williams, C. (2011). *Facebook and Twitter to help police track riots*. Retrieved 11 April 2016 from http://www.telegraph.co.uk/technology/social-media/8723038/Facebook-and-Twitter-to-help-police-track-riots.html

Willsher, K. (2009, February 7). French fighter planes grounded by computer virus. *The Telegraph*. Retrieved from http://www.telegraph.co.uk/news/worldnews/europe/france/4547649/French-fighter-planes-grounded-by-computer-virus.html

Winn, M. (1977). *The Plug-In Drug*. New York, NY: Viking Press.

Wong, W., & Brown, P. (2013). E-Bandits in Global Activism: WikiLeaks, Anonymous, and the Politics of No One. *American Political Science Association*, *11*(4), 1024.

World Economic Forum. (2015). *Global Risks 2015* (10th ed.). Geneva: World Economic Forum.

World Movement for Democracy (WMD). (2009). *Twitter Case Study*. Retrieved 17 June 2010 from: http://www.wmd.org/resources/whats-being-done/information-and-communication-technologies/case-study-twitter

Yonah, A. (2007). *Evolution of US counterterrorism policy*. Greenwood Publishing Group.

Young, K. S. (1999). The Evaluation and Treatment of Internet Addiction. In L. VandeCreek & T. Jackson (Eds.), Innovations in Clinical Practice: A Source Book 17 (pp. 19-31). Saratosa, FL: Professional Resource Press.

Young, K. S. (2004). Internet Addiction a New Clinical Phenomenon and its Consequences. *The American Behavioral Scientist*, *48*(4), 402–415. doi:10.1177/0002764204270278

Young, K. S., & Case, C. J. (2004). Internet Abuse in the Workplace: New Trends in Risk Management. *Cyberpsychology, Behavior, and Social Networking*, *7*(1), 105–111. doi:10.1089/109493104322820174 PMID:15006175

YouTube. (2015). *Statistics*. Retrieved 8 25, 2015, from YouTube: https://www.youtube.com/yt/press/en-GB/statistics.html

Zelikow, P. (2011). *The 9/11 commission report: Final report of the national commission on terrorist attacks upon the United States*. Government Printing Office.

Zetter, K. (2015, May 15). Feds say that banned researcher commandeered plane. *Wired*. Retrieved from http://www.wired.com/2015/05/feds-say-banned-researcher-commandeered-plane/

Zhang, H., Kuan Tan, H. B., Zhang, L., Lin, X., Wang, X., Zhang, C., & Mei, H. (2011). Checking enforcement of integrity constraints in database applications based on code patterns. *Journal of Systems and Software*, *84*(12), 2253–2264. doi:10.1016/j.jss.2011.06.044

Zimmerman, C. (2014). *Ten Strategies of a World-Class Cybersecurity Operations Centre*. The Mitre Corporation. Retrieved from https://www.mitre.org/sites/default/files/publications/pr-13-1028-mitre-10-strategies-cyber ops-center.pdf

Zizek, S. (2005). *The Matrix: or, the two sides of perversion*. Chicago: Open Court.

Zuckerberg, M. (2015, August 28). *Mark Zuckerberg*. Retrieved from Facebook: https://www.facebook.com/zuck

About the Contributors

Maximiliano E. Korstanje is editor in chief of International Journal of Safety and Security in Tourism (UP Argentina) and International Journal of Cyber Warfare and Terrorism(IGI-GlobalUS). Besides being Senior Researchers in the Department of Economics at University of Palermo, Argentina, he is a global affiliate of Tourism Crisis Management Institute (University of Florida US), Centre for Ethnicity and Racism Studies (University of Leeds), The Forge (University of Lancaster and University of Leeds UK) and The International Society for Philosopher, hosted in Sheffield UK. With more than 700 published papers and 25 books, Korstanje was awarded as Outstanding Reviewer 2012. International Journal of Disaster Resilience in the Built Environment. University of Salford, UK, Outstanding Reviewer 2013. Journal of Place Management and Development. Institute of Place, Manchester Metropolitan University, UK and Reviewer Certificate of Acknowledgement 2014. International Journal of Contemporary Hospitality Management (IJCHM), University of Central Florida, US. Now he co-edits almost 10 specialized journals in such themes as human rights, mobility, tourism and terrorism. Korstanje is subject to biographical records for Marquis Who`s Who in the World since 2009. He had nominated to 5 honorary doctorates for his contribution in the study of terrorism and mobilities. In 2015 he was awarded as Visiting Research Fellow at School of Sociology and Social Policy, University of Leeds, UK. In 2016 he served as Visiting Fellow at University of La Habana Cuba.

* * *

Nicholas Ayres is a PhD student at DeMontfort University within the Faculty of Technology with the DIGITS teams studying Dynamic software updates in antonymous vehicles. Prior to beginning his PhD program, Nicholas studied a Masters Degree in Cyber Security where he undertook research into the effects of a hypothetical cyberterrorist attack on a selected section of the general public and produced a final dissertation on the results as well a number of corresponding journal articles.

Matthew Crosston, Professor of Political Science, is the Miller Endowed Chair for Industrial and International Security and Director of the International Security and Intelligence Studies (ISIS) program at Bellevue University. He has authored two well-received books, several book chapters and over two dozen peer-reviewed articles in venues like the International Journal of Intelligence and Counterintelligence, Journal of Strategic Security, International Politics, Journal of Military and Strategic Affairs, Comparative Strategy, Journal of Conflict Transformation and Security, Strategic Studies Quarterly, Central Asian Survey, Journal of Global Analysis, and Democracy and Security. His research agenda continues to address counter-terrorism, intelligence analysis and education, failed democratization, global gender violence, and cyber war/ethics. His works have been translated into Russian, Arabic, Chinese, Hebrew, Spanish, Turkish, Farsi, Greek, and Uzbek. He has a BA from Colgate University, MA from the University of London, PhD from Brown University, and finished a Post-Doctoral Fellowship at the University of Toronto.

Marthie Grobler is a Principal Researcher at the Council for Scientific and Industrial Research (CSIR). She has worked at the CSIR since January 2008. She has a PhD Computer Science (Live Digital Forensics), and a MSc Computer Science (Information Security Governance), both from the University of Johannesburg. Her research focus is on cyber security, cyber psychology, strategic data management and incident management and response. She is co-editor of the now published ISO/IEC 27037, Guidelines for identification, collection and acquisition and preservation of digital evidence; and ISO/IEC 27035, Incident management. Marthie is an ISACA Certified Information Security Manager and is appointed as a visiting Professor at the University of Johannesburg, Academy for Computer Science and Software Engineering. She is currently supervising a number of post graduate students. Marthie is rated as a Y2 researcher with the National Research Foundation.

Pierre Jacobs has 15 years of consultation experience in the IT Security industry. He is currently with the Council for Scientific and Industrial Research (CSIR) where he is a member of various cybersecurity projects with a national impact and reach. He holds a MSc with specialisation in Information Security from Rhodes University, and is currently a PhD candidate at the University of Johannesburg.

Helge Janicke is heading De Montfort University's Software Technology Research Laboratory and the Cyber Security Centre. He is the Interim Head of School of Computer Science and Informatics. Dr. Janicke was awarded his PhD in Computer Science in 2007 and worked on Cyber Security with organisations such as QinetiQ, Ministry of Defence and General Dynamics UK as part of the DIF-DTC consortium. His interests are covering formal verification techniques and their application to

Cyber Security, SCADA and Industrial Control System Security as well as aspects of Cyber Warfare. He is working with Airbus Group and established DMU's Airbus Group Centre of Excellence in SCADA Cyber Security and Forensics Research in 2013. He is a general chair of the International Symposium on SCADA and Industrial Control Systems Cyber Security Research (ICS-CSR). He is Editor-in-Chief for the Journal of Security and Safety (EAI) and serves on the editorial board and as reviewer of international journals.

Yuriy Kostyuchenko, Ph.D., Associated Professor, is a Leading Research Scientist and leader of an Ad-hoc workgroup on disaster analysis and risk assessment of Scientific Centre for Aerospace Research of the Earth, National Academy of Sciences of Ukraine. He received an MS in Astrophysics and Stellar Astronomy from Moscow State University in 1993. After that he joined National Academy of Sciences of Ukraine as a Scientist. He received his PhD in Geophysics and Remote Sensing in 1997 from the Scientific Centre for Aerospace Research of the Earth of National Academy of Sciences of Ukraine. In 2011-2014 he served on position of Assistant Professor of the Geomorphology and Soil Science Department of the Faculty of Geography, Kiev University. Since 2000 he serves as Executive secretary of Ukrainian National Member Organization of International Institute for Applied Systems Analysis (IIASA). His research interests include remote sensing application for environmental and socio-ecological security analysis, risk analysis using non-parametric approaches. Yuriy V. Kostyuchenko is a member of Risk and Vulnerability Committee of American Society of Civil Engineers (ASCE), member of American Meteorological Society (AMS), American Statistical Society (ASA), and International Association for Promotion Geoethics (IAPG). He was honored by special award of Space Research Council of National Academy of Sciences of Ukraine in 2003 for fundamental research in remote sensing, and "Makarov" Medal of National Space Agency of Ukraine in 2012 for development of advanced technologies of remote sensing. Yuriy V. Kostyuchenko is author and co-author of more than 70 scientific papers and book chapters in fields of remote sensing, Earth sciences (ecology, climate change), engineering and social sciences.

Leandros Maglaras received the B.Sc. degree from Aristotle University of Thessaloniki, Greece in 1998, M.Sc. in Industrial Production and Management from University of Thessaly in 2004 and M.Sc. and PhD degrees in Electrical & Computer Engineering from University of Volos, in 2008 and 2014 respectively. He is currently a Lecturer in the School of Computer Science and Informatics at the De Montfort University, U.K. He has participated in various research programs investigating vehicular and ICT technologies (C4C-project.eu, reduction-project.eu), sustainable development (islepact.eu, Smilegov), cyber security (cockpitci.eu, fastpass-project.eu) and optimization and prediction of the dielectric behavior of air gaps (optithesi.webs.com). He is general (co-)Chair of INISCOM 2016, a new, annual, EAI and Springer sponsored conference on industrial networks and intelligent systems. He is an author of more than 45 papers in scientific magazines and conferences and is a senior member of IEEE.

Oğuzhan Menemencioğlu received BSc degree in Computer Education from Gazi University in 2006 and MSc Degree in Computer Engineering from Karabük University in 2011 where he is currently toward the PhD Degree under the supervision of Asst. Prof. Dr. İlhami M. Orak. He has been working as research assistant in Department of Computer Engineering since 2009. He has been working as administrator in Karabük University Computer Center about distance education infrastructure since 2010. His research interests include semantic web, database, web programming, ontology modelling, content analysis, multimedia databases, and security.

İlhami M. Orak graduated with a good degree from Computer and Control Engineering Department of İstanbul Technical University in 1989. He continued his further education in Automatic Control and Systems Engineering Department, Sheffield University, England. He finished his PhD thesis about "Parallel Algorithms for the Simulation of Sintering Process" in 1994. He participates in industrial projects committee of The Scientific and Technological Research Council of Turkey. He is an Assistant Prof. at Computer Engineering Department of Karabük University. He mainly interests programming in industrial systems, simulation, information security, and database.

Trishana Ramluckan has been an academic and researcher in the I.T field for the past 10 years. She is completing a PhD focussing on social media in crisis communication and disaster management at the School of Management, Information Systems & Technology and Public Governance, University of KwaZulu-Natal. In 2006, she graduated with an MA from the University of KwaZulu-Natal, South Africa, and holds a BA majoring in Politics, Legal Studies and Classics. She is a member of the IFIP working group on ICT Uses in Peace and War and serves as an academic advocate for ISACA South Africa.

Sayed Enayat Sayed Ally is an Information Systems and Technology researcher at the University of KwaZulu-Natal, South Africa and is currently working towards a PhD in Information Systems. He has completed his MCom in the area of Online Privacy and Facebook which sought to investigate the risk-attitude of students at University. He also holds a BCom with a double major in Information Systems. His research interests include Data Visualisation, Information Security and Data Analytics.

Brett van Niekerk graduated with his PhD in 2012 from the University of KwaZulu-Natal, and has completed two years of postdoctoral research into information operations, information warfare, and critical infrastructure protection. He is currently a Senior Security Analyst at Transnet and an Honorary Research Fellow at the University of KwaZulu-Natal. He also holds a BSc and a MSc in electronic engineering. He has over 10 journal articles and 15 conference presentations in the field. He is a member of the IFIP WG 9.10 on ICT in Peace and War, and in 2016 was appointed to the board of the ISACA South Africa Chapter. He has contributed to ISO/IEC standards on information security.

Shefali Virkar is post-doctoral researcher, having recently been awarded a D.Phil. in Politics from the University of Oxford(UK). Her doctoral research sought to explore the growing use of Information and Communication Technologies (ICTs) to promote better governance in the developing world, with special focus on the political and institutional impacts of ICTs on local public administration reform in India. Shefali holds an M.A. in Globalisation, Governance and Development from the University of Warwick, UK. Her Master's thesis analysed the concept of the Digital Divide in a globalising world, its impact developing countries and the ensuing policy implications. At Oxford, Shefali is a member of Keble College.

Maxim Yuschenko is Leading Engineer of Scientific Centre for Aerospace Research of the Earth, National Academy of Sciences of Ukraine. He received an MS in Geography and Education from Kiev National Pedagogical University in 2000. After that he joined National Academy of Sciences of Ukraine as an Engineer. His research interests include GIS and remote sensing application for environmental and socio-ecological security and risk analysis. Maxim Yuschenko is author and co-author of more than 30 scientific papers and book chapters in fields of Earth sciences, engineering and social sciences.

Index

A

C

D

U

V

W

www.ingramcontent.com/pod-product-compliance
Lightning Source LLC
LaVergne TN
LVHW061548290926
849162LV00004B/946
* 9 7 8 1 5 2 2 5 1 9 3 8 6 *